The Foundations of Newton's Alchemy

The Foundations of Newton's Alchemy

or

"*The Hunting of the Greene Lyon*"

BETTY JO TEETER DOBBS

Assistant Professor of History
Northwestern University, Evanston, Illinois

Cambridge University Press

CAMBRIDGE

LONDON . NEW YORK . MELBOURNE

Published by the Syndics of the Cambridge University Press
The Pitt Building, Trumpington Street, Cambridge CB2 1RP
Bentley House, 200 Euston Road, London NW1 2DB
32 East 57th Street, New York, NY 10022, USA
296 Beaconsfield Parade, Middle Park, Melbourne 3206, Australia

First published 1975

Printed in the United States of America

Library of Congress Cataloging in Publication Data
Dobbs, Betty Jo Teeter, 1930–
   The foundations of Newton's alchemy.
   Bibliography: p. 259
   Includes index.
   1. Newton, Isaac, Sir, 1642–1727. 2. Alchemy—History.
I. Title. II. Title: The hunting of the greene lyon.
QC16.N7D6   540'.1   74–31795
ISBN 0–521–20786–X

TO MARY GLADYS GREER TEETER, MY
MOTHER, WHO, FROM HER VANTAGE
POINT IN THE SMALL TOWNS OF ARKANSAS,
ALWAYS RECOGNIZED THAT THERE WAS
MORE TO BE LEARNED

Contents

CONTENTS

Plates

Preface

This study, since its original conception, has contracted in its projected coverage of Newton's own alchemy as it has expanded in its coverage of the tidal waves of the Hermetic modes of thought that engulfed his period. Only a tiny fraction of Newton's own alchemical experiments have been explored in the event, but it is hoped that they have been made more comprehensible by placing them in their matrix in the intellectual currents of the seventeenth century. What has resulted here is a composite, and the word "foundations" in the title may be taken in three separate ways. One sense in which the word is used in this work is that of origins, and the origins of Newton's alchemy in Restoration England are emphasized. Another sense is that of the experimental and scholarly bases Newton provided for himself in his alchemical studies, and those foundations are examined also for the earliest period of his work. The last sense in which the word is to be taken is that of supports, and alchemy is seen finally as comprising one of the pillars which supported the structure of Newton's mature science.

This book is, in a small way, a work of both intellectual history and the history of science. Perhaps that is appropriate, even necessary, in studies dealing with scientific thought in the seventeenth century. For modern science in its nascence surely bore the marks of the ancient womb of human thought in which it had had its long period of gestation. Not all of those antique thought patterns are acceptable today as valid approaches to the world of phenomena or as genuine and honest efforts at making that world comprehensible. Modern science, like an adolescent, denies its parentage.

But the rich complexities of the seventeenth century produced a "century of genius," and it will perhaps do no harm to recapture what we can of its intellectual atmosphere. Newton is the century's epitome: his great synthetic mind wove its multiple strands together to make a brilliant new tapestry. In methodology he integrated the empiricism of Bacon and the mathematics of the ancients; in physics he integrated Kepler's planetary motions with Galileo's terrestrial ones. In both cases he added his own inimitable flourishes, making his productions something wholly new. In his chemical studies the same pattern may be detected, for Newton strove to integrate alchemical and Hermetic ideas with the mechanical philosophies of his day. The result was somewhat less successful in the realm of chemistry proper than were the results in some of his other undertakings. But in fact his efforts to integrate alchemy and mechanism seem to have produced one of his best virtuoso performances: the creation of a new concept of force. Newton's alchemy is the historical link between Renaissance Hermeticism and the rational chemistry and mechanics of the eighteenth century.

Two research tools which have been invaluable in this study are Maurice P. Crosland, *Historical Studies in the Language of Chemistry* (Cambridge, Mass.: Harvard University Press, 1962) and J. W. Mellor, *A Comprehensive Treatise on Inorganic and Theoretical Chemistry* (16 vols.; New York: John Wiley & Sons, 1960). They have been used freely throughout for innumerable problems in the translation and interpretation of seventeenth-century chemistry and alchemy and no further citation will be made to them. The symbols and names in Tables 1–5, drawn in part from those works, in part from various dictionaries and lexicons cited in the Bibliography, and in part from primary sources cited in the text, may be of some use to the reader.

TABLE 1 *Metals*

Symbol	Metal	Celestial analogue
☿	Mercury	Mercury
♄ or ♄	Lead	Saturn
♃	Tin	Jupiter, Jove
♀	Copper	Venus
♂	Iron	Mars
☽	Silver	*Luna*, Moon
☉	Gold	*Sol*, Sun

TABLE 2 *Antimony*

Symbol	Name and modern symbol
♁	Antimony ore, stibnite, Sb_2S_3
Regulus of ♁, ☽, and [in Keynes MS 55] ℞	Metallic antimony, Sb
✳ [in Keynes MS 55]	Star, star regulus, Sb
R ♂, Regulus of iron	Metallic antimony prepared by the use of iron, Sb

PREFACE

TABLE 3 *Other chemicals*

Symbol	Seventeenth-century name	Modern name and symbol
♁	Sulphur	Sulfur, S
Æ	*Aqua fortis*	Nitric acid, HNO_3
⊕	Vitriol	Any sulfate but probably usually that of iron or copper, $FeSO_4$ or $CuSO_4$
✳ [usually]	Sal ammoniac	Ammonium chloride, NH_4Cl
☉	Salt peter, nitre	Potassium nitrate, KNO_3
♇	Tartar	Potassium hydrogen tartrate, $KH(C_2H_2O_3)_2$
⊖	Salt	Sodium chloride, NaCl, or other salt
♃	Mercury sublimate	Mercuric chloride, $HgCl_2$
---	Sal alkali	Probably sodium or potassium hydroxide, NaOH or KOH

TABLE 4 *Measures, apparatus, and processes*

Symbol	Name
℥	Ounce
β, ss	$\frac{1}{2}$
+	*Crucibulum*, crucible
⩔	*Tigillum*, crucible, cupel
a̅a̅a̅	Amalgam, amalgamate
℞ [usually]	Take or prepare

xiii

Table 5 *Miscellaneous*

Symbol	Name
Υ, q̄. ē.	Quintessence
♌	*Leo*, lion

Where square brackets occur in quotations from Newton's manuscripts, they are Newton's own. The arrows up and down set off Newton's interlineations; the carets indicate probable readings supplied by the present investigator where the manuscript is damaged or otherwise illegible and where it has been necessary to interject an explanatory remark.

References are cited in full when they are introduced for the first time. Subsequent citations to the same work will utilize a shortened form but will include also a parenthetical direction to the full title, such as (1, n. 5), to direct the reader to Chapter 1, note 5, if he desires the fuller form.

Many libraries and their staffs have assisted with books, manuscripts, microfilms, and Xeroxes: the several branches of the library of the University of North Carolina at Chapel Hill; University Library, Cambridge; the libraries of Trinity College, King's College, Queens' College, and Gonville and Caius College, Cambridge; The Fitzwilliam Museum, Cambridge; The Bodleian Library, Oxford; The British Museum; Babson College; the libraries of Stanford University, Cornell University, Harvard University, Duke University, the University of Wisconsin, and the Yale Medical School; Burndy Library, Norwalk, Conn.; Mercantile Library, St. Louis, Mo.; and the Jewish National and University Library, Jerusalem. Special thanks are due to the University of Wisconsin for a copy of Newton's copy of Eirenaeus Philalethes' *Secrets Reveal'd* and for the use of a copy of the 1650 edition of Sendivogius' *New Light of Alchemy*, and to the Syndics of University Library, Cambridge, to the Provost and Fellows of King's College, Cambridge, to the Master and Fellows of Trinity College, Cambridge, to The Frederick E. Brasch Collection of Sir Isaac Newton and the History of Scientific Thought of Stanford University, to Babson College, the Bodleian Library, the Fitzwilliam Museum, the Burndy Library, and the Yale Medical School for permission to quote from or describe manuscript material in their possession, and to the editors of *Ambix* for permission to use material previously published in that journal, which comprises parts of Chapters 2 and 3.

The inclusion of the illustrations was made possible through the courtesy of the Syndics of the Fitzwilliam Museum, Cambridge; the Provost and Fellows of King's College, Cambridge; the Science Museum, London; and the Stadtbibliothek Vadiana, St. Gallen.

Grateful acknowledgment for support is made to the National Science Foundation, the Danforth Foundation, the North Atlantic Treaty Organization, and to my husband and children. Katherine Roan Dobbs has assisted with the study of the Newton Collection, Trinity College, Cambridge, Gladys Rebecca Dobbs with the preparation of the index, George Byron Dobbs II and Jean Frances Dobbs with more personal sacrifices and contributions than they perhaps realize, and Dan Byron Dobbs with that good base *sine qua non*. Michael R. McVaugh and Frederick O. Behrends have assisted with translations from the Latin. Michael McVaugh, Robert Siegfried, Henry Guerlac, Ronald Sterne Wilkinson, Richard S. Westfall, John M. Headley, Seymour H. Mauskopf, and William H. Brock have given much valuable advice and encouragement along the way: Professors McVaugh, Guerlac, Westfall, Headley, and Mauskopf, in particular, having read part or all of the work in manuscript, have eliminated many errors which would otherwise appear. Unfortunately, none of them may be given credit for the ones which remain. But a very special debt of gratitude is owing to Michael McVaugh, without whom this project could never have been conceived, much less completed.

B. J. T. DOBBS

30 September 1974

1. "The Hunting of the Greene Lyon"

Introduction

In the late 1670's or early 1680's, a Fellow of Trinity College, Cambridge, made extensive abstracts from an alchemical poem entitled "The Hunting of the Greene Lyon." At the end of his abstract he made some notes.

> The ⁊ script ↙ conteins ye Regimen of ye work in common gold after ye Pher ☿ is made. Jn ye following notes you have the consent of it with other authors & elucidation thereby.[1]

The "work" referred to was the Great Work of alchemy and the man writing was Isaac Newton. How he came to be engaged in that peculiar activity has been the subject of considerable wonder and no little controversy. It is the purpose of the present work to attempt a relief of both wonder and controversy by establishing the historical context and foundations of Newton's alchemy.

Although alchemy is now considered a delusion, a mystical pursuit, or at best a pseudo-science, that was not always necessarily the case. In the seventeenth century quite different attitudes towards it were possible. In this study, after a brief biographical sketch of Newton's life and achievements, an historiographical survey of writings on Newton's alchemy, and a preliminary examination of his alchemical papers, some of the larger questions concerning alchemy in general will be undertaken. In succeeding chapters alchemy will be discussed from the viewpoint of analytical psychology, and a survey of its little-understood developments in the seventeenth century will be made.

With something of the broad currents of seventeenth-century alchemy in hand and with some background information on the flood of alchemical literature of the period, it will then be possible to turn to events in Cambridge as Newton came upon them and to approach the questions involved in the more immediate historical context of his work. It is hoped that a glimpse of that concrete setting Newton encountered in Cambridge will make his alchemical studies seem not only reasonable and natural but indeed almost inevitable.

Then Newton's own earliest alchemical studies and experiments will be considered. It will be seen that Newton attacked alchemical questions

---

[1] King's College, Cambridge, Keynes MS 20, f. 5r, hereinafter referred to as Keynes MS 20. Techniques utilized in the dating of the Newtonian manuscripts used in this study are discussed in Appendix D. For an explanation of transcription notation on manuscript materials and of alchemical symbols used, see the Preface.

1

with quite definite chemical processes, working within a fairly widespread alchemical tradition of the period. In the event, Newton achieved some alchemical success, preparing a "philosophical mercury" which he took to be preliminary to "ye Regimen of ye work in common gold."

That success, about 1675 or perhaps a little after, colored a great many of his later scientific ideas. Although it is beyond the scope of the present study to attempt a detailed analysis of Newton's alchemical studies after 1675, some places in his later scientific writings in which alchemical ideas are reflected will be indicated.

Biographical sketch

Isaac Newton was born on Christmas Day, 1642, the premature, posthumous, and only child of an illiterate yeoman farmer of Lincolnshire.[2] Not really expected at first to live – he was later to remark that at his birth he was so small that he might have been put into a quart mug – he survived war, revolution, plague, and the seventeenth-century pharmacopoeia to the age of eighty-four, to be buried in Westminster Abbey, idolized by his countrymen and admired by the world.

His genius appeared more mechanical than intellectual at first: as a boy he constructed water clocks, windmills, kites, and sundials, and cleverly used the force of the wind to enable himself to outjump the other boys. But nurtured by neighboring village schools and the King's School at Grantham his intellectual prowess and his enormous power of concentration slowly became apparent. Recalled from school by his mother to learn the art of farming, he spent his time under the hedges with his books and his calculations, to the utter neglect of the life of his ancestors. Eventually a maternal uncle, a Cambridge man himself, intervened to have him returned to the school at Grantham to be prepared for Cambridge, and Isaac went up to that venerable seat of learning in 1661, entering Trinity College. He was aged eighteen, a little older than most entering students and probably less well prepared than many, but evidently with all his faculties ready to flower. Stimulated by the new Cartesian ferment in physics, philosophy, and mathematics, by Kepler's optics and laws of planetary motion, by Galileo's mechanics, and by the work in mathematics and optics of his own teacher Isaac Barrow, the young Newton

[2] Lesser biographies of Newton are legion; a number are noted in the Bibliography. The three fullest and most serious biographical studies of Newton in English are: (1) David Brewster, *Memoirs of the Life, Writings, and Discoveries of Sir Isaac Newton* (2 vols.; Edinburgh: Thomas Constable and Co.; Boston: Little, Brown, and Co., 1855, hereinafter referred to as Brewster, *Memoirs*; (2) Louis Trenchard More, *Isaac Newton. A Biography* (London: Constable and Co.; 1934; New York: Charles Scribner's Sons, 1934; New York: Dover Publications, 1962), hereinafter referred to as More, *Newton*; (3) Frank E. Manuel, *A Portrait of Isaac Newton* (Cambridge, Mass.: The Belknap Press of Harvard University Press, 1968), hereinafter referred to as Manuel, *Portrait*. The following sketch is indebted to all but especially to More, *Newton*.

was soon to tackle – and solve – many of the physical and mathematical questions which engaged his contemporaries. And through all his subsequent work the niceness of his mechanical aptitude and the fine-grained quality of his physical comprehension, that early mechanical genius of his childhood, kept steady pace with his developing intellectual genius and strengthened it.

In January 1664/65 Newton took his Bachelor of Arts degree but in the summer of 1665 was forced to retire to his home at Woolsthorpe as the University was closed because of an outbreak of the plague. It remained closed most of the time until the spring of 1667 and Newton spent the period at his mother's manor, a fact which would hardly bear notice in this summary biography had that period not proved to be his *annus mirabilis*, the marvelous year in which he invented his "fluxions" (later to be developed into the calculus), discovered white light to be compounded of all the distinctly colored rays of the spectrum, and found out a mathematical law of gravity. The gradual development and unfolding to the world, throughout subsequent years, of the productions of that brief period were to establish his reputation upon the granite foundation it still enjoys.

In 1667, however, he merely returned to Cambridge, quietly proceeded Master of Arts, was elected to a Trinity Fellowship, and settled down. Only Isaac Barrow seems to have had an inkling of what was going on in Newton's mind, as Newton had shown him one of the mathematical papers from the Woolsthorpe period. Barrow immediately put it into circulation among interested mathematicians and in 1669 resigned his Lucasian Chair of Mathematics in Newton's favor.

In 1672 Newton disclosed some of the optical discoveries of the Woolsthorpe period to the Royal Society and was in consequence elected Fellow of that group. But it was not until 1684 that the full extent of his gravitational studies came to light. By that time a number of Fellows of the Royal Society had come to the conclusion that the centripetal force of the sun which acted upon the planets must be inversely proportional to the square of the intervening distance, but they could not prove it. Edmund Halley, later Astronomer-Royal, knowing from earlier exchanges that Newton had some ideas on the subject, made the journey to Cambridge to ask him about it. Halley

> at once indicated the object of his visit by asking Newton what would be the curve described by the planets on the supposition that gravity diminished as the square of the distance. Newton immediately answered, *an ellipse.* Struck with joy and amazement, Halley asked him how he knew it? Why, replied he, I have calculated it . . . .[3]

---

[3] From a manuscript by John Conduitt, who married Newton's niece and lived in Newton's London residence for many years, quoted in More, *Newton*, p. 299. Conduitt prepared the manuscript with a view to writing a "Life" of Newton, but never did so:

Newton had lost his papers on it but at Halley's urging prepared his proofs again and proceeded to work up a course of lectures on planetary motion, *De motu corporum*, and another *On the System of the World*. Due almost entirely to Halley's insistence these eventually became the *Philosophiae naturalis principia mathematica*, in English the *Mathematical Principles of Natural Philosophy*, commonly called now simply the *Principia* and said to be the greatest work of science ever published.[4]

Newton's life of retiring scholarship ended in 1696 with his appointment as Warden of the Mint just as the great recoinage of William III's reign got underway. Charles Montague, later Lord Halifax, who had known and admired Newton at Cambridge, was then Chancellor of the Exchequer. He had already been engaged for some time in his reorganization of the nation's finances, establishing the Bank of England and founding the national debt to finance William's wars. In 1695 Montague was empowered by William to prepare a bill for recoinage also, as a part of the total effort at the reform and stabilization of the monetary system, for coins in circulation were far below face value from clipping and counterfeiting. In January 1695/96 the bill was passed, in February the recoinage was begun, and in March, Newton, through Montague's efforts, arrived to take over the post of Warden.

At that time the principal duties of the Warden were of a legal nature and Newton was expected to oversee the detection and prosecution of counterfeiters and clippers of coins. He superintended that work in detail and also dealt from time to time with other infractions, such as conflicts between Mint personnel and the Tower garrison. In 1699 he was translated to the Mastership of the Mint, which position he held until his death. The latter was the general administrative post of the organization. Newton handled it with his accustomed thoroughness, straightening out

his papers on the subject now comprise a part of the Keynes Collection, King's College, Cambridge.

[4] Isaac Newton, *Philosophiae naturalis principia mathematica* (Londini: Jussu *Societatis Regiae* ac Typis *Josephi Streater*. Prostat apud plures Bibliopolas, 1687). Of the many subsequent editions, the most convenient modern one, and the one used primarily in the present study, is Isaac Newton, *Sir Isaac Newton's Mathematical Principles of Natural Philosophy and his System of the World. Translated into English by Andrew Motte in 1729. The translations revised, and supplied with an historical and explanatory appendix, by Florian Cajori* (2 vols.; Berkeley and Los Angeles: University of California Press, 1966), hereinafter referred to as Newton, *Principia*. Recently a definitive variorum edition has been prepared: Isaac Newton, *Isaac Newton's Philosophiae naturalis principia mathematica. The Third edition (1726) with Variant Readings. Assembled and Edited by Alexandre Koyré and I. Bernard Cohen with the Assistance of Anne Whitman* (2 vols.; Cambridge, Mass.: Harvard University Press, 1972), hereinafter referred to as Koyré and Cohen, *Newton's Principia*. Page references in the present study will be given for both the English of Mott and Cajori and the Latin of the Koyré and Cohen edition. A wealth of detail on the preparation of the first three editions of the *Principia* and on their contemporary reception appears in a companion volume to the variorum edition: I. Bernard Cohen, *Introduction to Newton's 'Principia'* (Cambridge, Mass.: Harvard University Press, 1971).

4

the chaos of records and accounts left by his predecessors and, when questions of procedure arose, searching out all the precedents as far back as Elizabethan times. The Mastership of the Mint cushioned Newton's later years with a financial security he had never known before, and the impecunious lad from Lincolnshire, who had entered Cambridge as a subsizar working for his keep, died a wealthy man.[5]

Although Newton's duties as Warden and as Master did not require it, he soon made himself familiar with the technical processes involved in minting. He learned the technique of the assaying of fine gold and silver, with which he had apparently been unfamiliar in 1696, and served sometimes as his own assayer even when the regular assay work was done by the Crown Assayer.[6] As John Conduitt observed, Newton "had frequent opportunities of employing his skill in mathematics and chemistry, particularly in his... assays of foreign coins...."[7] He seems in general to have been most conscientious in his attention to business, and Craig relates one humorous occasion when Newton's mathematical abilities gave aid and comfort to the Crown in an unexpected way. Parliament had passed such an involuted Tonnage and Poundage Act that the Treasury was at a loss as to how to apply it. It involved a percentage duty calculated on net price realized, less the amount of tax, and no one could quite cope with the computations. In 1703 the Treasury sent the problem to Newton and he, after two examinations of the Act, several arithmetical and algebraic approaches, and three draft replies, reduced the question to the point where only one sum needed to be done by the Customs Officer.[8]

For many years Newton never intended to publish any of his discoveries. His early disclosures to the Royal Society concerning his work in optical matters had generated controversies which caused him to forswear further publication and it had taken all Halley's tact and energy to obtain the manuscript of the *Principia* from him and to see it through the press. When additional controversies arose over the *Principia*, Newton became

[5] John Craig, *Newton at the Mint* (Cambridge: At the University Press, 1946), hereinafter referred to as Craig, *Mint*.

[6] *Ibid.*, p. 122.

[7] John Conduitt, "Memoirs of Sir Isaac Newton, sent by Mr. Conduitt to Monsieur Fontenelle, in 1727," in Edmund Turnor, *Collections for the History of the Town and Soke of Grantham. Containing Authentic Memoirs of Sir Isaac Newton, Now First Published From the Original MSS. in the Possession of the Earl of Portsmouth* (London: Printed for William Miller, Albemarle Street, by W. Bulmer and Co. Cleveland-Row, St. James's, 1806), p. 162, the larger work hereinafter referred to as Turnor, *Collections*. See also "Sir Isaac Newton's Memorial on the State of the Gold and Silver Coin," prepared for the Commissioners of the Treasury, dated Mint Office, September 21, 1717, in *Observations on the Subjects Treated Of in Dr Smith's Inquiry into the Nature and Causes of the Wealth of Nations by David Buchanan [1817] with Appendices: Sir Isaac Newton's Memorial on the Gold and Silver Coins 1717 And Other Documents* (The Adam Smith Library; Reprints of Economic Classics; New York: Augustus M. Kelley, 1966), Appendix, pp. 1–7.

[8] Craig, *Mint*, pp. 62–63 (1, n. 5). Craig's account is based on Newton MSS III, 441 and 446, of those "Papers relating to the Mint by Sir I. Newton," now at the Royal Mint.

even more reluctant to publish. But as the duties of the Mastership of the Mint allowed him the leisure, the death of old enemies and the kindly pressure of friends combined to dissuade Newton from his earlier reticence. The year 1704 saw the publication of his optical papers in a systematic form, two mathematical papers being included also in the same volume.[9] Second, third, and fourth editions of the *Opticks* were prepared by Newton with new material (1706, 1721, and the posthumous edition of 1730), as also second and third editions of the *Principia* (1713, 1726). More mathematical papers appeared (*Arithmetica universalis*, 1707, *De analysi* and *Methodus differentialis*, 1711) and a number of less well-known items.

Honours accumulated for the aging Newton: in 1699 he was elected as one of the first eight foreign associates of the French Academy of Sciences; in 1703 he became President of his own country's Royal Society, to which office he was re-elected annually until his death; in 1705 he was knighted by Queen Anne. Awestruck disciples saw to it that he became the virtual dictator of science while he yet lived, and he continued in that role so completely even after death that the sciences developed in the eighteenth and nineteenth centuries all came to be framed according to Newtonian models.

The Newtonian world-view, indeed, developed almost wholly on the basis of his successes in mathematics and physical science, so subtly and deeply colored the thoughts of succeeding generations that the fuller seventeenth-century context in which Newton's thought had developed was lost to view.[10] Thus it became a curious anomaly – and one to be explained away – that Newton's studies in astronomy, optics, and mathematics only occupied a small portion of his time. In fact most of his great powers were poured out upon church history, theology, "the chronology of ancient kingdoms," prophecy, and alchemy.

The historiography of Newton's alchemy

All of these, which now seem such obscure and unlikely studies for a great natural philosopher to pursue, were of vital interest in the seventeenth century. Then uncertainties in creed and dogma were fought out on intel-

[9] Isaac Newton, *Opticks: or, a Treatise of the Reflexions, Refractions, Inflexions and Colours of Light. Also Two Treatises of the Species and Magnitude of Curvilinear Figures* (London: Printed for Sam. Smith, and Benj. Walford, Printers to the Royal Society, at the *Prince's Arms* in St. *Paul*'s Church-yard, 1704). A convenient modern edition of the optical material is Isaac Newton, *Opticks, or A Treatise of the Reflections, Refractions, Inflections & Colours of Light,* foreword by Albert Einstein, introd. by Sir Edmund Whittaker, preface by I. Bernard Cohen, analytical table of contents by Duane H. D. Roller (based on the 4th edn., London, 1730; New York: Dover Publications, 1952), hereinafter referred to as Newton, *Opticks.*
[10] Gerd Buchdahl, *The Image of Newton and Locke in the Age of Reason* (Newman History

lectual as well as physical battlefields, and an age which still held largely to the literal truth of Scripture sought signs of the fulfillment of prophecy in contemporary events. When the question was agitated, whether the world was decaying towards its final end from a pristine Golden Age or was entering upon a joyful expansion of human knowledge and skill,[11] then it became important to know the exact relationships which held between the ancient kingdoms of the world. Newton struggled with these questions no more and no less than his contemporaries.

Twentieth-century scholarship is slowly restoring the lost context of Newton's thought. McLachlan has studied some of Newton's theological works[12] and Manuel some of his ideas on "chronology."[13] But Newton's alchemical studies have not yet received a thorough explication; the many attempts at explanation have long see-sawed between rationalistic and mystical interpretations and none of them has been wholly convincing.

Confusion over Newton's approach to alchemy arose in the very earliest period after his death, when biographical materials were being collected from those who had known the great man personally. At that time John Conduitt wrote to Dr. Humphrey Newton of Grantham, a man who had served Sir Isaac as amanuensis and laboratory assistant from 1685 until 1690. Humphrey Newton sent back his personal reminiscences, which painted a graphic picture of Sir Isaac in the heat of battle with the elements.

> He very rarely went to bed till *two* or *three* of the clock, sometimes not until *five* or *six*, lying about *four* or *five* hours, especially at spring and fall of the leaf, at which times he used to employ about six weeks in his elaboratory, the fire scarcely going out either night or day; he sitting up one night and I another, till he had finished his chemical experiments, in the performances of which he was the most accurate, strict, exact. What his aim might be I was not able to penetrate into, but his pains, his diligence at these set times made me think he aimed at something beyond the reach of human art and industry.... On the left end of the garden was his elaboratory, near the east end of the

and Philosophy of Science Series, no. 6, gen. ed., M. A. Hoskin; London and New York: Sheed and Ward, 1961).

[11] Richard Foster Jones, *Ancients and Moderns. A Study of the Rise of the Scientific Movement in Seventeenth-Century England* (2nd edn.; St. Louis: Washington University Studies, 1961), esp. pp. 23–40.

[12] Herbert McLachlan, *Religious Opinions of Milton, Locke, and Newton* (Publications of the University of Manchester, no. 276, Theological Series, no. 6; Manchester: Manchester University Press, 1941); Isaac Newton, *Sir Isaac Newton: Theological Manuscripts*, selected and ed., with introd., by Herbert McLachlan (Liverpool: At the University Press, 1950), hereinafter referred to as Newton, *Theological MSS*.

[13] Frank E. Manuel, *Isaac Newton, Historian* (Cambridge, Mass.: The Belknap Press of Harvard University Press, 1963), hereinafter referred to as Manuel, *Historian*.

chapel, where he at these set times employed himself in with a great deal of satisfaction and delight. Nothing extraordinary, as I can remember, happened in making his experiments; which, if there did, he was of so sedate and even temper, that I could not in the least discover it.[14]

About a month later Humphrey Newton sent Conduitt a second letter, recalling additional details.

About 6 weeks at spring, and 6 at the fall, the fire in the elaboratory scarcely went out, which was well furnished with chemical materials as bodies, receivers, heads, crucibles, etc., which was [sic] made very little use of, the crucibles excepted, in which he fused his metals; he would sometimes, tho' very seldom, look into an old mouldy book which lay in his elaboratory, I think it was titled *Agricola de Metallis*, the transmuting of metals being his chief design, for which purpose antimony was a great ingredient.... His brick furnaces, *pro re nata*, he made and altered himself without troubling a bricklayer.[15]

Humphrey Newton presented a portrait of Isaac as the Great Experimenter in these reminiscences to Conduitt, consistent with what he said also to William Stukeley, who was collecting materials for a "Life" of Newton a little later. Stukeley wrote,

Dr Newton of Grantham...was assistant to him [Sir Isaac], particularly in his chymical operations, which he pursu'd many years. He often admir'd Sir Isaacs patience in his experiments, how scrupulously nice he was in weighing his materials, and that his fires were almost perpetual.[16]

But Stukeley was one of the awestruck disciples and already by the mid-eighteenth century he was attempting to rescue Sir Isaac's reputation from the taint of alchemy. Stukeley knew a little more than Humphrey Newton did about the intellectual production that might have resulted from Sir Isaac's laboratory work and was not averse to "explaining" more about it than he actually knew. Stukeley said:

He [Sir Isaac] wrote likewise an intire work on chymistry, explaining the principles of matter, and elementary components, from that abstruse art; on experimental and mathematical proof. He had himself a good opinion of this work; but the MS. was unluckily burnt in the

---

[14] Humphrey Newton to John Conduitt, January 17, 1727/28, Keynes MS 135, King's College, Cambridge, as quoted in More, *Newton*, pp. 247–48 (1, n. 2), and in Brewster, *Memoirs*, II, 93–94 (1, n. 2).

[15] Humphrey Newton to John Conduitt, Feb. 14, 1727/28, Keynes MS 135, King's College, Cambridge, as quoted in More, *Newton*, p. 249 (1, n. 2), and in Brewster, *Memoirs*, II, 95–97 (1, n. 2).

[16] William Stukeley, *Memoirs of Sir Isaac Newton's Life by William Stukeley, M.D., F.R.S. 1752 Being some Account of his Family and Chiefly of the Junior Part of his Life*, ed. by A. Hastings White (London: Taylor and Francis, 1936), p. 56, hereinafter referred to as Stukeley, *Memoirs*.

laboratory, which casually took fire. He never could undertake it again, a loss not to be sufficiently regretted....

As to chymistry in general, we may very well presume Sir Isaac, from his long and constant application to that pyrotechnical amusement, had made very important discoverys in this branch of philosophy, which had need enough of his masterly skill, to rescue it from superstition, from vanity and imposture, and from the fond inquiry of alchymy and transmutation. By this means Sir Isaac carried his inquiry very far downwards into the ultimate component parts of matter, as well as upwards towards the boundless regions of space....[17]

Now Stukeley got it almost right, recognizing as he did Newton's strong experimental program as well as his intellectual commitment to inquiring into "the ultimate component parts of matter." But Stukeley, a young man when Newton died, had had his thinking colored by the eighteenth century's reaction against alchemy, and he called Sir Isaac's work "chymistry" and said it would have rescued that field from "alchymy and transmutation." In contrast to that, it will be the position of the present study that it was precisely *by* the route of alchemy and transmutation that Sir Isaac expected to elucidate "the ultimate component parts of matter."

But even before Stukeley's death in the eighteenth century there arose the ghost of Newton's predilection for the mystical component of alchemy, a ghost which has been adequately laid only in the present century if at all. A Rev. Mr. William Law (1686–1761) first made the suggestion that Jacob Boehme had inspired much of Newton's thought in natural philosophy.

The illustrious Sir *Isaac Newton* when he wrote his *Principia*, and publish'd to the World his great Doctrine of *Attraction*, and those *Laws of Nature* by which the *Planets* began, and continue to move in their Orbits, could have told the World, that the *true and infallible* Ground of what he there advanced, was to be found in the *teutonick Theosopher*, in his *three first Properties of Eternal Nature*; he could have told them, that he had been a *diligent Reader* of that wonderful Author, that he had made large extracts out of him ....[18]

Jacob Boehme (1575–1624), or Behmen as he was usually called by the English, was a Christian mystic, poorly educated, a journeyman shoemaker in Prussia. Having acquired some knowledge of alchemy during

---

[17] *Ibid.*, pp. 59–60.

[18] William Law, *An Appeal To all that Doubt, or Disbelieve The Truths of the Gospel, Whether they be Deists, Arians, Socinians, or Nominal Christians. In which, the true Grounds and Reasons of the whole Christian Faith and Life are plainly and fully demonstrated. To which are added, Some Animadversions upon Dr. Trap's late Reply* (London: Printed for W. Innys, at the West-End of St. *Paul's*, 1742), p. 314.

his apprenticeship, he described his later ecstatic religious experiences in a kind of transfigured Hermetic terminology.[19] If Newton had indeed found the key to his natural philosophy in Boehme, it would require rather radical revision of current understanding of Newton's methodology, but happily this is not true. Stephen Hobhouse, a student of William Law, has established that Law only had an untrustworthy oral tradition for his statements, and has shown furthermore that Boehme's thought stands in complete contradistinction to Newton's.[20] The evidence against Law's position is clearly summarized by McLachlan who further notes the absence of any large extracts from Boehme in the Newton theological manuscripts.[21] And the present writer can testify likewise to Boehme's absence from the alchemical manuscripts.

But at the time the first major biography of Newton was undertaken around the middle of the nineteenth century by Sir David Brewster, no one had yet searched through Newton's papers to establish the fact that Boehme was really not one of his favorite authors. Furthermore, Brewster had seen among Newton's papers a number of items which seemed quite as bad to him: an autograph transcript of *The Metamorphosis of the Planets* by John de Monte Snyders which ran to sixty-two quarto pages, page after page of alchemical verse from Thomas Norton's *Ordinall of Alchimy* and Basilius Valentinus' *Mystery of the Microcosm*, and a copy of Eirenaeus Philalethes' *Secrets Reveal'd, or an Open Entrance to the Shut Palace of the King*, heavily annotated in Newton's own hand.[22]

There were other alchemical papers too, many, many of them. Their extent was indeed hardly indicated by Brewster, for they comprised about 650 000 words, almost all in Newton's hand.[23] Although Brewster was hagiographic in his approach to Newton's biography, he was too honest a researcher to pretend he had not seen that "damaging" evidence, and he was consequently forced into the position of distinguishing the alchemy of Newton – and of his great contemporaries John Locke (1632–1704) and

[19] Arthur Edward Waite, *Alchemists Through the Ages. Lives of the Famous Alchemistical Philosophers from the year 850 to the close of the 18th century, together with a Study of the Principles and Practice of Alchemy, including a Bibliography of Alchemical and Hermetic Philosophy*, introd. by Paul M. Allen (Reprint of the London edn. of 1888; Blauvelt, N.Y.: Rudolf Steiner Publications, 1970), pp. 161–66, hereinafter referred to as Waite, *Lives*.

[20] William Law, *Selected Mystical Writings of William Law edited with Notes and Twenty-four Studies in the Mystical Theology of William Law and Jacob Boehme and an Enquiry into the Influence of Jacob Boehme on Isaac Newton by Stephen Hobhouse*, foreword by Aldous Huxley (2nd edn., revised; New York and London: Harper and Brothers, 1948), pp. 397–422.

[21] Herbert McLachlan, "Introduction," in Newton, *Theological MSS*, pp. 20–21 (1, n. 12).

[22] Brewster, *Memoirs*, II, 371–72 (1, n. 2).

[23] Newton's alchemical papers are listed in Appendix A and the collection will be discussed below. The estimated number of words was made at the time of their sale in 1936 and was published in the unpaginated "Foreword" to the sale catalogue cited below at note 31.

10

Robert Boyle (1627–91), who were also interested in alchemy – from common alchemy, "a process . . . commencing in fraud and terminating in mysticism . . . ."

The alchemy of Boyle, Newton, and Locke cannot be thus characterized. The ambition neither of wealth nor of praise prompted their studies, and we may safely say that a love of truth alone, a desire to make new discoveries in chemistry, and a wish to test the extraordinary pretensions of their predecessors and their contemporaries were the only motives by which they were actuated.[24]

A neat rationalization of the problem, one might think, and a sensible one which Brewster might have accepted wholeheartedly. But those alchemical manuscripts he had seen were bothering him; they marred his idealized picture of Newton terribly, and he immediately burst out in complete exasperation.

In so far as Newton's inquiries were limited to the transmutation and multiplication of metals, and even to the discovery of the universal tincture, we may find some apology for his researches; but we cannot understand how a mind of such power, and so nobly occupied with the abstractions of geometry, and the study of the material world, could stoop to be even the copyist of the most contemptible alchemical poetry, and the annotator of a work, the obvious production of a fool and a knave.[25]

Those alchemical papers were Brewster's real nemesis, and he had to recognize that his straightforward rational interpretation was not quite adequate, although he felt that it should have been.

Louis Trenchard More, on the other hand, even though he was in many respects more thorough than Brewster, does not seem to have examined those papers closely when he undertook the next major biography of Newton in the 1930's, and consequently he did not have quite the same problem of interpretation that Brewster had had. Neither does More seem to have studied the long series of chemical laboratory notes at which Brewster had looked. For although the notes are a part of the Portsmouth Collection, University Library, Cambridge, which collection More did use, the laboratory notes were for a time mislaid in the University Archives, from which they emerged to rejoin the Portsmouth papers only in the 1950's, and More probably relied on the summary description of them in the catalogue of the Portsmouth Collection. Thus More neglected the primary sources for a study of Newton's alchemy and was able in good conscience to suggest a severance of it from Newton's serious work in natural philosophy. He referred to Humphrey Newton's recollections as an "interesting description of Newton's method of relaxation when

[24] Brewster, *Memoirs*, ii, 374 (1, n. 2).
[25] *Ibid.*, ii, 374–75.

fatigued with the composition of the Principia"[26] and went on to characterize Newton's alchemy as mystical, accepting uncritically the influence of Jacob Boehme upon him.

There was a mystical strain in his [Newton's] character which has been quite overlooked. It showed itself not only in his persistent reading of the esoteric formulae of the alchemists, but also in his sympathy for the philosophy of the Cambridge Platonists and in his extended interpretations of the prophecies of the Books of Daniel and of the Revelation. Nor did his enquiry stop at these bounds, there is evidence that he studied the writings of Jacob Boehme and became, more or less, a follower of the mystical shoemaker.[27]

Then More compounded his mis-interpretation by suggesting that Newton would really have had no objection at all to making gold and that that was more his motive for studying alchemy than was disinterested scientific research.

There is a vast deal of nonsense talked about science for science [sic] sake, just as there is about art for art's sake. Newton, Boyle, and Locke desired to apply their knowledge to obtain practical results, and would have welcomed power and wealth as men of science do today. Francis Bacon esteemed science only for its fruits. And it is a commentary on human blindness for anyone in this age, absorbed as it is in the pursuit of mechanical power and with its ideals confused with material welfare, complacently to apologise for Newton's practical tastes lest his scientific reputation be smirched.[28]

With More, Humphrey Newton's evidence showing Isaac Newton as the Great Experimenter has gone completely out of focus. More did not at all mind accepting Newton's alchemical interests as real, but he did not feel much obligation to integrate them with the rest of Newton's science. For More, Newton's alchemy might have been recreational, mystical, or practical, but it could safely be passed over lightly when one came to discuss his "real" science.

Not long after More wrote, however, an event occurred which made it impossible simply to dismiss Newton's alchemy in a summary fashion ever again, for in 1936 Newton's alchemical manuscripts were sold at auction and many of them became available for serious study. The manuscripts had led a very quiet existence up to that time. When Thomas Pellet examined Newton's papers after his death in 1727, he marked the alchemical papers as "not fit to be printed" and they were put back in their boxes. Nor were they included in the Opera omnia of Newton prepared by Samuel Horsley in the eighteenth century,[29] and except for the mention of some of them

[26] More, Newton, p. 158 (1, n. 2).   [27] Ibid., pp. 158–59.   [28] Ibid., p. 161.
[29] George J. Gray, A Bibliography of the Works of Sir Isaac Newton, Together with a List of Books illustrating his Works, with Notes (2nd edn., Cambridge: Bowes and Bowes, 1907),

by Brewster, noted above, they remained largely unknown, passing to descendants of Catherine Barton Conduitt, Newton's niece. In 1936 the family decided to sell the alchemical, theological, and other papers still in its possession, the scientific papers having already been given to the University Library, Cambridge, where they comprise the Portsmouth Collection.[30]

The 1936 sale was handled by Sotheby and Company of London. For the sale Mr. John Taylor of Sotheby's prepared a descriptive catalogue, still the best published survey of the papers and for some of the manuscripts now the only available description.[31] The manuscripts were dispersed at the sale, but John Maynard Keynes, regretting the loss of such a national treasure, undertook to re-collect as many as he could. Eventually he was able to acquire between one-third and one-half of them, and these he left at his death to King's College, Cambridge, where they now comprise the Keynes Collection.[32]

Before his death Lord Keynes himself examined all the alchemical manuscripts he had acquired, some fifty-seven of the original one hundred and twenty-one lots on alchemy in the sale, and wrote a paper on them for the Royal Society's Newton Tercentenary Celebrations of 1946. In his paper Keynes reached some startling conclusions.

> Newton was not the first of the age of reason. He was the last of the magicians, the last of the Babylonians and Sumerians, the last great mind which looked out on the visible and intellectual world with the same eyes as those who began to build our intellectual inheritance rather less than 10,000 years ago.[33]

pp. 1–2; Isaac Newton, *Isaaci Newtoni opera quae exstant omnia. Commentariis illustrabat Samuel Horsley, LL.D., R.S.S., Reverendo admodum in Christo patri Roberto episcopo Londinensi a sacris* (5 vols., Londini, Excudebat Joannes Nichols, 1779–85).

[30] *A Catalogue of the Portsmouth Collection of Books and Papers written by or belonging to Sir Isaac Newton, the scientific portion of which has been presented by the Earl of Portsmouth to the University of Cambridge. Drawn up by the Syndicate appointed the 6th November, 1872* (Cambridge: At the University Press, 1888). The alchemical manuscripts were returned to the family as being non-scientific. A full scholarly history of the vicissitudes undergone by Newton's papers has recently been offered by Derek T. Whiteside in Isaac Newton, *The Mathematical Papers of Isaac Newton*, ed. by Derek T. Whiteside with the assistance in publication of M. A. Hoskin (6 vols., cont.; Cambridge: At the University Press, 1967–), "General Introduction," I, xv–xxxvi, hereinafter referred to as Newton, *Mathematical Papers*.

[31] *Catalogue of the Newton Papers sold by order of The Viscount Lymington to whom they have descended from Catherine Conduitt, Viscountess Lymington, Great-niece of Sir Isaac Newton* (London: Sotheby and Co., 1936). Peter Spargo is currently preparing a new annotated edition of this catalogue, showing as far as possible the present locations of the papers.

[32] The Sotheby lot numbers of the manuscripts obtained by Lord Keynes are given in A. N. L. Munby, "The Keynes Collection of the works of Sir Isaac Newton at King's College, Cambridge," *Notes and Records of the Royal Society of London* 10 (1952), 40–50, hereinafter referred to as Munby, "Keynes Collection."

[33] John Maynard Keynes, 1st Baron Keynes, "Newton, the Man," in *The Royal Society Newton Tercentenary Celebrations 15–19 July 1946* (Cambridge: At the University Press, 1947), pp. 27–34, quotation from p. 27, hereinafter referred to as Keynes, "Newton, the Man." The larger work is hereinafter referred to as *Royal Society NTC*.

Now Keynes was the first scholar since Brewster to attempt to come to grips with Newton's alchemical manuscripts, and his statements created considerable interest. He elaborated his views in the following manner.

Why do I call him a magician? Because he looked on the whole universe and all that is in it *as a riddle*, as a secret which could be read by applying pure thought to certain evidence, certain mystic clues which God had laid about the world to allow a sort of philosopher's treasure hunt to the esoteric brotherhood. He believed that these clues were to be found partly in the evidence of the heavens and in the constitution of elements (and that is what gives the false suggestion of his being an experimental natural philosopher), but also partly in certain papers and traditions handed down by the brethren in an unbroken chain back to the original cryptic revelation in Babylonia. He regarded the universe as a cryptogram set by the Almighty....[34]

There is much to recommend Keynes' view, for, as will be seen in subsequent chapters, Newton did have a strong belief in the *prisca sapientia*, an original wisdom or knowledge in the ancients which had been mostly lost to mankind. That belief was a part of Newton's Renaissance heritage and was still a part of the seventeenth-century's intellectual furniture. In a sense, perhaps Newton did believe that the secrets known to the ancients could be decoded "by applying pure thought," and he did establish his own technique for rational determination of their "significations." But Newton did not stop with that technique or even rely on it completely, for he also endeavored to validate conclusions with external, objective events. In the case of alchemy, that meant experimental verification. The experimental work was for Newton a crucial and absolutely essential step, and the experimental work was what Keynes did not take into account in his dramatic pronouncement. Completely missing from Keynes' new perspective was that clear description by Humphrey Newton of Sir Isaac's meticulous experiments.

After the Keynes Collection became available for study, three of the manuscripts were published in whole or in part, with explanatory articles attached: Keynes MS 38 by Taylor,[35] Keynes MS 26 by Geoghegan,[36] and Keynes MS 27 by Churchill.[37] The writers of these articles made a start towards integrating Newton's alchemy with the rest of his work.

[34] Keynes, "Newton, the Man," p. 29 (1, n. 33).

[35] Keynes MS 38 (part), Sotheby Lot 58, in Frank Sherwood Taylor, "An alchemical work of Sir Isaac Newton, " *Ambix* **5** (1956), 59–84.

[36] Keynes MS 26, Sotheby Lot 45, in D. Geoghegan, "Some indications of Newton's attitude towards alchemy," *Ambix* **6** (1957), 102–06.

[37] Keynes MS 27, Sotheby Lot 30, in Mary S. Churchill, " *The Seven Chapters*, with explanatory notes," *Chymia* **12** (1967), 29–57.

Taylor found him to be "in the fullest sense an alchemist" but observed that he seemed to be seeking principles which could be realized in the laboratory.

Churchill went a step further, pointing out places in "The Seven Chapters" in which Newton reduced the ancient esoteric terminology to a more chemical form. But Churchill rather spoiled her effect by insisting, without adequate evidence, that Newton was also interested in alchemy because of its partial congruence with his heretical religious beliefs.

It does seem to be quite true, as Churchill also suggests, that Newton found the alchemists to be exemplars of *prisci theologi* because of the exhortations to morality and piety which appear so frequently in the older alchemy. Newton thought "piety and righteousness" were the "fundamental and immutable" part of religion, as he made explicit in a theological manuscript entitled "A Short Scheme of the True Religion."

> Religion is partly fundamental and immutable, partly circumstantial and mutable. The first was the religion of Adam, Enoch, Noah, Abraham, Moses, Christ and all the saints, and consists of two parts, our duty towards God and our duty towards man, or piety and righteousness....[38]

And Newton in old age told Conduitt that "They who search after the Philosopher's Stone by their own rules [are] obliged to a strict & religious life."[39] But Churchill extrapolated her argument much past the place where it could be supported by similarly pious attitudes on the part of Newton and the alchemists, saying that Newton's anti-Trinitarianism made him view the old adepts as fellow spirits liable to prosecutions for heresy as he was, and indeed that he saw them as early protestants against the Romanism he disliked. To suggest then as she does that Newton read alchemy *for the purpose* of drawing out its religious content and thereby obtaining insight into ancient religion or into primitive Christianity is to miss the point of Newton's alchemy and of Newton's religion. The alchemical manuscripts do not deal with religious matters. It is true that Newton copied the old moral exhortations into his texts whenever he was making full transcripts, but when he was extracting or making brief notes such injunctions were generally omitted. On the other hand, in religious studies Newton was much more conventional than it is easy for a product of the twentieth-century's non-literalism and non-religion to understand. The sources Newton turned to for the study of religion were the Bible and the Church Fathers, nothing so out-of-the-way as alchemy: McLachlan has observed that basically Newton accepted the authority of Scripture

---

[38] Isaac Newton, "A Short Scheme of the True Religion," in Newton, *Theological MSS*, pp. 48–53, quotation from p. 48 (1, n. 12).

[39] Keynes MS 130, King's College, Cambridge, a Conduitt notebook, quoted in Manuel, *Portrait*, p. 173 (1, n. 2).

although he believed it should be interpreted by reason and never by Roman Catholic tradition.[40]

About the same time that the alchemical manuscripts were first being studied, a series of papers began to appear which were based largely on an entirely different set of materials, on the few chemical items published by Newton and on the chemical experiments in the Portsmouth Collection.[41] In general the writers of those papers took the attitude that Newton did have a consistent chemical philosophy, one which involved a corpuscularian theory of matter coupled with a concept of attractive forces acting over short distances.

The only one of that series of articles which dealt with Newton's chemical experiments in detail was the pioneering study of Boas and Hall. Those authors reviewed much of the earlier literature and concluded with previous writers that Newton was basically a scientific chemist rather than an alchemical one, in spite of the fact that he mentioned the "Oak," the "Green Lion," and the "Star" in his laboratory notes. They saw in most of Newton's experiments an attempt to get at the roots of the material structure of matter by destroying the fixity of the metals, by "opening" them, to use the seventeenth-century term, and noted that Newton

> seems to have felt that the profoundly esoteric terminology of alchemy was only an extension of the superficially esoteric language of ordinary chemistry; and similarly that the more recondite experimentation of the former was but an extension of the better-known and clearer experimentation of the latter.[42]

So far, so good, although that position fails to take fully into account the very real belief which Newton held, that the ancients knew *all* the

---

[40] Herbert McLachlan, "Introduction," in Newton, *Theological MSS*, pp. 17–19 (1, n. 12).

[41] (1) Lyman C. Newall, "Newton's Work in Alchemy and Chemistry," in *Sir Isaac Newton, 1727–1927. A Bicentenary Evaluation of His Work. A Series of Papers Prepared under the Auspices of The History of Science Society in collaboration with The American Astronomical Society, The American Mathematical Society, The American Physical Society, The Mathematical Association of America and Various Other Organizations* (Baltimore: The Williams & Wilkins Co., 1928), pp. 203–55; (2) Douglas McKie, "Some notes on Newton's chemical philosophy written upon the occasion of the tercentenary of his birth," *Philosophical Magazine* **33** (1942), 847–70; (3) S. I. Vavilov, "Newton and the Atomic Theory," in *Royal Society NTC* (1, n. 33), pp. 43–55, hereinafter referred to as Vavilov, "Newton and the Atomic Theory"; (4) R. J. Forbes, "Was Newton an alchemist?" *Chymia* **2** (1949), 27–36; (5) Thomas S. Kuhn, "Newton's '31st Query' and the degradation of gold," *Isis* **42** (1951), 296–98; (6) Marie Boas (Hall), "Newton and the theory of chemical solution," *Isis* **43** (1952), 123; (7) Thomas S. Kuhn, "Reply to Marie Boas (Hall)," *Isis* **43** (1952), 123–24; (8) Thomas S. Kuhn, "The independence of density and pore-size in Newton's theory of matter," *Isis* **43** (1952), 364–65, hereinafter referred to as Kuhn, "Density and pore-size"; (9) Joshua C. Gregory, "The Newtonian hierarchic system of particles," *Archives internationales d'histoire des sciences* **7** (1954), 243–47, hereinafter referred to as Gregory, "Hierarchic system"; (10) Marie Boas (Hall) and A. Rupert Hall, "Newton's chemical experiments," *Archives internationales d'histoire des sciences* **11** (1958), 113–52, hereinafter referred to as Boas and Hall, "Experiments."

[42] Boas and Hall, "Experiments," p. 151 (1, n. 41).

secrets at one time. More unfortunate were some of the following statements of Boas and Hall.

> Newton was not in any admissible sense of the word an alchemist; there is no evidence that any of his processes are of the kind necessarily preliminary to the Great Work....
>
> Newton's experiments read like those of a rational, experimental scientist at a time when alchemy could not be discounted – when, one might say, transmutability was a far less vulnerable concept than the chemical immutability of the modern element. . . . [S]ome dent on the smiling, passive resistance of iron, copper, lead and tin to all assaults upon their integrity, seems to have been Newton's object.[43]

Boas and Hall had readily admitted their inability to determine the rationale for Newton's experiments: as they pointed out, the experimental notes are empirical records and the theoretical dimension is almost totally missing. Furthermore, many of the symbols Newton used seem to have been of his own devising, and to attempt to determine what sort of book on chemistry Newton might have produced on the basis of these notes would be, they said, somewhat like attempting to reproduce the finished *Opticks* from his bare optical experiments. However, as succeeding chapters will demonstrate, the task is not entirely hopeless, because of the extensive alchemical manuscripts that parallel the experimental notes chronologically.

A comparison between the experimental notes and the alchemical manuscripts was not undertaken by Boas and Hall. On the contrary, they chose to ignore that possibility. Even though Lord Keynes had suggested that there appeared to be a few manuscripts in his collection in which Newton presented his own ideas on alchemy,[44] Boas and Hall made the sweeping statement, apparently on the basis of the two manuscripts published by Taylor and Geoghegan, that

> Newton never wrote an alchemical treatise; nor did he ever declare his opinion of alchemy, except to comment on a few well-publicized alchemical processes.[45]

Starting from that position, they of course were unable to find the rationale for any of Newton's experiments, for alchemical theories are far from obvious to modern historians or modern chemists. Boas and Hall, who are among the most eminent modern elucidators of seventeenth-century corpuscularianism, nevertheless recognized that within the framework of the seventeenth-century mechanical philosophies transmutation was considered quite possible. For transmutation, all that was thought to be required was a mechanical rearrangement of the minute particles

[43] *Ibid.*, pp. 151–52.

[44] Private communications of Keynes, quoted in Munby, "Keynes Collection," pp. 42–43 (1, n. 32).

[45] Boas and Hall, "Experiments," p. 114 (1, n. 41).

of one catholic and universal matter, as will be considered in more detail in other chapters. Boas and Hall seem to have assumed that something of that nature was in Newton's mind and that he was trying in a rather straightforward way to rearrange the minute particles of the metals with which he worked. It is the position of the present study, however, that Newton's techniques for attempting transmutation were not so straightforward as all that. Rather they really were the techniques of alchemy, and no matter how carefully Newton may have thought about the changes in terms of corpuscular rearrangements he nevertheless expected them to take place in processes traditionally associated with the Great Work of alchemy. It will be a substantial part of the present study to show that the rationales behind many of Newton's early chemical experiments were quite explicitly alchemical.

Frank Manuel, Newton's next major biographer, reacting against the excessive rationalism of the Boas and Hall position, managed in many respects to present a more balanced view of Newton's alchemy. Drawing upon several of the Keynes manuscripts which had not previously been studied and upon some of Newton's alchemical books, Manuel went on to strengthen his *Portrait* with considerable new material which evoked something of the large interest in alchemy during Newton's lifetime. Manuel called attention to Boyle's alchemical publications, to the fact that the King's physician Edmund Dickinson wrote an alchemical treatise, to the groups of "adventurers" in London who were attempting transmutation, to the extraordinary general upsurge in alchemical publication and translation, to the alchemical content of the correspondence between Newton and the brilliant young continental mathematician Fatio de Dullier, and many other items of great interest.[46] Manuel wrote as an intellectual historian, however, not as an historian of science, and showed not the slightest interest in the content of Newton's alchemy, i.e., in the experimental material utilized by Boas and Hall. While Manuel should perhaps not be criticized for that – after all it was never his intention to attempt a technical study of any of Newton's science – the resulting portrait does lack something of the solid dependability of Newton the Great Experimenter. Manuel also is a little too ready to believe that Newton perhaps found psychological relief for his fantasies of aggression in the roasting of his metals, and he also suggested that Newton might have been excessively interested in an elixir of life because of his hypochondria. The evidence for either of those notions seems to be essentially lacking: one can only wonder what has happened in Manuel's hands to Humphrey Newton's old employer, the man of "sedate and even temper," and to the majestic brow and calm good health of so many other portraits of Sir Isaac Newton.

[46] Manuel, *Portrait*, pp. 160–205 (1, n. 2).

Following Manuel's lead in ignoring the experimental aspect of Newton's alchemy, a more recent commentator brings the problem back almost full-circle to the discredited position of William Law. Frances Yates, after carefully establishing the influence of John Dee, the sixteenth-century English alchemist, upon the mystical Rosicrucian movement of the seventeenth century, offers the wholly unwarranted suggestion that Newton was attracted to Rosicrucianism in his turn.

As a deeply religious man, like Dee, Newton was profoundly occupied by the search for One, for the One God, and for the divine Unity in nature. Newton's marvellous physical and mathematical explorations of nature had not entirely satisfied him. Perhaps he entertained, or half-entertained, a hope that the "Rosicrucian" alchemical way through nature might lead him even higher.[47]

Yates based her suggestion on the fact that Newton had read works by Michael Maier and Elias Ashmole, both of whom certainly were influenced by Rosicrucianism, and on the fact that Newton was also familiar with the Rosicrucian manifestos. That Newton studied those items is not to be denied for a moment, but Yates was apparently unaware that Newton also read virtually everything else alchemical that had ever been published and a good many things that had not. He had read Greek alchemists, Arabic alchemists, the alchemists of the medieval Latin West, of the Renaissance, and of his own period. He had read Aristotelian alchemists, medical alchemists, Neoplatonic alchemists, and mechanical alchemists. He had read the most mystical and the most pragmatic. The fact that he had read Rosicrucian writers simply does not speak to the question of what kind of alchemist he was. Fortunately, Yates prefaced her discussion of Newton's alchemy with the admission that she had not examined his unpublished papers, so perhaps readers of her stimulating book will be cautioned by that and her error will not have as long a life as that of William Law has had.

There have been two quite recent comments on Newton's alchemical interests, however, which coincide, each in its own way, rather more closely with the results of the present study, even though neither considers the details of the problem of Newton's alchemical experimentation. Both appear, appropriately, in a publication in honor of that historian of science who has done more than any other to emphasize the importance of the total intellectual environment in scientific advances, Walter Pagel.[48]

[47] Frances A. Yates, *The Rosicrucian Enlightenment* (London and Boston: Routledge & Kegan Paul, 1972), pp. 193–205, quotation from pp. 201–02, hereinafter referred to as Yates, *Enlightenment*.

[48] *Science, Medicine and Society in the Renaissance. Essays to honor Walter Pagel*, ed. by Allen G. Debus (2 vols.; New York: Science History Publications, 1972), hereinafter referred to as Debus, *SMS*.

The first of these important essays, by Rattansi, points up the difficulties of accepting the heavily rationalistic position of Boas and Hall and brings in new manuscript material to demonstrate that Newton used some alchemical beliefs quite matter-of-factly. Rattansi also draws attention to the concept of the "universal spirit" as it appears in the writings of Newton's contemporary, Elias Ashmole, and in works attributed to the fabulous Hermes Trismegistus, both of which writers Newton certainly read. The theme of the "universal spirit" will be one of the most important ones for the rest of this present work, as will another point emphasized by Rattansi, the doctrine of the *prisca sapientia* to which Newton adhered.[49]

The other recent essay which offers a fundamentally new insight into Newton's alchemical work, by Westfall, pinpoints some of the similarities between the ancient Hermetic tradition and the new mechanical philosophies in the seventeenth century. Far from being completely antithetical, as modern writers have tended to assume, those two approaches to the nature of the universe were capable of subtle interactions. Hermetic modes of thought could easily be expressed in terms of mechanical, particulate systems. And furthermore, according to Westfall, Newton actually incorporated the Hermetic concept of "active principles" into his mechanical philosophy in a most important way, modifying orthodox mechanism and allowing it to rise above itself to the more sophisticated level of modern science.[50] As will appear in subsequent chapters, the present study reinforces Westfall's view of Newton's mature science and provides a larger context for it.

Documents and techniques of the present study

As will already have become apparent, the primary sources for a study of Newton's alchemy are the alchemical manuscripts and the experimental laboratory notes. In addition there are several items in Newton's published works and scattered throughout his other unpublished papers which bear on the question of his alchemy and of his general chemical philosophy. No attempt will be made to discuss the last-named group here, but they will be introduced and examined in Chapter 6.

Newton's alchemical papers were dispersed in 1936, as noted above, but the Sotheby sale catalogue has preserved some description of them. For convenient reference, these alchemical materials are all listed in Appendix A. The list is based on the Sotheby sale catalogue and contains

[49] P. M. Rattansi, "Newton's Alchemical Studies," in Debus, *SMS*, II, 167–82 (1, n. 48), hereinafter referred to as Rattansi, "Newton's Alchemical Studies."

[50] Richard S. Westfall, "Newton and the Hermetic Tradition," in Debus, *SMS*, II, 183–98 (1, n. 48), hereinafter referred to as Westfall, "Newton and the Hermetic Tradition."

the Sotheby descriptions and lot numbers and the location of the manuscripts used in the present study. No other Newtonian alchemical manuscripts are known to be extant. This study is based on those in the Keynes
Collection, supplemented by several of the others, in all comprising
seventy-six of the original one hundred twenty-one lots in the alchemical
section of the Sotheby sale. Since the location of twenty-eight of the original
number is unknown at the present time, and since three lots are in the
hands of a private collector, the seventy-six lots represent a total of eighty-
four percent of those available for study.

It has long been recognized that most of these Newtonian alchemical
manuscripts are transcripts of the texts of other authors, but that not all
of them are. Keynes thought there were four classifications into which his
collection might be divided: (1) the more or less complete exact transcripts of others' works, the largest group; (2) translations of others'
writings, probably done by Newton himself; (3) a group of summaries,
indexes, and comparisons; (4) a very small group of Newton's own contributions.[51] It is the last group, of course, which it is of most importance
to distinguish, as they are the ones which give some clues to his experimental program.

Keynes himself never had the opportunity to work through his manuscripts in detail and effect that suggested classification, and in fact no one
else has ever done it either in any systematic way, although some few individual manuscripts have been classified by other investigators. In consequence a part of the present study has been to classify the manuscripts
insofar as possible, and some of the results of that undertaking are presented in Chapter 5. An extensive bibliographical knowledge of alchemical
literature was required for this task, and various approaches have been
utilized, such as modern general and specialized bibliographies, seventeenth-century publishers' lists, and the catalogues of seventeenth-century
libraries. Many of these bibliographical sources are listed in Chapter 3,
where the vast outpouring of alchemical literature during Newton's period
is surveyed.

Fortunately, Newton was a remarkably conscientious note-taker and
transcriber. He followed his texts word for word when he had undertaken
to prepare full transcriptions of them, and, when he was making brief
notes or abstracts, he customarily adhered to his author's phraseology to
a great extent. Moreover, he usually included either title or author, frequently both, and ordinarily gave chapter or page references and sometimes even noted the exact line on the page. Since most of the alchemical
works in question appeared during the course of the century in several
editions and translations, these Newtonian habits have made the task of
locating his exact sources a great deal easier.

[51] Munby, "Keynes Collection," pp. 42–43 (1, n. 32).

21

Keynes had thought that, in the main, Newton made use of a group of writers published by the London bookseller William Cooper in the period 1668–88, but that has not proved to be the case. Newton's range was much wider than that. Probably he did not read German, however, and his French seems to have been rather sketchy. So the large alchemical literature in German was never explored by him and his excursions into the French literature were very limited. The latter perhaps occurred mostly during his period of friendship with Fatio de Dullier when Fatio, an accomplished linguist, assisted him with French readings.[52] Those limitations of language were not really substantial, however, as everything of importance was generally available in Latin or English, frequently in both, and through all the Latin and English materials – from late sixteenth-century to early eighteenth-century editions – Newton ranged freely.

Partly because of Newton's very extensive coverage of the literature and partly because of the fact that much of what he used had been through several editions before he began to study alchemy, the location of the exact edition of an alchemical work used by him is usually not very helpful in dating the manuscript in question. Only in cases when he used a work published by a contemporary during his lifetime is it sometimes possible to determine the earliest possible date for a manuscript by a study of Newton's sources. But location of the sources with precision does make it possible to decide which manuscripts may be included in that all-important fourth classification comprised of those containing Newton's own ideas on alchemy. Keynes seems to have held somewhat oversimplified views on the fourth group, as he did on the sources for the first group. For while it is true that only a few of the manuscripts are Newtonian from beginning to end, quite a few do contain Newtonian notes at the end or scattered through the texts, where brief ones were sometimes placed by Newton in square brackets. In some of the manuscripts Newton drew a line down the centers of his pages and placed his author's text on the left and his own commentaries on the right in a parallel column.

No item has yet appeared which is definitely assignable to Keynes' second group, that of translations probably by Newton. All of the manuscripts except for one in French are in Latin or English. Newton seems to have used Latin and English with equal ease, and, when the text from which he was working was in Latin, he simply used Latin for his transcript or notes, conversely when the original was in English. The language of the

---

[52] Cf. Fatio de Dullier to Isaac Newton, Feb. 24, 1689/90, in Isaac Newton, *The Correspondence of Isaac Newton*, ed. by H. W. Turnbull, J. F. Scott, A. R. Hall and Laura Tilling (5 vols., cont.; Cambridge: Published for the Royal Society at the University Press, 1959–), III, 390–91, hereinafter referred to as Newton, *Correspondence*. Fatio was soon to receive a copy of Huygens' *Traité de la lumière* and wanted to know whether to send it on to Newton in Cambridge, saying, "It beeing writ in French you may perhaps choose rather to read it here with me."

original continues even to be reflected in Newton's summary anthologies, where he frequently flips from English to Latin or from Latin to English as he changes from one author to another without a break in the text or any noticeable change in handwriting.

The third group suggested by Keynes, that of summaries, indexes, and comparisons of other authors, is fairly easily distinguishable but does not at the present writing seem to be of much value in the interpretation of Newton's alchemical thought. For example, an entry in an index, such as the alchemical term "Azoth," may be followed by numerous precise references to places where the term appears in the alchemical literature, but no definition of the term is given. Presumably Newton was attempting to synthesize a definition of it from its various usages, but was never able to do so. As for the manuscripts which consist of summaries and comparison, they seem usually to be short extracts, from texts of which Newton sometimes already had complete transcripts, thrown together in an attempt to elucidate some obscure point or process. It may be that future investigations will establish their meaning for Newton, but that effort lies beyond the scope of the present study.

In addition to the problem of classifying the manuscripts, there is the problem of dating them. Newton only dated one or two of them himself, and only a few contain any useful internal material from which tentative dates may be derived. Occasional external clues appear in his correspondence or notebooks, but the heaviest reliance in the present attempt at dating has had to fall on changes in Newton's handwriting.

Newton scholars have long recognized some major changes in Newton's flowing adult script over his long life and have used them to date various materials. Nothing systematic has been published on the techniques involved, but Professor R. S. Westfall kindly shared his personal experiences with Newton's manuscripts with the present writer.[53] In addition to Westfall's suggestions, the laboratory notes themselves have proved to be extremely useful in learning to recognize the variations. The laboratory notes, although interspersed in places with reading notes and out-of-place material, do form in general a chronological sequence. And while by no means all of them bear dates, enough of them were dated by Newton himself to provide an excellent framework, which framework is consistent with the little external evidence that is available. The most obvious variations in Newton's handwriting in the laboratory notes are detailed in Appendix D, where a tentative periodization for Newton's alchemical studies is also offered.

The laboratory notes are in one calf-bound notebook and on several loose sheets in University Library, Cambridge, Portsmouth Collection MS Add. 3975 and MS Add. 3973, respectively. Like the alchemical

---

[53] Private communication, April 16, 1971.

23

manuscripts, they were originally part of the papers that passed to Catherine Barton Conduitt's descendants and came into the family of the Earls of Portsmouth. In 1872, when the family decided to donate the purely scientific papers to the University of Cambridge, the laboratory notes were included with the scientific papers because of their experimental nature. Their alchemical character was not noted at the time and has been partially denied in the only published study of them.[54]

But with the alchemical manuscripts classified and dated at least approximately, it is possible for the first time to draw Newton's own alchemical ideas out of the mass of transcripts and to begin the comparison of them with the empirical laboratory records for each period. The process of comparison allows each to throw light on the other, and in Chapter 5 of this study the alchemical manuscripts and the laboratory notes for the early period of Newton's alchemy will be examined in detail.

[54] Boas and Hall, "Experiments" (1, n. 41).

2 Conceptual Background for
Seventeenth-Century Alchemy

Introduction

There was an enormous efflorescence of alchemical study during the seventeenth century, and large issues of a religious and philosophical nature inevitably arise when one considers general trends in alchemy in that period. No adequate history of the impact of alchemy on other aspects of European intellectual life exists, and it is extremely difficult to disentangle the tangled skein of occult studies during the seventeenth century and to determine their influences.

For many years only the elements of seventeenth-century thought which led into eighteenth-century rationalism were considered important, and the scientific revolution was treated as an abrupt break with occultism and as an escape from superstition and animism. In recent years the poverty of that point of view has become increasingly apparent. Allen Debus has been particularly influential in arguing that an enlarged understanding of "pseudo-science" is positively required for dealing with the seventeenth century in general and with its scientific revolution in particular.[1]

The task is so enormous, however, that it can only be undertaken piecemeal in the present stage of scholarship. With the caveat that it is necessarily inadequate both in scope and in depth, a preliminary sketch of the history of alchemy in the seventeenth century will be undertaken in Chapter 3.

That will be aimed, of course, at providing a more meaningful background for Newton's activities, and the present chapter is merely ancillary to that purpose. The complexity of the problems involved seem to make imperative some prior discussion of alchemical concepts and some attempt to provide a reasonably solid point of departure from which to embark upon the survey of seventeenth-century alchemy. So although some historical variations in alchemical thought and practice will be indicated here, the primary function of this chapter is to offer a schematic model of the older alchemy, one that will perhaps make all of alchemy more understandable to non-occultists, and to indicate very briefly some of the theoretical constructs that played an important role in seventeenth-century alchemical thought.

[1] See Allen G. Debus, "Alchemy and the historian of science," *History of Science* **6** (1967), 128–38, and the literature cited there.

25

There has always been something of an historical problem in the very existence of alchemy. Its evident appeal to generation after generation of adepts is inaccessible to the modern critical intellect, and most books on the subject do little to elucidate the grounds of the fascination it once exerted, even though its relations to innumerable scientific, technical, religious, and philosophical currents have been carefully explored.[2] In recent years, however, the insights of twentieth-century analytical psychology as applied to alchemy by C. G. Jung have come to provide a really promising approach to the problem, allowing as they do for an understanding of the many factors in alchemy which are not only obscure but patently irrational. Jung was not very historically minded, as will be discussed below, but his views offer a comprehensive and comprehensible model of alchemy as a field of human endeavor. Thus Jung's understanding of alchemy in general may usefully be explored first and historical variations examined against the background that his framework provides.

The nature of the older alchemy and its soteriological function: the Jungian model

Jung came to his study of alchemical literature through the realization that the dreams of his patients were often rich in the archaic symbolism of alchemy even though the patients were totally unacquainted with alchemical subject matter. This discovery led him into an exhaustive study

---

[2] Works on the origins and development of alchemy of the greatest importance include: (1) Marcellin Pierre Eugene Berthelot, *Les origines de l'alchemie* (Paris: Georges Steinheil, 1885), hereinafter referred to as Berthelot, *Les origines*; (2) Marcellin Pierre Eugene Berthelot, *Introduction à l'étude de la chimie des anciens et du moyen âge* (Paris: Georges Steinheil, 1889), hereinafter referred to as Berthelot, *Introduction*; (3) Mircea Eliade, *The Forge and the Crucible*, trans. by Stephen Corrin (Harper Torchbooks; New York and Evanston: Harper & Row, 1971), hereinafter referred to as Eliade, *Forge and Crucible*; (4) Eric John Holmyard, *Alchemy* (Harmondsworth: Penguin Books, 1968), hereinafter referred to as Holmyard, *Alchemy*; (5) Arthur John Hopkins, *Alchemy, Child of Greek Philosophy* (New York: Columbia University Press, 1934); (6) Hermann Franz Moritz Kopp, *Die Alchemie in Älterer und Neuerer Zeit. Ein Beitrag zur Kulturgeschichte* (2 vols.; Heidelberg: Carl Winter's Universitätsbuchhandlung, 1886), hereinafter referred to as Kopp, *Die Alchemie*; (7) Henry M. Leicester, *The Historical Background of Chemistry* (reprint of the 1956 edn.; New York: Dover Publications, 1971), hereinafter referred to as Leicester, *Historical Background*; (8) Jack Lindsay, *The Origins of Alchemy in Graeco-Roman Egypt* (New York: Barnes & Noble, 1970); (9) John Read, *Prelude to Chemistry: An Outline of Alchemy, its Literature and Relationships* (London: G. Bell and Sons, 1936; New York: The Macmillan Co., 1937), hereinafter referred to as Read, *Prelude to Chemistry*; (10) Herbert Stanley Redgrove, *Alchemy, Ancient and Modern. . .*(2nd and revised edn.; London: William Rider, 1922), hereinafter referred to as Redgrove, *Alchemy*; (11) Karl Christoph Schmieder, *Geschichte der Alchemie* (Halle: Verlag der Buchhandlung des Waisenhauses, 1832), hereinafter referred to as Schmieder, *Geschichte*; (12) John Maxson Stillman, *The Story of Alchemy and Early Chemistry* (reprint of the 1924 edn.; New York: Dover Publications, 1960), hereinafter referred to as Stillman, *Story of Alchemy*; (13) Frank Sherwood Taylor, *The Alchemists, Founders of Modern Chemistry* (New York: Henry Schuman, 1949), hereinafter referred to as Taylor, *The Alchemists*.

of the corpus of alchemical literature, both published and manuscript, which occupied him for many years, and the fruit of which was a long series of lectures, articles, and books, now readily available in his *Collected Works*.[3] These publications contain a wealth of detail on alchemical writings and writers which has as yet been little exploited by historians of science. However, their value was long ago recognized by that eminent exponent of non-scientific motives in the history of science, Walter Pagel,[4] and also by the historian of alchemy, Gerard Heym.[5]

Since Jung's views are not as widely known as they might be, what follows may serve as an introduction to them; it is in no way intended, however, as a brief for all the details of his psychology or for all of his interpretation of alchemy. The desire here is to emphasize a general viewpoint which seems especially valuable, namely: *the older alchemy never was, and never was intended to be, solely a study of matter for its own sake*. For the true adept alchemy was a way of life, a great work which absorbed all his mental and material resources, but it was never a rational branch of natural philosophy.

Until the seventeenth century, alchemy had always been composed of two inextricable parts: (1) a secret knowledge or understanding and (2) the labor at the furnace. According to Jung, these two sides of alchemy really were inextricable, for the secret knowledge about transformation was in reality an unconscious or semi-conscious understanding of certain psychological changes *internal* to the adept. Since he was unaware of their true nature, however, the alchemist projected the process of change upon matter, projected, that is, in the psychological sense, which meant that he "saw" the processes taking place *externally*. Now this sort of psychological

---

[3] Carl Gustav Jung, *The Collected Works of C. G. Jung*, ed. by Herbert Read, Michael Fordham, Gerhard Adler and William McGuire; trans. by R. F. C. Hull, Leopold Stein in collaboration with Diana Riviere, and H. G. Baynes (17 vols. in 18 cont.; Bollingen Series xx; New York: Pantheon Books; Princeton: Princeton University Press; London: Routledge & Kegan Paul, 1953–); hereinafter referred to as Jung, *Works*. The volumes most important for the study of Jung's views on alchemy are: vol. 9, pt. 1: *The Archetypes and the Collective Unconscious*; vol. 9, pt. 2. *Aion: Researches into the Phenomenology of the Self*; vol. 12, *Psychology and Alchemy*; vol. 13, *Alchemical Studies*; vol. 14, *Mysterium coniunctionis: An Inquiry into the Separation and Synthesis of Psychic Opposites in Alchemy*. Not in the *Collected Works*, but prepared as a companion work to Jung's vol. 14, is *Aurora consurgens: A Document Attributed to Thomas Aquinas on the Problem of Opposites in Alchemy*, ed., with commentary, by Marie-Louis von Franz; trans. by R. F. C. Hull and A. S. B. Glover (Bollingen Series LXXVII; New York: Pantheon Books, 1966), hereinafter rereferred to as von Franz, *Aurora consurgens*. A convenient introduction to Jung's psychology is Frieda Fordham, *An Introduction to Jung's Psychology* (A Pelican Book; Harmondsworth: Penguin Books, 1956).

[4] Walter Pagel, "Jung's views on alchemy," *Isis* **39** (1948), 44–48, which reviews Jung's *Paracelsica* and *Psychologie und Alchemie*, hereinafter referred to as Pagel, "Jung's views."

[5] Gerard Heym, reviews of (1) *Paracelsica*, *Ambix* **2** (1938–46), 196–98; (2) *Psychologie und Alchemie*, *Ambix* **3** (1948–49), 64–67; (3) *Mysterium coniunctionis*, *Ambix* **6** (1957–58), 47–51.

projection presupposes a certain lack of self-awareness and also an amorphous, ambiguous, external medium which the psyche can act upon and structure in its own image. A rational and too-detailed knowledge of matter precludes its use for this purpose as then matter has its own structure, so to speak, and cannot have one shaped for it by the alchemist. Nevertheless, the use of matter as the external medium for projection, as well as the acts actually performed in the laboratory, were absolutely essential for the functioning of alchemy. It was only that the knowledge of matter had to be kept vague. Thus the alchemist necessarily used obscure, symbolic, and irrational terminology for his materials: the "Greene Lyon,"[6] a special but undefined "earth,"[7] "'our' water,"[8] "Diana's doves,"[9] etc. Equally important was the lack of self-awareness. Too much

[6] *The Hunting of the Greene Lyon. Written by the Viccar of Malden* in *Theatrum Chemicum Britannicum, Containing Severall Poeticall Pieces of our Famous English Philosophers, who have written the Hermetique Mysteries in their owne Ancient Language. Faithfully Collected into one Volume, with Annotations thereon, by Elias Ashmole, Esq. The First Part* (London: Printed by *J.* Grismond for Nath. Brooke, at the Angel in *Cornhill,* 1652), p. 279:

> This *Lyon* maketh the *Sun* sith soone
> To be joyned to hys Sister the *Moone*:
> By way of wedding a wonderous thing,
> Thys *Lyon* should cause hem to begett a King:
> And tis as strange that thys Kings food,
> Can be nothing but thys *Lyons* Blood;
> And tis as true that thys ys none other,
> Than ys it the Kings Father and Mother.

Ashmole's single volume collection, the second part of which never appeared, is hereinafter referred to as Ashmole, *TCB.*

[7] *Pearce the Black Monke upon the Elixir, ibid.,* p. 269:

> Take Erth of Erth, Erths Moder,
> And Watur of Erth yt ys no oder,
> And Fier of Erth that beryth the pryse,
> But of that Erth louke thow be wyse,
> The trew *Elixer* yf thow wylt make,
> Erth owte of Erth looke that thow take....

[8] Michael Sendivogius, *A new Light of Alchymie, taken out of the fountaine of Nature and Manuall Experience. To which is added a Treatise of Sulphur; also Nine Books of the Nature of Things written by Paracelsus, viz., of the Generations, Growths, Conservations, Life, Death, Renewing, Transmutation, Separation, Signatures of Natural Things. Also a Chymical Dictionary explaining hard places and words met withall in the writings of Paracelsus and other obscure Authors. All which are faithfully translated out of the Latin into the English tongue by J. F. [John French]* M. D. (London: Printed by Richard Cotes, for Thomas Williams at the Bible in Little-Britain, 1650), p. 44: "O our Water! O our Mercury! O our Salt-nitre abiding in the sea of the world! O our Vegetable! O our Sulphur fixed, and volatill! O our *Caput Mortuum,* or dead head, or feces of our Sea! Our Water that wets not our hands, without which no mortall can live, and without which nothing grows, or is generated in the whole world! And these are the Epithites of *Hermes* his bird, which never is at rest." Hereinafter referred to as Sendivogius, *New Light.*

[9] Eirenaeus Philalethes, *Ripley Reviv'd: or an Exposition upon Sir George Ripley's Hermetico-Poetical Works. Containing the plainest and most excellent Discoveries of the most hidden Secrets of the Ancient Philosophers, that were ever yet Published. Written by Eirenaeus Philalethes an Englishman, stiling himself Citizen of the World* (London: Printed by *Tho. Ratcliff* and *Nat. Thompson,* for *William Cooper* at the *Pelican* in *Little-Britain,* 1678), pp. 24–25: ". . .our

insight into his own psyche would impair the alchemist's projection of his psychological processes onto matter. Thus the older alchemy comprised a delicate balance of ignorance, and an overemphasis on either its material or its psychological side would seriously impair its vitality.

It may be noted, leaving Jung aside for a moment, that alternative overemphases on first one side and then the other side of alchemy do seem to have occurred about the beginning of the seventeenth century. The result was an irreversible disintegration of the older alchemy. The overemphasis on the psychological side probably occurred first, as will be discussed in the next chapter. The reaction of more rationally minded intellects to the extreme mysticism of that movement then resulted in an overemphasis on the material aspects of alchemy. In the sequence the intertwined halves were split apart and the psychological side, loosened from its firm moorings in the laboratory, degenerated into theosophy and the so-called spiritual chemistry. Conversely, the laboratory side was released from the exigencies of a psychological progression, and, joining itself to the burgeoning scientific movement, it became a rational study of matter for its own sake.

The very fact of this split, however, and also the fact of the enormous upsurge of interest in alchemy during the seventeenth century, raise additional historical problems which seem to be related to the soteriological function of the older alchemy and to the religious qualities inherent in it. Before an attempt is made to grapple with these questions historically, it will be well to probe a little more deeply into Jung's views. Jung's scholarship is thorough, not to say overwhelming, and he has amply demonstrated the affinity of alchemical concepts with the major religions of the world as well as with his own psychological theories, so that even though one may not agree with every detail of his analyses, there hardly seems room to doubt the general correctness of his stance.

In addition to the sex drive and the power attributed to man by the other great names in analytical psychology, Freud and Adler, Jung postulated a drive towards wholeness and maturity inherent in the human psyche. He called the actualization of this drive the process of individuation and found it to occur both spontaneously and under clinical supervision, especially in persons of middle years and beyond.

Jung's theories on individuation are particularly applicable to so-called "normal" individuals of middle age who are generally reasonably well

Diana hath a wood, for in the first days of the Stone, our Body after it is whitened grows vegetably. In this wood are at the last found two Doves; for about the end of three weeks the Soul of the *Mercury* ascends with the Soul of the dissolved Gold; these are infolded in the everlasting Arms of *Venus*, for in this season the confections are all tincted with a pure green colour; These Doves are circulated seaven times, for in seaven is perfection...." This volume is hereinafter referred to as Philalethes, *Ripley Reviv'd*, and a fuller discussion of its contents will appear subsequently.

adapted to their social and physical environments and already have personalities which are successfully integrated around certain psychic functions. The public face of such a personality, those aspects which the individual shows to the world and for which he receives social approval, Jung calls the "persona," a name taken from the masks worn by the actors of antiquity. The "persona" includes those traits of character and behavior which the individual has emphasized and developed and become conscious of in the course of making his social adjustments – traits which help him hold a job, find a marriage partner, act as a parent, etc. The development of the "persona," however, necessarily left other potential characteristics repressed or underdeveloped, according to Jung, and these potentialities are all still present in the individual's unconscious. In general a repressed or underdeveloped trait will be the inverse of one which is flourishing openly in the "persona." For example, a person who presents a benign and gentle aspect to the world will yet have buried aggressive tendencies in his unconscious, whereas the converse might be true in another person.

For a personality to become more whole, more individuated, it must somehow accommodate both conscious and unconscious areas to each other and form a new center which Jung calls the "self." In the "self" many pairs of psychic opposites strike a new balance. For Jung it was precisely this process of individuation which was being unconsciously acted out by the practitioners of alchemy. The following details continue his argument.

Before the maturation process begins in the psyche, the conscious "persona" or mask-like part of the personality is able to accept only the "light" or socially desirable member of each pair of opposites as belonging to the personality, the "dark" or unwanted members having been repressed into the personal unconscious at an earlier stage of psychic development. Here they constitute what Jung calls the "shadow." The first step in individuation is facing the "shadow" and incorporating it together with the "persona" in the burgeoning new center. The first step in alchemy is similarly an evocation of a "blackness," followed by its transformation into something new.

Numerous symbolic expressions for the step occur in alchemy: ideas of putrefaction, death, and blackness; processes of reducing the matter to the state of *nigredo*. Since in a sense the "persona" must kill its own bright image of itself, must die and sink into the blackness of the "shadow" in order to be born again, these are peculiarly and dramatically appropriate symbols. The alchemists found one image in the words of Jesus: "Except a corn of wheat fall into the ground and die, it abideth alone: but if it die, it bringeth forth much fruit."[10] Sir George Ripley referred to those

---

[10] John 12: 24 (St. James).

words,[11] as did Bernard Trevisan.[12] Basilius Valentinus represented the process pictorially with grave and resurrection in his Eighth Key.[13] Michael Maier said simply, and with striking parallelism to Jung's terminology: "The sun and his shadow complete the work."[14] When it is recalled that the blackening, dying phase occurred early in the process of alchemy as it does in Jung's individuation, it seems entirely possible that Jung and the several alchemists were each describing a stage in the process of human development in his own way.

The "shadow," however, comprises only the personal unconscious. Even if it is successfully incorporated into the "self," the individual must still face the deeper realms of unconsciousness where the psychic functions,

[11] John Jacob Manget, *Bibliotheca chemica curiosa, seu Rerum ad alchemiam pertinentium thesaurus instructissimus: Quo non tantùm artis auriferae, Ac Scriptorum in ea Nobiliorum Historia traditur; Lapidis Veritas Argumentis & Experimentis innumeris, immò & Juris Consultorum Judiciis evincitur; Termini obscuriores explicantur; Cautiones contra Impostores, & Difficultates in Tinctura Universali conficienda occurrentes, declarantur: Verùm etiam tractatus omnes virorum Celebriorum qui in Magno sudarunt Elixyre, quìque ab ipso Hermete, ut dicitur, Trismegisto, ad nostra usque Tempora de Chrysopoea scripserunt, cum praecipuis suis Commentariis, concinno Ordine dispositi exhibentur. Ad quorum omnium Illustrationem additae sunt quamplurimae Figurae aeneae* (2 vols.; Genevae: Sumpt. Chouet, G. De Tournes, Cramer, Perachon, Ritter, & S. De Tournes, 1702), II, 280, hereinafter referred to as Manget, *BCC*. The passage occurs in the Fifth Gate, *De putrefactione*, of Sir George Ripley's *Liber duodecim portarum*: "Nam corpora omnia naturaliter alterari non possunt sine ipsa, ut Christus est testis, dicens: Nisi granum tritici moriatur in terra, augmentum nullum faciet, sic nisi putrefiat materia, nullo modo potest recte alterari, nec possunt elementa naturaliter dividi, nec eorum absoluta esse conjunctio."

[12] *The Answer of Bernardus Trevisanus, to the Epistle of Thomas of Bononia, Physician to King Charles the Eighth,* in *Aurifontina Chymica: or, A Collection of Fourteen small Treatises Concerning the First Matter of Philosophers, For the discovery of their (hitherto so much concealed) Mercury. Which many have studiously endeavoured to Hide, but these to make Manifest, for the benefit of Mankind in general,* trans. by John Frederick Houpreght (London: Printed for William Cooper, at the *Pelican* in *Little-Britain,* 1680), pp. 200–01: "Notwithstanding this dissolution makes several colours appear, because the *species* remain as it were dead, yet their intrinsical proportion is permanent and entire. So the Lord in the Gospel speaks by way of similitude of Vegetables. *Unless a grain of corn fallen on the earth do dye, it abides alone; but if it dye, it brings forth much fruit:* Therefore this alterative corruption hides forms, perfects natures, keeps proportions, and changes colours from the beginning to the end...."

[13] *Musaeum hermeticum reformatum et amplificatum,* introd. by Karl R. H. Frick, introduction translated by C. A. Burland (Graz: Akademische Druck-u. Verlagsanstalt, 1970), p. 409, hereinafter referred to as *MHRA*. This volume is a reprint of the edition of 1678: *Musaeum Hermeticum reformatum et amplificatum, omnes sophospagyricae artis discipulos fidelissimè erudiens, quo pacto Summan illa veraque lapidis philosophici Medicina, qua res omnes qualemcunque defectum patientes, instaurantur, inveniri & haberi queat. Continens tractatus chemicos XXI. Praestantissimos, quorum Nomina et Seriem versa pagella indicabit.* (Francofurti; Apud Hermannum à Sande, 1678). The tract by Basilius Valentinus referred to here is entitled *Practica cum duodecim clavibus et appendice, De magno lapide antiquorum sapientum.* The text for the Eighth Key begins: "Omnis caro sive humana, sive animalium sui augmentationem aut propagationem nullam producere potest, nisi id fiat principio per putrefactionem. . . ," pp. 409–10.

[14] Michael Maier, *Atalanta fugiens hoc est emblemata nova de secretis naturae chymica Authore Michaele Majero,* herausgegeben, mit Nachwort, von Lucas Heinrich Wüthrich (Faksimile-Druck der Oppenheimer Originalausgabe von 1618 mit 52 Stichen von Matthaeus Merien d. Ä.: Kassel und Basel: Im Bäsenreiter-Verlag: 1964), p. 189, Emblema XLV: "Sol & ejus umbra perficiunt opus."

31

occurring still in pairs, can only be known through archetypal images common to all mankind. Here Jung has done some of his most important work: in identifying the great mysterious symbols of alchemy as psychic images. These all-pervasive symbols occur throughout the literature of alchemy: Flamel's dragons – one winged and one wingless:[15] the pair of lions – one red and one green – on the Ripley "Scrowles"; [16] the king and queen of Basilius Valentinus' Sixth Key;[17] the dog and wolf struggling in Lambsprinck's Fifth Figure;[18] and so forth. Before Jung's work few people were even willing to hazard a guess as to their meaning and no one was ever able to put forth an explanation of them which became generally accepted. Clearly these picturesque symbols have nothing to do with chemical realities or with rational theories of transmutation. Viewed as poetic clothing for unconscious psychic processes, however, they become somewhat comprehensible. Furthermore, it becomes clear why the symbols varied so dramatically from one alchemical writer to the next: their form was determined in each case by individualized psychic contents.

If and when the psyche found itself able to fuse these archetypal opposites into its new "self," symbols of wholeness and unity appeared, such

[15] *Tractatus brevis, sive svmmarivm philosophicum, Nicolai Flamelli* in *MHRA*, pp. 172–73 (2, n. 13): "Dicta ista duo spermata veteres sapientes duos per Dracones vel serpentes praefigurarunt, quorum unus alas, alter autem nullas habuit. Draco sine alis sulphur est, numquam ab igne avolans. Alatus serpens argentum vivum est, quod à vento transportatur. . . ."

[16] Fitzwilliam Museum, Cambridge, Fitzwilliam MS 276*, *George Ripley, Canon of Bridlington, The Emblematical Scroll.* A number of these manuscript "scrowles" are known to have survived: see Rossell Hope Robbins, "Alchemical texts in Middle English verse: corrigenda and addenda," *Ambix* **13** (1966), 62–73. All of the "scrowles" are rolls, Fitzwilliam MS 276* being over twenty feet long and nearly three feet across, and consist of elaborate alchemical illustrations done in colored inks, with groups of couplets describing the drawings. In Fitzwilliam MS 276* the standing red and green lions lean on either side of a furnace, warming their paws at the fire (Plate 1). It was Robbins' opinion that these curious scrolls must at times have served as advertisements in alchemists' shops. The verses from one of them were published by Ashmole in 1652 (Ashmole, *TCB*, pp. 375–79) (2, n. 6); selected pictures from others appear in C. A. Burland, *The Arts of the Alchemists* (New York: The Macmillan Co., 1968), pp. 81–83 and facing p. 40. Burland's volume is hereinafter referred to as Burland, *Arts.*

[17] *Practica cum duodecim clavibus et appendice, De magno lapide antiquorum sapientum, Scripta & relicta à Basilio Valentino Germ. Benedictini Ordinis Monacho* in *MHRA*, p. 405 (2, n. 13): "Mas absque faemina pro dimidiato habetur corpore, & faemina absque viro similiter dimidii corporis vicem obtinet: nam singuli per se nullum fructum edere possunt. . . ."

[18] *Lambsprinck Nobilis Germani Philosophi Antiqvi Libellvs de lapide philosophico, E Germanico versu Latinè redditus, per Nicolaum Barnaudum Delphinatem Medicum, hujus scientiae studiosissimum* in *MHRA*, p. 351 (2, n. 13): "Lupus & Canis sunt in una domo / Postremo tamen ex his unum fit." The complicated, symbolic Lambsprinck engravings with their realistic landscape backgrounds are thought to have been executed by the famous alchemical engraver Matthaus Merien, or by one of his school. Merien also did the engravings for the books by Michael Maier and Robert Fludd: see *MHRA*, introd., pp. xx–xxi and xxxvii–xxxix (2, n. 13). The Lambsprinck pictures have been reproduced in Burland, *Arts*, pp. 145–52 (2, n. 16).

THE REDE LIONE ✦

THE GRENE LIONE

The movth of colerke be wware

THE MOVTH OF COLERKƐ

Here is the laſt of the Red, and the beginning to put a waye ỹ ded ỹ clere vine,

Plate 1. RED AND GREEN "LYONS" FROM A RIPLEY "SCROWLE".

According to the Jungian analysis of the older alchemy, this pair of lions is a symbolic representation of paired psychic functions. (Reproduced by permission of the Syndics of the Fitzwilliam Museum, Cambridge.)

as the ourobouros,[19] the mandala – a pictorial representation in a circular or other balanced form[20] – and, of course, the philosopher's stone.

Jung found that the encounter with deeply hidden psychic material had subjectively a strongly numinous quality for the adepts. Psychologically speaking, the effect of achieving a new level of psychic integration through the exercises of alchemy was roughly equivalent to a religious mystic's experience of direct union with God. Jung himself was very careful to point out that his researches were strictly empirical and spoke not at all to the question of the metaphysical reality, if any, towards which this sort of experience points, but he had no doubt that the true adept searched for and found some level of religious satisfaction in alchemy. The literature of alchemy seems to justify his views.

In order not to belabor the point at too great a length, a single example will be cited: a table of philosophical correspondencies compiled by B. a Portù Aquitanius whom Jung has tentatively identified as Bernhardus Georgius Penotus (c. 1525–1620).[21] One of the sets of correspondencies in this table directly related all the Christian sacraments to alchemical processes: extreme unction to putrefaction, ordination to distillation, contrition or repentance to calcination, marriage to coagulation, baptism to solution, confirmation to sublimation, and as might be expected the *Mysterium Altaris*, or the Mass, to transmutation.[22] Jung has devoted considerable space also to what he calls the *Christus-lapis* parallel and the interested reader may find ample entrance into the alchemical literature in his references.[23]

In summary, it may be said then that the older alchemy served large psychological and religious needs for the adepts, although this was not clearly understood by them, but it definitely did not attempt a rational elucidation of matter in its chemical relationships as that would have been inimicable to its other functions. Only one more point of a concomitant nature need be made before turning to historical developments. That is the rather specific point of the doctrine of relationships between matter and

---

[19] Lambsprinck's Sixth Figure, *MHRA*, p. 353 (2, n. 13), shows the dragon biting its own tail with the accompanying text: "Hoc verè est magnum miraculum & cita fraus / In venenoso Dracone summam medicinam inesse." The ourobouros is one of the oldest of alchemical symbols. One from ancient Greece, bearing the text, "One is all, and by it all, and to it all, and if one does not contain all, all is nought," has been reproduced in John Read, *Through Alchemy to Chemistry. A Procession of Ideas & Personalities* (Harper Torchbooks; New York and Evanston: Harper & Row, 1963), p. 25. See also Berthelot, *Les origines*, p. 62 and Planche 1, facing p. 64 (2, n. 2), and Berthelot, *Introduction*, p. 18 (2, n. 2).

[20] Jung, *Works*, vol. 9, pt. 1: *The Archetypes and the Collective Unconscious*, pp. 355–90, and vols. 9, pt. 2, 12, 13, and 14, *passim* (2, n. 3).

[21] Jung, *Works*, vol. 12: *Psychology and Alchemy*, p. 408 (2, n. 3).

[22] In the seventeenth century the Penotus table became associated with Ripley's *Twelve Gates* and was sometimes printed with it, as it is in Manget, *BCC*, II, 275 (2, n. 11).

[23] For religious ideas in alchemy, see esp. Jung *Works*, vol. 12: *Psychology and Alchemy*, pp. 225–472 (2, n. 3).

spirit, which was to have many ramifications in the seventeenth century.

One way in which the alchemists strove to express the reconciliation of the psychic opposites which they had projected upon matter was in the interconvertibility and final union of the elements. Earth, air, fire, and water were to be successively changed one into another in a wheeling movement or a great circulation. Another related form of expression, and one of greater historical interest, fixed upon the opposites of body and spirit for its symbolism. Body was to be changed into spirit and spirit into body until each joined with the other to form a new and more valuable entity.

Matter–spirit relationships

Some of the most ancient alchemical authorities who took the position that the elements could and should be interconverted are conveniently summarized in the text of *Aurora consurgens*, in the Fourth Parable of that document, which is entitled "Of the Philosophic Faith, Which Consisteth in the Number Three." The modern editor of *Aurora consurgens*, M.-L. von Franz, has argued that this alchemical treatise was probably really composed in part by St. Thomas Aquinas, who perhaps at an advanced age had a mystical experience which he could only express in the language of alchemy. If it was written by Aquinas, the text dates from the thirteenth century; certainly it is no later than the fifteenth. A few excerpts follow: it may be noted that passages from the Bible are included, indicating a spiritual understanding of the alchemical process on the part of the author.

> And Hermes: Thou shalt separate the dense from the subtle, the earth from the fire. Alphidius: Earth is liquefied and turned into water, water is liquefied and turned into air, air is liquefied and turned into fire, fire is liquefied and turned into glorified earth.... And Isaias: The Spirit of the Lord lifted me up.... And [in] the *Turba*: Make bodies incorporeal and the fixed volatile; but all these things are brought about and fulfilled by our spirit.... And to Naaman (the Syrian) was it said: Go and wash seven times in the Jordan and thou shalt be clean. For there is one baptism for the remission of sins, as the Creed and the Prophet bear witness...which the philosophers set forth in these words: Distil seven times and thou hast set it apart from the corrupting humidity.[24]

Consider also the following passage from the fifteenth-century Canon of Bridlington, Sir George Ripley, who was famous as an alchemist even in his lifetime.

> ...[W]e do sublimation for three reasons. The first is that the body may be made spiritual, the second is that the spirit may be made

[24] von Franz, *Aurora consurgens*, pp. 92–99 (2, n. 3).

corporeal, and be fixed with it [the body] and become consubstantial [with it]. The third reason is that it may be purged from original impurity and that its sulphureous saltness may be diminished in it, by which it is infected.[25]

Ripley and Aquinas were using "body" and "spirit" in these passages much the same way that other alchemists used their red and green lions or their winged and wingless dragons. "Body "and "spirit," however, are the bearers of fairly obvious physical characteristics, so that even though this pair of opposites is still suitably vague and so can perform its symbolic function in the older alchemy of which Ripley and Aquinas were a part, at the same time these particular opposites are more readily translated to chemical referents than some of the other symbols. Early rationalizers of alchemy found an entrée in the sort of terminology Ripley and Aquinas were using and came to emphasize changes between "body" and "spirit."

Furthermore, with the spread of Renaissance Neoplatonism from its late fifteenth-century center in Florence, ideas about the relationships and intermediate stages between matter and spirit received a new emphasis, for such relationships were fundamental to Neoplatonism. Indeed, the whole complex of materials promulgated by the Florentine Academy, which included the Hermetic Corpus and the Cabbala and which has been so brilliantly discussed by Yates,[26] served to emphasize matter–spirit relationships. Renaissance alchemy took on a new and rather distinctive appearance as a result of Neoplatonic ideas. The work of Paracelsus and, later, van Helmont further developed the same line of thought. Both of these authors went on to postulate intermediate entities which acted as physiological directing agents in matter–spirit conversions.[27] And as the present writer has noted, the doctrine of matter–spirit relationships received a peculiarly chemical expression at the hands of Nicolas le Fèvre, instructor in chemistry at the Jardin Royale in Paris in the 1650's.[28]

Also a Neoplatonic natural philosophy which also emphasized matter–spirit relationships came in the seventeenth century to offer a clean and

[25] Manget, *BCC*, II, 283 (2, n. 11). From the Eighth Gate, *De sublimatione*, of Sir George Ripley's *Liber duodecim portarum*: "Sublimationem autem facimus tribus de causis, prima est, ut corpus fiat spirituale, secunda ut spiritus fiat corporalis, & fixetur cum eo, & consubstantialis fiat; tertia causa est, ut ab originali immunditia purgetur, & ejus salsedo sulphurea diminuatur in eo, qua est infectum."

[26] Frances A. Yates, *Giordano Bruno and the Hermetic Tradition* (Chicago: The University of Chicago Press; London: Routledge & Kegan Paul; Toronto: The University of Toronto Press, 1964), hereinafter referred to as Yates, *Bruno*.

[27] Walter Pagel, *Paracelsus. An Introduction to Philosophical Medicine in the Era of the Renaissance* (Basel and New York: S. Karger, 1958), and Walter Pagel, "The religious and philosophical aspects of van Helmont's science and medicine," *Bulletin of the History of Medicine Supplements* 1–4 (1943–45), no. 2 (Baltimore: The Johns Hopkins Press, 1944).

[28] Betty Jo Dobbs, "Studies in the natural philosophy of Sir Kenelm Digby. Part I," *Ambix* 18 (1971), 1–25, esp. pp. 22–23, hereinafter referred to as Dobbs, "Digby. Part I."

well-developed general alternative to decaying Aristotelianism and rising mechanism. Although after the decline of the Florentine Academy, Neo-platonism was frequently associated with the barbarous mixture of Latin and German which Paracelsus employed or with doubtful alchemical or medical practices, those were not the only sources of it available to the European intellectual community at mid-seventeenth-century. Sometime early in the century Jean d'Espagnet (1564–1637), a French magistrate with Hermetic inclinations, retired to a private life for the express purpose of working towards a "restoration" of physics. The book he produced, *Enchiridion physicae restitutae*, was a literary masterpiece, and by 1651 it was available in Latin, French, and English. Not really alchemical at all, although frequent quotations from Hermes Trismegistus and "the Philos-ophers" are included, it is a rational statement of Neoplatonic physical theory. As d'Espagnet's majestic periods roll on through canon after canon, one is allowed a glimpse of the tranquility, harmony, and beauty found in the universe by the Neoplatonists. Some passages from d'Espagnet's work will demonstrate the contemporary strength of the Neoplatonic philo-sophical background upon which seventeenth-century natural philosophers might draw.

[B]efore the creation of the Universe he [God] was a book rowld up in himself giving light onely to himself; but, as it were, travailing with the birth of the world, he unfolded himself, and that work which lay hid in the womb of his own mind, was manifested by extending it to view, and so brought forth the *Idaeal-world*, as it were in the transcript of that divine Original, into an actual and material world....

[The world is framed] so that the extreams of the whole work by a secret bond, have a fast coherence between themselves through insensible *mediums*, and all things do freely combine in an obedience to their Supream Ruler....[29]

The "Universal Spirit," or the "Soul of the World," takes second place only to God and is centered in the sun.

[The second universal cause attending on the first is] the spirit of the Universe, or the quickening virtue of that light created in the

[29] [Jean d'Espagnet], *Enchyridion Physicae Restitutae; or, The Summary of Physicks Re-covered. Wherein the true Harmony of Nature is explained, and many Errours of the Ancient Philoso-phers, by Canons and certain Demonstrations, are clearly evidenced and evinced* (London: Printed for *W. Bentley*, and are to be sold by *W. Sheares* at the *Bible*, and *Robert Tutchein* at the *Phenix*, in the New-Rents in S. *Pauls* Church-Yard, 1651), pp. 2–3, hereinafter referred to as d'Espagnet, *Enchiridion*. The book had originally appeared in Latin and although the date of its first edition has not been definitely established, it was published at least by 1623 and had had more than one Latin edition before mid-century. The first French edition seems to have appeared the same year as the first English, i.e. 1651. A new French translation has now quite recently been published: *La philosophie naturelle restituée*, introd. and trans. by J. Lefebure Desagues (Paris: E. P. Denoël, 1972), not seen. On d'Espagnet's life, see *Dictionnaire de biographie française* Sous la direction de J. Balteau...(13 vols, cont.; Paris: Librairie Letouzey et Ané, 1933–   ), **12**, 1941.

beginning, and contracted into the body of the Sun, and endowed with a hidden foecundity.[30]

D'Espagnet continues his paean on the sun, characterizing its "middle Nature" and its work of mediation as the "Joyner" of the "Extreams" of the universe.

It was not an improbable assertion of some of the Philosophers, *That the soul of the World was in the Sun, and the Sun in the Centre of the whole.* For the consideration of equity and nature seem to require, that the body of the Sun should have an equal distance from the fountain and rise of created Light, to wit, the *Empyrean Heaven,* and from the dark Centre the Earth, which are the extreams of the whole Fabrick, whereby this lamp of the world, as a middle Nature and Joyner of both Extreams, might have its scite in the middle, that it may the more commodiously receive the rich treasuries of all powers from the chief Spring, and upon a like distance convey them to things below.[31]

Air is the medium through which the powers are conveyed; it is an air far removed conceptually from an atomistic atmosphere, much vaster in extent and richer in function.

The whole Air is the Heaven, the floor of the World, Natures sieve, through which the virtues and influences of other bodies are transmitted: a middle nature in that it knits all the scattered natures of the Universe together...impatient of vacuity...the easiest receiver of almost all qualities and effects, yet the constant retainer of none: a borderer upon the spiritual nature....[32]

And "through the universal Regions of the Air," d'Espagnet asserted, the Spirit of the Universe "doth extend it self perpetually . . . pouring out all gifts for generation and life, through all the bodies of the Universe."[33]

The Neoplatonists understood that certain earthly substances with specific "Magneticck" virtues were required to draw the "Universal Spirit" to earth and cause it to differentiate into all the manifold forms of matter which its universality encompassed. The comments of the Neoplatonic chemist Nicolas le Fèvre, who performed chemical demonstrations at the Jardin Royale in Paris for about ten years during the 1650's, elucidate that supposed process in some detail.

And as this Spirit is universal, so can he not be specificated but by the means of particular Ferments, which do print in it the Character and Idea of mixt bodies.... That when we say, that this Spirit is specificated in such or such a Matrix, that we understand nothing else, but that this Spirit is imbodied in such or such a Compound, according

---

[30] d'Espagnet, *Enchyridion,* p. 5 (2, n. 29).   [31] *Ibid.,* pp. 19–20.
[32] *Ibid.,* pp. 51–52.   [33] *Ibid.,* p. 135.

to the different Idea it hath received, by the means of particular ferment. . . .[34]

Thus the "Magnet" which draws to itself a portion of the "Universal Spirit" is also the matrix in which the "Universal Spirit" assumes a "specificated" form. The specific form of the matter so produced from spirit is determined by the "Idea" carried by the matrix.

One further selection from d'Espagnet will point up the existence of a sort of intermediate "Magnet" thought to be continuously present in the air. In d'Espagnet's opinion the Sun and the "rest of the superior Natures [the stars and planets], taking care of the inferiour [the earth],"

> do instil by continual breathings enlivening spirits, as so many trilling rivulets from their most clear and pure Fountains: But the Vapours [the water vapor of the air] being thin, and so swimming in the Air, or else bound up into a Cloud, do most eagerly suck in that spiritual *Nectar* [from the heavenly bodies], and attract it to them by a Magnetick virtue, and having received it, they grow big, and being impregnated and quickened with that ingendering seed, as being delivered of their burden, do freely fall down back into the lap of the Earth in some Dew, hoar Frost, Rain, or some other nature. . . .[35]

Thus the water which came to earth was the bearer of the heavenly virtues of the "Universal Spirit," and certain crucial astronomical configurations or times of the zodiac, such as the equinoxes, were considered to be more efficacious than others in allowing for a plentiful supply of "spiritual *Nectar*" to be absorbed by the water.[36]

It has seemed well to emphasize here the importance of the type of thinking which regarded matter and spirit as interconvertible and the prevalence of that sort of thinking not only in alchemy but also in medicine, chemistry, and natural philosophy, for it was to become the focal point later for many arguments. Descartes argued in favor of a split between matter and spirit, but his ideas were not fully accepted in all circles, as will be seen in Chapters 4, 5, and 6. Descartes' writings on matter–spirit relationships, indeed, were subjected to extensive criticism by some of the people who most influenced Newton, and Newton himself became thoroughly engaged with the problem.

[34] Nicolas le Fèvre, *A Compleat Body of Chymistry: Wherein is contained whatsoever is necessary for the attaining to the Curious Knowledge of this Art; Comprehending in General the whole Practice thereof; and Teaching the most exact Preparation of Animals, Vegetables and Minerals, so as to preserve their Essential Vertues. Laid open in two books, and Dedicated to the Use of all Apothecaries, &c. By Nicasius le Febure, Royal Professor in Chymistry to His Majesty of England, and Apothecary in Ordinary to His Honourable Household. Fellow of the Royal Society. Rendred into English, by P. D. C. Esq; one of the Gentlemen of His Majesties Privy-Chamber. Corrected and amended; with the Additions of the late French Copy* (2 vols. in 1; London; Printed for *O. Pulleyn* Junior, and are to be sold by *John Wright* at the Sign of the *Globe* in *Little-Brittain*, 1670), 1, 16, hereinafter referred to as le Fèvre, *Chymistry*.

[35] d'Espagnet, *Enchyridion*, pp. 99–100 (2, n. 29).

[36] le Fèvre, *Chymistry*, 1, 107–08 (2, n. 34).

Variations in the older alchemy

One of the main difficulties in accepting Jung's psychological analysis of alchemy in its entirety lies in his basically a-historical approach to the problem.[37] Jung in fact drew his alchemical materials from all periods without regard to the historical relationships which might be involved. On a fairly typical page, chosen at random for an example, Jung relies on definitions from Pliny and Dioscorides, from Martin Ruland of the early seventeenth century, and from the nineteenth-century historian of alchemy, M. Berthelot. References to the seventeenth-century writings of Mylius and Sendivogius are jumbled with comments from the Arabic pesudo-Aristotle of the twelfth or early thirteenth century, with a passage from the sixteenth-century mystic Henry Khunrath, with mention of the *Aurora consurgens* which was possibly written in part by St. Thomas Aquinas, and with a quotation from the tenth-century Arabic *Turba philosophorum*.[38]

There is perhaps something to be said for treating alchemy as an historical constant in that manner since a great deal of what was written in philosophical alchemy did simply repeat and abstract earlier writers. Nevertheless, there are detectable historical variations which ought to be explored and which Jung's approach obscures. To leave the discussion with Jung's a-historical analysis, in which everything is dealt with as psychology and symbolism, is to fail to do justice to alchemy as a precursor to science, as Pagel has noted.[39]

For the purposes of a discussion of alchemy which is to be focused on seventeenth-century Europe, the most important historical variation to be noted here is that between medieval and Renaissance alchemy. In the medieval period in western Europe there was much less emphasis on the psychological and religious factors which Jung has treated and more on practical chemical processes and rational theories. The medieval West seems first to have absorbed and used a few technical collections of recipes in which theories on the origins and transmutations of metals are not to be found. Then the great period of translations from the Arabic provided the West with the more mystical alchemy of Egypt, Greece, and the Arabic world in the eleventh and twelfth centuries, as well as with fresh technical knowledge. But when the encyclopedists of the twelfth and

[37] Much of the following section has resulted from the stimulating criticisms of the Third Medieval Science Colloquium, Raleigh–Durham–Chapel Hill, N.C., May 18–20, 1973, especially the comments of John Murdoch, Edward Grant, Francis Nichols, Brian Stock, and John Riddle.

[38] Jung, *Works*, vol. 9, pt. 2, *Aion. Researches into the Phenomenology of the Self*, p. 156 (2, n. 3).

[39] Pagel, "Jung's views," p. 48 (2, n. 4).

thirteenth centuries approached their task of organizing and digesting the new materials, they seem to have been tempered with a great deal of Aristotelian rationalism and to have omitted most of the mystical elements from the received Arabic, Greek, and Egyptian alchemy. They presented ideas on transmutation, certainly, but usually treated it as a naturally occurring and lawful physical phenomenon.[40]

An example of the sort of treatise frequently produced then during the High Middle Ages may be drawn from the pseudo-Jabir, usually known as Geber, who wrote about 1310.[41] Geber, who is thought to have been a Spanish alchemist familiar with practical metallurgy, offered a rational theory of the composition of metals based on Aristotelianism and on the Arabic theory of sulfur–mercury composition. He also provided many descriptions of substances, processes, and furnaces, which, although perhaps not immediately transparent to modern readers, are certainly not mystical. Here is the beginning of his discussion " *Of the Nature of* Sol, *or* Gold."

> Now of *Bodies*, We will more amply declare the intimate *Nature* of them. And first of *Sol*, but afterward of *Luna*, and then of all the other, according as shall be thought expedient, with their *Probations*, which are acquired by *Experiment.*
>
> *Sol* is created of the most subtile *Substance* of *Argentvive*, and of most clear *Fixture*; and of a small *Substance* of *Sulphur* clean, and of pure *Redness*, fixed, clear, and changed from its own *Nature,* tinging that. And because there happens a *Diversity* in the *Colours* of that *Sulphur*, the *Citrinity* (or *Yellowness*) of *Gold* must needs have a like *Diversity*. For some is more intense, other less in *Yellowness*. That *Gold* is of the most subtile *Substance* of *Argentvive* is most evident, because *Argentvive* easily retains it. For *Argentvive* retains not any *Thing*, that is not of its own *Nature*. And that it hath the clear and clean *Substance* of that, is manifest by its splendid and radiant *Brightness*, manifesting it self not only in the *Day*, but also in the *Night*. And that it hath a fixed *Substance* void of all burning *Sulphureity*, is evident by every *Operation* of it in *Fire*: for it is neither diminished nor inflamed.[42]

Geber's work in its naturalistic approach to the nature of metals stands in strong contrast to the mystical literature usually cited by Jung, and the fact that there were such rational alchemical treatises written during the Middle Ages should be kept in mind as a corrective to his psychological

[40] Leicester, *Historical Background*, pp. 74–83 (2, n. 2); Stillman, *Story of Alchemy*, pp. 184–272 (2, n. 2).

[41] Leicester, *Historical Background*, pp. 85–86 (2, n. 2); Stillman, *Story of Alchemy*, pp. 277–86 (2, n. 2).

[42] Geber, *The Works of Geber Englished by Richard Russell, 1678*, ed. with introd. by E. J. Holmyard (a new edition; London and Toronto: J. M. Dent & Sons; New York: E. P. Dutton & Co., 1928), pp. 129–30.

theories. It appears that Jung may have been rather selective in his choice of alchemists for intensive study. Nevertheless, the sort of alchemy with which Jung dealt did exist in all periods and in the fifteenth, sixteenth, and seventeenth centuries it grew in volume. How that came about may be suggested by a discussion of the more historical psychological approach of Erich Fromm.

Fromm, an analytical psychologist whose work is more recent than most of Jung's, has examined some of the social, political, and economic developments in western Europe since the Middle Ages in terms of the insights gained from depth psychology. Although his work does not deal with alchemy *per se*, some of his ideas seem to offer a possible explanation of the increased interest in mystical alchemy in the Renaissance.

Fromm suggests, in brief, that the breakdown of the highly structured social groups of the Middle Ages left the individual with more freedom to develop his own potentialities. Fromm uses the term "individuation" somewhat in the same way Jung does, although without many of Jung's technical details, and argues that it was only when more individuals emerged from their close ties to the natural and social world of the medieval period that large-scale individuation became possible. Presumably the process might have occurred in rare individuals in any age and place and perhaps occurred more in some cultures other than the medieval one – Fromm's discussion only encompasses the transition from the Middle Ages to the Renaissance in western Europe. But for that time and place Fromm argues, along the lines of Jacob Burckhardt's famous thesis on *The Civilization of the Renaissance in Italy*, that in the medieval period man was only conscious of himself as a member of some general category such as race, people, party, family, or corporation. Only with the progressive destruction of the medieval social structure did it become possible for men to conceive of themselves as individuals free to develop as far as their own potentialities might allow – be it economically, socially, politically, or spiritually.[43]

It seems reasonable to assume then, on the basis of Fromm's analysis, that perhaps individuation was actually taking place more and more frequently in the period after the fourteenth century. If that was true, it would explain the rising interest in a more spiritual variety of alchemy, in which a process of psychological growth, imperfectly understood by the individual involved, might be projected upon the unresisting materials in the furnace.

In addition, by the sixteenth century, religious tensions probably served to increase interest in the natural soteriological functions of alchemy, further enlarging its attraction. But for an exploration of that aspect of

[43] Erich Fromm, *Escape from Freedom* (New York: Holt, Rinehart and Winston, 1941), esp. pp. 24–63.

alchemy's historical development, it is necessary to await the turn to more particular events in Chapter 3. Here first may be considered briefly those relationships between alchemy and the new mechanical philosophies of the seventeenth century which seem to some extent to be a continuation of the more rational and non-mystical medieval alchemical tradition.

Alchemy and mechanical philosophies

Writers on the development of rational chemistry generally point to the seventeenth century as a time of great diversity and uneven quality in chemical literature. Many technical chemical treatises were circulating, chemical medicines were receiving much attention, and of course the alchemists were still being published. These disparate facets of chemical studies might be united in some persons, but there was no single body of knowledge or doctrine that might be called chemistry. Although there was one series of works published which on the face of it looks to be a cumulative scientific tradition, close analysis reveals little theoretical agreement in it. The series was that of the textbooks issued by holders of the chair of chemistry at the Jardin Royale in Paris: Davidson (1635), de Clave (1646), Arnaud (1656), Barlet (1657), Lefèvre, le Fèvre, or le Febure (1660), Glaser (1663), Thibaut (1667), de Tressel (1671), Lémèry (1675). Metzger, in her analyses of these works, had continually to point out their individualistic character and noted that each author wrote as if he were completely isolated from the others.[44] Following the terminology of Kuhn,[45] the period is usually considered "pre-paradigmatic" for chemical studies.

Yet as a matter of fact an alchemical paradigm was still widely accepted. Cast in its most general chemical form, leaving aside its more mystical potentials for a moment, it could be reduced to a belief in metallic transmutation, and rested usually on Aristotelian or Arabic theory derived from the rational alchemical treatises of the Middle Ages. There were not very many who were ready totally to deny the possibility of transmutation at the beginning of the century, however variously they might express their ideas about it. Metals grew in the bowels of the earth until they reached a state of maturity in gold, just as plants grew on the surface of the earth until they reached their final form. A sort of organic growth and maturation in the great womb of Mother Earth was a belief of great antiquity,[46] a belief hardly questioned as yet. In the early seventeenth

[44] Hélène Metzger, *Les doctrines chimiques en France du début du XVIIe à la fin du XVIIIe siècle* (Paris: Les Presses Universitaires de France, 1923), pp. 17–97, hereinafter referred to as Metzger, *Les doctrines*.

[45] Thomas S. Kuhn, *The Structure of Scientific Revolutions* (Phoenix Books; Chicago and London: The University of Chicago Press, 1964), pp. 10–22, hereinafter referred to as Kuhn, *Revolutions*.

[46] Eliade, *Forge and Crucible* (2, n. 2).

century, arguments did not yet turn very much on the question of whether metals in fact grew in the earth; rather men were more likely to argue whether art might assist nature in her movement towards perfection and whether those alchemists who claimed to have done so already were honest men, fools, or imposters.

The advent of mechanical philosophies in the seventeenth century is usually supposed to have sounded the death knell of alchemy. Following Metzger's pioneer work in the area,[47] that thesis has been rather generally accepted. However, the single example of Boerhaave is sufficient to cast grave doubt on the proposition.

Herman Boerhaave (1664–1734) is usually considered to be the first great rational chemist, imbued with the Newtonian philosophy, a thorough-going experimentalist and careful empiricist. Indeed he was all of those things, but he still believed in transmutation also. In his lectures on chemistry he set out the principles of alchemy as he had synthesized them from the works of the alchemists and from Boyle and Homberg.

All metals must first be mercury e'er they be gold....

Metals...appear transmutable into one another....

The imperfect metals consist of impure mercury, and imperfect sulphur....

If this matter can either be changed, or be perfectly purg'd away, any one metal were convertible into any other; the most impure, into the purest.[48]

Boerhaave presented his ideas on alchemy at the end of a long section on metals in which massive amounts of empirical data are given for each metal. His alchemical "corollaries" are drawn, he thinks, from the metals' "several apparent properties," all in good Newtonian form. Boerhaave furthermore stated in his title that he was laying down the art of chemistry on "Mechanical Principles" and his corpuscularianism is everywhere apparent. For example, he explains the gain in specific gravity when mercury is "fixed" into gold by the addition of the heavier particles of fire or light. When particles of fire or light insinuate themselves into the pores between the particles of the mercury, then the weight is increased although the bulk is not; thus the specific gravity is increased.[49]

Undoubtedly mechanical philosophies did alter traditional approaches

[47] Metzger, *Les doctrines*, esp. pp. 229–77 and 421–68 (2, n. 44).

[48] Herman Boerhaave, *A New Method of Chemistry: including the Theory and Practice of that art: Laid down on Mechanical Principles, and accomodated to the Uses of Life. The whole making a Clear and Rational System of Chemical Philosophy. To which is prefix'd A Critical History of Chemistry and Chemists, From the Origin of the Art to the present Time.* trans. by P. Shaw and E. Chambers (London: Printed for J. Osborn and T. Longman at the *Ship* in Paternoster-Row, 1727), pp. 99–104, quotations from pp. 102–03, hereinafter referred to as Boerhaave, *Chemistry*.

[49] *Ibid.*, p. 101. Boerhaave accounts for the increased weight when lead is calcined in the same way.

to chemistry a great deal, but they simply cannot be said to have killed alchemy. Some of the best of the new mechanical philosophers adhered to theories of transmutation, re-thinking alchemical ideas in mechanical terms.

That that should have been the case may appear somewhat paradoxical. At first glance any mechanical philosophy appears to be diametrically opposed to the organic conceptions of alchemy, such as the putrefaction or maturation and perfection of matter. Organic transformations or alchemical ones might easily be treated in terms of mechanical rearrangements of parts, however; Sir Kenelm Digby (1603–65), one of the earliest of the mechanical philosophers, provided an example in his mechanical treatment of putrefaction in his treatise *Of Bodies* in 1644.[50] Boerhaave provides another when he argues for the possibility of transmutation being effected by the philosopher's stone.

> Now 'tis certain there is gold in every mass of lead: had we then a body which would so agitate all the parts of lead, as to burn every thing out but mercury and gold; and had we some fixing sulphur to coagulate the matter remaining: would not it be gold?...The philosopher's stone is held a fix'd subtle, concentrated fire [which does these things]....[51]

That Boyle's views on alchemical change were somewhat similar has been suggested by Ihde[52] and More,[53] and some of Boyle's particular ideas on transmutation will be explored in more detail in Chapter 6.

Seventeenth-century ideas about alchemical change based on the mechanical rearrangements of parts need an additional word for their understanding, as modern views of chemical change are quite dissimilar. Modern chemistry is predicated upon distinct chemical elements, each with its own unique type of particle which remains of constant mass at the energy levels involved in chemical (i.e., non-nuclear) reactions. It was otherwise in the seventeenth century. No one was even clear that mass was a fundamental property of matter until the time of Newton;[54] even then it was not immediately apparent that the ultimate chemical particles might *differ* in mass. Newton himself seems to have held that the smallest parts were inertially homogeneous.[55]

[50] Dobbs, "Digby. Part 1", 24–25 (2, n. 28).

[51] Boerhaave, *Chemistry*, p. 217 (2, n. 48).

[52] Aaron J. Ihde, "Alchemy in reverse: Robert Boyle on the degradation of gold," *Chymia* 9 (1964), 47–57, hereinafter referred to as Ihde, "Alchemy in reverse."

[53] Louis Trenchard More, "Boyle as alchemist," *Journal of the History of Ideas* 2 (1941), 61–76, hereinafter referred to as More, "Boyle as alchemist."

[54] Hélène Metzger, *Newton, Stahl, Boerhaave et la doctrine chimique* (Paris: Librairie Félix Alcan, 1930), pp. 20–33.

[55] Arnold Thackray, *Atoms and Powers. An Essay on Newtonian Matter-Theory and the Development of Chemistry* (Harvard Monographs in the History of Science; Cambridge, Mass.: Harvard University Press, 1970), pp. 12–24, hereinafter referred to as Thackray, *Atoms and Powers*.

In the mechanical philosophies of the seventeenth century in fact the common assumption was that all the ultimate particles were of the same type of matter. John Harris, summing up the principles attributed to the mechanical philosophers in his *Lexicon Technicum* in 1704, made the following observation.

> They suppose, that there is but one Catholick or Universal Matter, which is an extended, impenetrable, and divisible Substance, common to all Bodies, and capable of all Forms. . . .

> They suppose also, that these differently siz'd and shap'd Particles may have as different *Orders* and *Positions, Situations* or *Postures*: From whence great variety may arise in the composition of Bodies.[56]

Harris was following an exposition originally presented by Boyle, but few corpuscularians of the period would have disagreed with this formulation about the ultimate particles of matter.[57] The primitive particles might differ in magnitude and figure, as did the letters of the alphabet; larger units, like words, were formed by the combinations of the primitive particles in different orders, groups, and positions. The alphabet analogy was quite commonly drawn upon to explain chemical changes. Yet however the particles might differ in size, shape, and arrangement, they were all made from the same basic substance.

The ultimate substance which formed the letters of the alphabet of matter being one and universal, it is apparent that mechanical philosophies did nothing to strike at the root of alchemy. Indeed, if anything, they encouraged the concept of transmutability and gave it a new rationale by emphasizing that new substances were formed by the rearrangements of the minute particles of one catholic matter.

Perhaps then instead of the prevailing tendency towards mechanism it was rather the accumulation of negative evidence that finally caused a significant decline in the popularity of the concept of transmutability. Indeed there is evidence, some of which is offered in Chapter 3, that the process of the rejection of the claims of the old alchemists may have taken place in two stages.

[56] John Harris, *Lexicon Technicum or an Universal English Dictionary of Arts and Sciences* (2 vols.; Facsimile reprints of the London editions of 1704 [vol. I] and 1710 [vol. II]; The Sources of Science, no. 28; New York and London: Johnson Reprint Corp., 1966), I, article "Corpuscular Philosophy," the larger work hereinafter referred to as Harris, *Lexicon Technicum*.

[57] One notable exception to this was John Woodward, who came much closer to a Daltonian conception. See *ibid.*, I, article "Matter." Woodward, better known as a pioneer geologist, presented his ideas in *An Essay toward a Natural History of the Earth: and Terrestrial Bodies, Especially Minerals: As also of the Sea, Rivers, and Springs. With an Account of the Universal Deluge: And of the Effects that it had upon the Earth* (London: Printed for *Ric. Wilkin* at the *Kings-Head* in St. *Paul's* Church-yard, 1695), pp. 229–30. For the history of that work, see Melvin E. Jahn, "A bibliographic history of John Woodward's *An Essay toward a natural history of the Earth*," *The Journal of the Society for the Bibliography of Natural History* 6 (1972), 181–213.

The first of those stages may be called the clarification and chemicalization of alchemical thought and practice. This step seems to have served to provide a body of literature which offered examples of transmutations in rational terminology, examples which were in fact presented operationally. The second stage, which apparently lasted on into the eighteenth century, occurred when such operations came to be attempted again and new generations of experimentalists found the earlier reports of transmutations to be inaccurate.

The first stage of the process of rejection probably took place in the generation immediately preceding the time of Newton's alchemical studies, i.e., in the 1640's, 1650's and 1660's. One of the prime representatives of the movement towards the clarification of alchemy was that *virtuoso* and "compleat Gentleman," Sir Kenelm Digby.[58] Due to the peculiar and romantic circumstances of his life, Digby spent a great deal of his time in travel and apparently collected alchemical recipes and processes wherever he went. He was as thoroughly convinced of the "fact" of metallic transmutation as he was of the general correctness of Aristotle and of the possibility of sympathetic medical treatments, but in his work he rationalized the language of alchemy and simplified the processes. His own favorite recipe was fully operational; for the most part those he collected from others were also. Many of them may even be translated into twentieth-century terminology, and some representative ones chosen from his book of *Secrets* have been presented with their seventeenth- and twentieth-century descriptions in parallel by the present writer.[59]

Since the work of Digby and his contemporaries established the immediate historical context for Newton's alchemy, it must be examined in some detail, and it is now time to turn to some of the specific developments in alchemical studies in the seventeenth century, both mystical and practical.

[58] For biographical references and an analytical treatment of Digby's methodology, his theory of matter, and his general approach to natural philosophy, see Dobbs, "Digby. Part I" (?, n. ?8). For Digby's relations with the alchemical circles of his day, see Betty Jo Dobbs, "Studies in the natural philosophy of Sir Kenelm Digby. Part II. Digby and alchemy," *Ambix* **20** (1973), 143–63, hereinafter referred to as Dobbs, "Digby. Part II."

[59] Betty Jo Dobbs, "Studies in the natural philosophy of Sir Kenelm Digby. Part III. Digby's experimental alchemy – The Book of *Secrets*," *Ambix* **21** (1974), 1–28, hereinafter referred to as Dobbs, "Digby. Part III."

3 Seventeenth-Century Alchemy: a Few of its Internal Developments and their Relations to Religion, Philosophy, and Natural Philosophy

Alchemy and dogmatic religion

With the concept of the sotoriological function of alchemy in hand, it becomes less of an historical problem to consider the enormous interest in alchemy in the late sixteenth and early seventeenth centuries. Of course interest in alchemy was stimulated through the publications of the Florentine Neoplatonists. Alchemy also must certainly have received encouragement through the spreading use of chemical medicines by Paracelsian physicians,[1] Paracelsian medicine being itself a movement grounded in Renaissance Neoplatonism. But if Jung is right, the interest in alchemy may also have been a reflection of the religious needs of the age.

The movements of the Reformation and Counter-Reformation swept through the different countries of Europe during the period in question, and it would seem that those more general religious surges may have been related in a negative way to the great flowering of alchemy. Religious doctrines had generated hot disputes in the religious wars first in the Germanies and then in France in the sixteenth century, as they were to do in the whole Empire and then in England in the seventeenth century. Men were growing weary of such arguments. As Paracelsus had once characterized Luther and the pope as two whores discussing chastity,[2] so more and more of Europe's intelligentsia turned aside from theological disputation, finding it irrelevant and socially divisive.

Alchemy, with its built-in potential for the satisfaction of religious needs, may have seemed most attractive at first, but all branches of natural philosophy and mathematics were soon to benefit. Kepler, for example, at a crucial point in his career, found that his conscience would not tolerate the rigidities of Lutheran orthodoxy and that he could not pursue his original intention of becoming a Lutheran theologian. He then turned to his fruitful association with Tycho Brahe and satisfied his religious yearnings by showing forth the glories of God in the harmonies of the heavens.[3]

[1] Allen G. Debus, *The English Paracelsians* (New York: Franklin Watts, 1966).

[2] Quoted in Henry M. Pachter, *Paracelsus: Magic into Science* (New York: Henry Schuman, 1951), p. 267.

[3] Max Caspar, *Kepler*, trans. and ed. by C. Doris Hellman (London and New York: Abelard-Schuman, 1959), pp. 48–52, 98–99, 374–76.

English naturalists and natural philosophers seem to have found a similar religious solace in demonstrating God's wisdom and providence in the intricate forms of plants and animals and in the whole frame of creation; their work was part of a movement towards a low-keyed, non-dogmatic Christianity which reached its climax with men such as Boyle, Ray, and Newton.[4]

The upsurge in alchemical publication

But in the first instance the movement away from dogmatic theology seems to have led many to alchemical studies, and during the second half of the sixteenth century publishers began to issue collections of medieval and Renaissance alchemy. These works had never ceased to circulate in manuscript, of course,[5] but they were now made readily available to a wide public for the first time, and the steady increase in the volume of alchemical publication presumably reflects a considerable interest on the part of the public.

The extent and variety of alchemical publication from about 1560 until the end of the seventeenth century has been noted in a general way by several writers, but a more detailed survey will serve to emphasize the importance of alchemy for seventeenth-century thought. In the present study two of the great alchemical libraries gathered by men of the seventeenth century have been utilized for the purpose of making such a survey: Newton's own library and that of the colonial American John Winthrop, Jr., and his descendants.

The Winthrop library was scattered in the early part of the nineteenth century when Francis Bayard Winthrop gave selections from it to various institutions.[6] However, R. S. Wilkinson has recently traced many of the volumes and has published a catalogue of those on alchemy, chemistry, and medicine.[7] His catalogue lists 275 items, which is a larger number of alchemical-chemical-medical works than has survived in the Newton Collection.

A list of Newton's books was compiled after his death in 1727 by John

[4] Richard S. Westfall, *Science and Religion in Seventeenth-Century England* (Yale Historical Publications, David Horne, ed., Miscellany 67; New Haven: Yale University Press, 1958).

[5] Lynn Thorndike, *History of Magic and Experimental Science* (8 vols.; New York: Columbia University Press, 1941–58), v, 532–49 discusses alchemy in the first half of the sixteenth century, and v, 679–95 presents an alchemical bibliography of 1572. Hereinafter referred to as Thorndike, *Magic*.

[6] Ronald Sterne Wilkinson, "The alchemical library of John Winthrop, Jr. (1606–1676) and his descendants in colonial America. Parts i–iii," *Ambix* 11 (1963), 33–51.

[7] Ronald Sterne Wilkinson, "The alchemical library of John Winthrop, Jr. (1606–1676) and his descendants in colonial America. Part iv. The catalogue of books," *Ambix* 13 (1966), 139–86.

Huggins, which list is now in the British Museum, but which is unfortunately incomplete, containing for example an entry for "Three dozen of small chymical books" but not specifying them. Huggins himself bought Newton's library almost immediately and it seems later to have passed largely intact to the Rev. Dr. James Musgrave, after which it dropped almost completely out of sight until the present century. In 1920 part of it was sold at auction and dispersed, but a substantial portion of it was rediscovered elsewhere, purchased by the Pilgrim Trust, and presented as a gift to Trinity College, Cambridge. There it is now housed in honor in a separate bay in that portion of the library of which Sir Christopher Wren was the architect, along with a bust of Newton by Roubilliac, Newton's death mask, and his walking stick. Information about its history and re-discovery and about a number of items of interest which are known to have been scattered along the way has been presented by de Villamil,[8] who has also published the Musgrave Catalogue supplemented by the Huggins' list, by Zeitlinger,[9] by Feisenberger,[10] and by Spargo.[11] At the present writing, J. R. Harrison is engaged in preparing what will be the definitive work on the subject, as far as possible reconstructing the content of the entire library from manuscript sources and giving the present location of surviving volumes.

The Newton Collection now at Trinity contains a total of 109 items of an alchemical, chemical, or medical nature.[12] By far the largest number of these books are alchemical, as are most of those in the Winthrop Library Catalogue. Of the 109 items only eight are duplicates of the larger number in the Winthrop Catalogue, as given by Wilkinson. There is great overlap of alchemical authors acquired by Newton and by the Winthrops in many instances nevertheless. For example, Newton had five of the twenty-five works known to have been published by Michael Maier; the Winthrops had four other ones. In several cases Newton and the Winthrops had different editions of the same work. For example, Newton had the 1673 Geneva edition of Nathan Albineus' *Bibliotheca chemica contracta*; the Winthrops had one from the same city issued in 1652.

Much of the difference in the two collections arises from the fact that

---

[8] Richard de Villamil, *Newton: the Man* (London: G. D. Knox, n. d. [1931]).

[9] *Library of Sir Isaac Newton. Presentation by the Pilgrim Trust to Trinity College, Cambridge, 30 October 1943. Address of Presentation by the Right Hon. Lord Macmillan, G.C.V.O., Chairman of the Pilgrim Trust, and of Acceptance by George Macauley Trevelyan, O.M., Master of Trinity College* (Cambridge: Printed at the University Press, 1944), "Appendix" by H. Zeitlinger, "Newton's Library and its Discovery."

[10] H. A. Feisenberger, "The libraries of Newton, Hooke, and Boyle," *Notes and Records of the Royal Society of London* **21** (1966), 42–55.

[11] P. E. Spargo, "Newton's Library," *Endeavour* **31** (1972), 29–33.

[12] Newton's alchemical-chemical-medical books have never had a separate detailed publication devoted to them. The above statement is based on the present writer's study of the collection itself.

Newton lived for fifty years after John Winthrop, Jr., died. Wait Winthrop, in the next generation, inherited something of his father's interest in alchemy but he was never much of a book collector. Consequently, only a few alchemical titles were added to the Winthrop Collection after John, Jr., died in 1676, whereas Newton continued to add to his on into the eighteenth century. Taken together, however, the two libraries give a vivid reflection of the fascination alchemical studies exerted throughout the seventeenth century. While it is not intended to suggest here that either Newton or the Winthrops pursued alchemy for religious reasons, their books did cover the whole range of alchemical and chemical literature available, from sixteenth-century editions of Paracelsus[13] and Raymund Lull[14] to Frederick Slare's 1715 *Experiments and Observations upon Oriental and Other Bezoar-Stones, Which Prove them to be of no Use in Physick.*[15]

A number of fairly broad statements about sixteenth- and seventeenth-century alchemical literature may be ventured, largely on the basis of the Winthrop and Newton libaries. By 1678 most of the important literature of alchemy was available in print. Much of it was in one of several large collected editions. The *Ars chemica* (Strassburg, 1566) and the *Artis auriferae* (Basle, 1572, 1593, and 1610) had set the trend. The *Theatrum chemicum*, originally published in Ursel in three volumes in 1602, had grown to its final massive form of six volumes (Strassburg, 1659–61), and the *Musaeum Hermeticum*, which had been limited to nine major treatises in 1625, had been "reformed and amplified" to include twenty-one (Frankfurt, 1678). Only one other really major collection of alchemical materials was assembled before Newton's death but later than 1678: the *Bibliotheca chemica curiosa* edited by J. J. Manget (Geneva, 1702).

Avicenna, Morienus, Albertus Magnus, Thomas Aquinas, Roger Bacon, Raymund Lull, and the pseudo-Arnold of Villanova were among the most popular of the older authors published and re-published in the late sixteenth and early seventeenth centuries, ranking with the fabled Hermes Trismegistus. The anonymous *Turba philosophorum* was printed often, frequently with one or more commentaries. Nicolas Flamel, Isaac of Holland, Bernard Trevisan, and Denis Zachaire were fourteenth-, fifteenth-, and sixteenth-century figures considered by the compilers to have considerable merit. Sir George Ripley, who died about 1490, received a great deal of attention: an English edition in 1591,[16] a separate Latin

---

[13] Trinity College NQ. 9. 170 (1, 2, 3) and Winthrop Catalogue 198–203.
[14] Trinity College NQ. 16. 37 and NQ. 16. 133 and Winthrop Catalogue 170–171.
[15] Trinity College NQ. 9. 60.
[16] George Ripley, *The Compound of Alchymy. Or, the ancient hidden Art of Archemie: Conteining the right and perfectest meanes to make the Philosophers Stone, Aurum potabile, with other excellent Experiments. Divided into twelve Gates, First written by the learned and rare Philosopher of our Nation George Ripley, sometime Chanon of Bridlington in Yorkeshire: and Dedicated to K. Edward the 4. Whereunto is adioyned his Epistle to the King, his Vision, his Wheele, and other his Workes, neuer before published: with certaine briefe Additions of other notable Writers concerning*

Opera omnia in 1649,[17] inclusion in Ashmole's *Theatrum chemicum britannicum* in 1652,[18] and a re-working by the last great philosophical alchemist of the seventeenth century, the anonymous Eirenaeus Philalethes.[19]

In addition to the new editions of the older alchemists, there were several works of importance which were original to the seventeenth century. Michael Sendivogius, who had once been resident at the court of that great collector of alchemists, the Holy Roman Emperor Rudolph II, was enormously popular. Standard bibliographic works[20] reveal around fifty editions of works in which his tracts appeared either alone or collected with other items, all before the century was over.[21] The books of Basilius Valentinus, falsely ascribed in the seventeenth century to an earlier period but now established as contemporary works,[22] were published in numerous translations and editions.[23] A flurry of interest shortly after the mid-point

the same. *Set foorth by Raph Rabbards Gentleman, studious and expert in Archemicall Artes* (London: Imprinted by Thomas Orwin, 1591).

[17] George Ripley, *Georgii Riplaei Canonici Angli Opera omnia Chemica, quotquot hactenus visa sunt, quorum aliqua jam primum in lucem prodeunt, aliqua MS. exemplarium collatione à mendis & lacunis repurgata, atque integritati restituta sunt. Horvm Seriem pagina post praefationem prima monstrabit* (Cassellis: Typis Jacobi Gentschii, Impensis Sebaldi Kohlers, 1649), hereinafter referred to as Ripley, *Opera omnia.*

[18] Ashmole, *TCB*, pp. 107–93, 374–96 (2, n. 6).

[19] Philalethes, *Ripley Reviv'd* (2, n. 9) contained the following five tracts: (1) *An Exposition upon Sir George Ripley's Epistle*, (2) *An Exposition upon Sir George Ripley's Preface*, (3) *An Exposition upon the First Six Gates of Sir George Ripley's Compound of Alchymie*, (4) *A Breviary of Alchemy; Or A Commentary upon Sir George Ripley's Recapitulation*, (5) *An Exposition upon Sir George Ripley's Vision.*

[20] Bibliographic sources for the present writer's survey of the Sendivogian tracts include: (1) *Alchemy and the Occult. A Catalogue of Books and Manuscripts from the Collection of Paul and Mary Mellon given to Yale University Library*, compiled by Ian Macphail... (2 vols.; New Haven: Yale University Library, 1968); (2) *Bibliotheca Alchemica et Chemica: An Annotated Catalogue of . . . the Library of Denis I. Duveen* (London: E. Weil, 1949). (3) Henry Carrington Bolton, *A Select Bibliography of Chemistry, 1492–1892* (Smithsonian Miscellaneous Collections, vol. 36; Washington: Smithsonian Institution, 1893); (4) *British Museum General Catalogue of Printed Books*...(263 vols.; London: Trustees of the British Museum, 1965–66); (5) Albert L. Caillet, *Manuel bibliographique des sciences psychiques ou occultes*...(3 vols.; Paris: Lucien Dorbon, 1912); (6) *Catalogue général des livres imprimés de la Bibliotheque Nationale. Auteurs* (217 vols.; Paris: Imprimerie Nationale, 1897); (7) *Chemical, Medicinal and Pharmaceutical Books Printed before 1800 In the Collections of the University of Wisconsin Libraries*, ed. by John Neu...(Madison and Milwaukee: The University of Wisconsin Press, 1965); (8) John Ferguson, *Bibliotheca Chemica: A Catalogue of the Alchemical, Chemical and Pharmaceutical Books in the Collection of the Late James Young*... (2 vols.; Glasgow: James Maclehose and Sons, 1906); (9) J. R. Partington, *A History of Chemistry* (4 vols.); London: Macmillan & Co.; New York: St. Martin's Press, 1961–70), hereinafter referred to as Partington, *History.*

[21] The Sendivogian tracts were four in number: (1) *Novum lumen chymicum* (sometimes attributed to Alexander Seton), (2) *Aenigma philosophicum*, (3) *Dialogus Mercurii, Alchymistae et Naturae*, and (4) *Tractatus de sulphure*. The *Lettres philosophiques*, which only began to appear in the 1690's, are spurious and have not been included in the total of fifty editions.

[22] Read, *Prelude to Chemistry*, pp. 183–84 (2, n. 2).

[23] The tracts attributed to Basilius Valentinus include (1) *Concerning the Great Stone of the Ancients*, (2) *Concerning Natural and Supernatural Things*, (3) *The Secret Generation of Planets*

of the century attached to the writings of Eirenaeus Philalethes.[24] Philalethes has often been identified as George Starkey, who studied alchemy
with John Winthrop, Jr., and who had access for a while to the library
discussed above.[25]

Philalethes was the last contemporary alchemist of importance to appear
in print, but the publishing of alchemical works did not immediately
abate. Around seventy less important treatises of an alchemical nature,
almost all of them published in the seventeenth century, are in the
Newton Collection. However, enough has been said to make further detail
superfluous and to make it necessary to attack the historical problems
involved in this vast pan-European publishing venture and the equally
great general interest it reflected.

Alchemy, mechanism and reform

For even if the primary impulse towards the study of alchemy had initially
been religious, as has been suggested here, that impulse became radically
different for some of those who later became interested in alchemical
studies so that an entirely new phenomenon made its appearance as the
century wore on. The change that came about was early reflected in the
attitude of Marin Mersenne. Mersenne clearly perceived the dangers to
orthodox religion inherent in the alchemical movement, and his attack
on the occultists and the battle which followed not only signaled a change
in the course of alchemical studies but presaged a new approach in
philosophy and natural philosophy as well.

The events which roused Mersenne's ire had begun to take shape in
Germany in the later sixteenth century. There a school of mysticism arose
made up of alchemists, cabbalists, and astrologers compounded from
the disciples of Ficinus, Pico della Mirandola, Reuchlin, Agrippa von

---

and Metals, (4) *The Triumphal Chariot of Antimony*, (5) *The Last Will and Testament: A
Practick Treatise Together with the XII. Keys and Appendix of the Great Stone of the Ancient
Philosophers*, and (6) *Azoth*. They have been discussed in Read, *Prelude to Chemistry*, pp.
184–211 (2, n. 2).

[24] In addition to the five tracts which comprised *Ripley Reviv'd*, eight others appeared
in print: (1) *Introitus apertus*, (2) *De metallorum metamorphosi*, (3) *Brevis Manuductio ad
Rubinum Coelestem*, (4) *Fons chemicae philosophiae*, (5) *Experiments for the Preparation of the
Sophick Mercury*, (6) *Enarratio methodica trium gebri medicinarum*, (7) *Vade mecum philosophicum*,
(8) *The Secret of the Immortal Liquor called Alkahest*. All these Philalethes tracts circulated
in manuscript for some time before they were published and there were apparently yet
other tracts which were never printed and which have been lost, as will be discussed
subsequently.

[25] Ronal Sterne Wilkinson, "The problem of the identity of Eirenaeus Philalethes,"
Ambix **12** (1964), 24–43, hereinafter referred to as Wilkinson, "Identity." Wilkinson,
who earlier favored the thesis that Philalethes was Winthrop himself, has now come to a
tentative agreement with the more traditional position of identifying Philalethes as
Starkey. See Ronald Sterne Wilkinson, "Further thoughts on the identity of 'Eirenaeus
Philalethes,'" *Ambix* **19** (1972), 204–08.

Nettesheim, Trithemius, and Paracelsus. According to Waite, Henry Khunrath was the chief representative of this group and his *Ampitheatrum sapientiae aeternae* one of its principal publications. Many other books and pamphlets exalting the spiritual side of the alchemical and Renaissance Hermetic traditions appeared also, and the adherents of the movement claimed that their processes afforded the sublimity of infallible knowledge and wisdom through divine illumination.[26] Khunrath, for example, identified the Philosopher's Stone as Jesus Christ, the "Son of the Macrocosm," and thought that its discovery would reveal the true wholeness of the macrocosm just as Christ gave wholeness to the microcosm, man. His views have been shown by Montgomery to have achieved considerable influence in Lutheran circles.[27]

The English Hermeticist John Dee[28] has recently been shown by Yates to have been intimately related to this mystical movement in the Germanies, to have been in personal contact with Khunrath, and to have contributed to some of Khunrath's ideas in the *Amphitheatrum sapientiae aeternae*.[29] Although Dee had been nurtured in his Hermeticism by the same sort of literature which had been absorbed by Khunrath, the direct personal influence of Dee on Khunrath became a new factor in the movement and led the movement on to complicated relations involving Protestant England's political alignments with the rest of Europe and especially with Protestant princes in the Germanies.

This late-sixteenth-century mystical movement became involved with Protestant politics not because of any particular intellectual affinity of the one for the other but because of the fact that where Protestantism was dominant the long arm of the Inquisition could not reach. Therefore any movement which had detached itself from extreme orthodoxy might readily be attracted to Protestant areas and in Prague, under Rudolph II, Holy Roman Emperor until 1612, all sorts of occult studies flourished along with religious toleration.[30]

But the occultists, apparently finding it satisfactory to have a relative freedom to develop their ideas, soon became involved in an effort to maintain that freedom. As the Counter-Reformation gathered its forces

[26] Arthur Edward Waite, *The Real History of the Rosicrucians, Founded on Their Own Manifestoes, and on Facts and Documents Collected from the Writings of Initiated Brethren* (London: George Redway, 1887), pp. 27–33, hereinafter referred to as Waite, *Rosicrucians*.

[27] John Warwick Montgomery, "Cross, constellation, and crucible: Lutheran astrology and alchemy in the age of the Reformation," *Ambix* **11** (1963), 65–86.

[28] Charlotte Fell Smith, *John Dee (1527–1608)* (London: Constable & Co., 1909); Frances A. Yates, *Theatre of the World* (Chicago: The University of Chicago Press, 1969), pp. 1–41; hereinafter referred to as Yates, *Theatre*; Peter J. French, *John Dee. The World of an Elizabethan Magus* (London: Routledge & Kegan Paul, 1972).

[29] Yates, *Enlightenment*, pp. 37–40 (1, n. 47).

[30] *Ibid., passim*; Robert John Weston Evans, *Rudolf II and His World. A Study in Intellectual History 1576–1612* (Oxford: At the Clarendon Press, 1973), esp. pp. 196–242, hereinafter referred to as Evans, *Rudolf*.

and it became apparent that the arch-conservative Archduke Ferdinand of Styria was destined to marshal them for the fight against heresy, the Calvinist Prince Frederick, Elector Palatine, began to be groomed to enter the lists against Ferdinand. The chief architect of Protestant policy in this case was Christian of Anhalt, adviser to the Palatine court at Heidelberg and an occultist of long standing. Under his guidance an English marriage was arranged for Frederick, in the expectation that then England would side with the German Protestants in their coming struggle against the Catholic Hapsburgs, and English influences became strong at the Heidelberg court. Conversely, the Hermeticists and other occultists rallied behind Anhalt in the expectation that the bulwark of Protestant alliances which he was building would shelter reform, intellectual enlightenment, and liberal philosophy all across the board and that the shelter thus afforded would cause the dawn of a new and brighter day to break over northern Europe.[31]

About 1614 there appeared a manifesto thought to be a product of this group of occultists, or perhaps of a single member of it, entitled *Fama Fraternitatis; or, a Discovery of the Fraternity of the Most Laudable Order of the Rosy Cross.*[32] It was followed in 1615 by a *Confessio Fraternitatis R. C.*[33] Printed with the *Confessio* was a tract entitled *A Brief Consideration of the more Secret Philosophy*, which was based largely on John Dee's *Monas hieroglyphica*, demonstrating Dee's continuing influence upon the occult movement as it flowed into Rosicrucianism.[34] Both the *Fama* and the *Confessio* were addressed to the learned of Europe and offered a universal reformation under the auspices of the fraternity. The political references in the tracts, explored by Yates, make it clear that the universal reformation was to take place under the Elector Palatine as soon as he had successfully challenged the Hapsburg domination.[35]

There was a large public response to these tracts. Many people wrote letters which they had published in an attempt to contact the fraternity, the group – if it ever actually existed as a group – having somehow failed to include its mailing address in its manifestos. Others wrote small pamphlets claiming to have intimate knowledge of the fraternity or arguing its principles pro or con. No authentic response by the fraternity ever appeared, but the uproar increased. Charges of heresy and atheism were hurled; imposters arose to capitalize on the now famous name of the

[31] Yates, *Enlightenment*, pp. 1–40 (1, n. 47).

[32] Printed in Waite, *Rosicrucians*, pp. 65–84 (3, n. 26), and in Yates, *Enlightenment*, pp. 238–51 (1, n. 47).

[33] Printed in Waite, *Rosicrucians*, pp. 86–98 (3, n. 26), and in Yates, *Enlightenment*, pp. 251–60 (1, n. 47).

[34] Yates, *Enlightenment*, pp. 45–47 (1, n. 47). Dee's original work is available in C. H. Josten, "A translation of John Dee's 'Monas Hieroglyphica' (Antwerp, 1564), with an introduction and notes," *Ambix* 12 (1964), 84–221.

[35] Yates, *Enlightenment*, pp. 53–58 (1, n. 47).

Rosy-Cross. Libavius, although critical of the grounds of Rosicrucian knowledge, nevertheless advised everyone to join the order as he thought there was much wisdom to be gained by doing so.[36] The fraternity found two truly erudite defenders in Michael Maier[37] and Robert Fludd.[38] Yates has noted that all of the major works of Fludd and Maier were actually published in the Palatinate during Frederick V's reign, which fact lends credence to her suggestion of a relationship between the Rosicrucians and Anhalt's plan for Frederick's leadership of Protestantism.[39]

Mersenne, born in 1588, came to maturity at the height of the Rosicrucian controversy and was profoundly affected by it. To him it appeared clear that there was great danger to true religion in the rage for the occult and for the spiritual alchemy advocated by the Rosicrucians, and he undertook in some of his earliest works to demonstrate the peril. A true son of the Church, trained in the Jesuit school of La Fléche and later a member of the Minorite Order, Mersenne thought that the confounding of the natural and the supernatural in the occult philosophies was disastrous both to true religion and to true science. He attacked occultism in all its variety: sorcery, cabbalism, animism, astrology, and alchemy. It was basically on two grounds that he found these studies objectionable. One was that by attributing independent powers to stars, demons, natural spirits, a world-soul, etc., the occultists denied both God's power and man's freedom. The other was their lack of certainty; their schemes of causality were unverifiable. As a remedy to their lamentable influence, Mersenne came out strongly for traditional Christian faith and for a program of verity in the sciences based solidly on Aristotelian rationalism and causal relationships and on the demonstrations of astronomy, arithmetic, and geometry. Thus certainty in religion and certainty in science would accompany one another.[40]

[36] Waite, *Rosicrucians*, pp. 246–52 (3, n. 26).

[37] *Ibid.*, pp. 268–72; J. B. Craven, *Count Michael Maier, Doctor of Philosphy and of Medicine, Alchemist, Rosicrucian, Mystic, 1568–1622. Life and Writings* (Reprint of the Kirkwall edn. of 1910; London: Dawsons, 1968), hereinafter referred to as Craven, *Maier*; Evans, *Rudolf*, pp. 205–06 (3, n. 30).

[38] Waite, *Rosicrucians*, pp. 283–307 (3, n. 26); J. B. Craven, *Doctor Robert Fludd (Robertus de Fluctibus), The English Rosicrucian. Life and Writings* (Reprint of the Kirkwall edn. of 1902; n. p.: Occult Research Press, n. d. [196?]); Yates, *Theatre*, pp. 60–79 (3, n. 28). See also the following papers by Allen G. Debus: "Robert Fludd and the circulation of the blood," *Journal of the History of Medicine and Allied Sciences* 16 (1961), 374–93; "Robert Fludd and the use of Gilbert's *De Magnete* in the weapon–salve controversy," *Journal of the History of Medicine and Allied Sciences* 19 (1964), 389–417; "Renaissance chemistry and the work of Robert Fludd," *Ambix* 14 (1967), 42–59; "Harvey and Fludd: The irrational factor in the rational science of the seventeenth century," *Journal of the History of Biology* 3 (1970), 81–105.

[39] Yates, *Enlightenment*, pp. 70–90 (1, n. 47).

[40] Robert Lenoble, *Mersenne, ou la naissance du mécanisme* (Paris: Librairie philosophique J. Vrin, 1943), pp. 83–167, hereinafter referred to as Lenoble, *Mersenne*. The works in which Mersenne opened his skirmish with the occultists were *Quaestiones in Genesim* (Paris, 1623) and *La vérité des sciences* (Paris, 1625); they are discussed by Lenoble.

Alchemy was particularly horrendous to Mersenne for it offered a salvation without faith and constituted a sort of counter-church. However, he was far from suggesting that alchemical studies should be abandoned. His solution was rather that the alchemists should found an academy in which they could study the positive results of their experiences without mystery and without secrets.[41]

After Mersenne made public his firm stand against occultism, there were a number of important developments. Some of these are well known: Descartes undertook the elaboration of a full philosophical system in which matter and spirit were carefully kept apart and all science was to depend on a rational, mathematical approach to matter in motion. And Mersenne's friend, Pierre Gassendi, probably at Mersenne's suggestion, took up the public arguments against the Rosicrucians. In the course of his writings, Gassendi worked out a Christianization of ancient atomic doctrines which made them acceptable to the devout natural philosophers of the time. But Mersenne's suggestion that alchemy should be kept apart from religion and divested of its mystery fell finally on fertile soil of quite a different sort.

As mentioned in the above discussion, the Rosicrucians had been much involved with ideas of a general reformation. A document called *Die Reformation der ganzen weiten Welt*[42] appeared with the *Fama* in 1614 and was generally reprinted with the Rosicrucian tracts although it is not alchemical in nature. Despite the differences in style and surface content, however, *Die Reformation*, the *Fama*, and the *Confessio* carry the same larger message of the need for a new reformation, and the three documents were linked politically, as Yates has recently shown. *Die Reformation* was a translated extract from a larger work by an Italian anti-Hapsburg liberal published at Venice the year before. Venice had long been in conflict with the pope and dissatisfied with the state of religion, culture, and politics in Italy, and was watching affairs in the Palatinate with great interest.[43] Reform of all sorts, enlightenment, and alchemy were thus associated from the very beginning in Rosicrucianism.

The hope of a universal reformation and the mystical progressions of spiritual alchemy were perhaps best united in the important figure of J. V. Andreae. Andreae (1586–1654) is known for his *Reipublicae Christianopolitanae descriptio*, a utopia which shows traces of the influence of More's *Utopia* and Campanella's *Civitas solis* and in turn must have had its own effect on Bacon's *New Atlantis*.[44] Andreae's object was the educational and religious reform of society but he described certain ideal institutions,

[41] Lenoble, *Mersenne*, pp. 149–50 (3, n. 40). For other contemporary reactions in France to the Rosicrucian movement, see Yates, *Enlightenment*, pp. 103–17 (1, n. 47).

[42] Printed in Waite, *Rosicrucians*, pp. 36–63 (3, n. 26).

[43] Yates, *Enlightenment*, pp. 130–39 (1, n. 47).

[44] Johann Valentin Andreae, *Christianopolis, an Ideal State of the Seventeenth Century*, trans., with historical introd., by Felix Emil Held (New York: Oxford University Press

maintained in Christianopolis for the study of nature and the gathering of information about her, that were remarkably similar in function to the early Royal Society as it actually developed. About the "Laboratory" in his ideal republic, Andreae had the following to say, for example:

Here the properties of metals, minerals, and vegetables, and even the life of animals are examined, purified, increased, and united, for the use of the human race and in the interests of health. Here the sky and the earth are married together; divine mysteries impressed upon the land are discovered; here men learn to regulate fire, make use of air, value the water, and test earth. Here the ape of nature has wherewith it may play, while it emulates her principles and so by the traces of the large mechanism forms another, minute and most exquisite. Whatever has been dug out and extracted from the bowels of nature by the industry of the ancients, is here subjected to close examination, that we may know whether nature has been truly and faithfully opened to us. Truly that is a humane and generous undertaking, which all who are true human beings deservedly favor.[45]

However, Andreae's fame does not rest solely on his *Christianopolis*. Held attributed the *Fama*, the *Confessio*, and *Die Reformation* all to him.[46] Although actually *Die Reformation* was written by the Italian Traiano Boccalini[47] and the authorship of the *Fama* and the *Confessio* is still in doubt,[48] it seems to be a fact that Andreae wrote *The Chymical Marriage of Christian Rosencreutz*, as he himself claimed that tract in his autobiography.[49] The *Chymical Marriage* was one of the most popular of the so-called Rosicrucian documents and was frequently printed with the manifestos.[50] It

American Branch; London, Toronto, Melbourne, and Bombay: Humphrey Milford, 1916), Held, "Introduction," pp. 16–74, the larger work hereinafter referred to as Andreae, *Christianopolis*. See also Yates, *Enlightenment*, pp. 145–51 (1, n. 47), and John Warwick Montgomery, *Cross and Crucible. Johann Valentin Andreae (1586–1654), Phoenix of the Theologians. Vol. I. Andreae's Life, World-view, and the Relations with Rosicrucianism and Alchemy. Vol. II. The Chymische Hochzeit with Notes and Commentary* (International Archives of the History of Ideas, no. 55; The Hague: Martinus Nijhoff, 1973), hereinafter referred to as Montgomery, *Andreae I* and *Andreae II*, respectively.

[45] Andreae, *Christianopolis*, pp. 196–97 (3, n. 44).

[46] Held, "Introduction," in Andreae, *Christianopolis*, p. 126 (3, n. 44).

[47] Yates, *Enlightenment*, p. 133 (1, n. 47); Montgomery, *Andreae I*, p. 165 (3, n. 44).

[48] The question of Andreae's authorship has been treated exhaustively pro and con in Montgomery, *Andreae I*, pp. 160–231 (3, n. 44), but not necessarily settled.

[49] Waite, *Rosicrucians*, p. 226 (3, n. 26); Yates, *Enlightenment*, pp. 30–33 (1, n. 47); Montgomery, *Andreae I*, p. 171 (3, n. 44); Johann Valentin Andreae, *Ioannis Valentini Andreae Theologi Q. Württembergensis. Vita, ab ipso conscripta. Ex autographo, in Bibl. Guelferbytano recondito, adsumtis Codd. Stuttgartianis, Schorndorfiensi, Tubingensi, nunc primum edidit F. H. Rheinwald, Dr. Cum icone et chirographo Andreano* (Berolini: Apud Herm. Schultzium, 1849), p. 10.

[50] Printed in Waite, *Rosicrucians*, pp. 100–96 (3, n. 26); résumé in Yates, *Enlightenment*, pp. 60–64 (1, n. 47); facsimile reproduction of the seventeenth-century English translation, with notes and commentary, in Montgomery, *Andreae II*, pp. 288–487 (3, n. 44).

is a highly allegorical rendition of the alchemical process, and since Andreae indeed wrote it, there can be no doubt of his great interest in the more mystical side of alchemy.

But whether Andreae really wrote all the tracts sometimes attributed to him is less important for the present discussion than the unquestioned fact that from the second decade of the seventeenth century writings on spiritual alchemy were associated with discussions of general reform. Contemporaries found the tracts here discussed printed with each other and often associated with Andreae's name, and these writings, however strange and ill-assorted they now appear, and however closely linked to evanescent contemporary political events they were, stood at the beginning of a long series of efforts for educational, religious, and scientific reform in Protestant countries.

The most important of the other reformers, men who all had some debt to Andreae or to each other, were Francis Bacon (1561–1626), Jan Comenius (1592–1670), and Samuel Hartlib (c. 1600–62).[51] In recent years the importance of archaic, alchemical modes of thought in the basic philosophies of these reformers has been receiving considerable attention, and Armytage has pointed up the relationship existing between early science and utopian thought, noting the involvement of alchemists and Rosicrucians with both.[52]

As for Bacon, there can be no doubt that he was familiar with alchemical literature. Although he criticized the alchemists on some occasions and significantly modified their ideas at other times, still many of the tenets of his natural philosophy derive from them, and he presented his own suggestions for the maturation of other metals into gold.[53] As West has pointed out, Bacon sounds remarkably like Paracelsus when he speaks on experiment and on the "Light of Nature,"[54] and Yates has clearly established the influence of the Rosicrucian manifestos upon Bacon's *New Atlantis*.[55]

The reputation of Comenius has always been highest in educational circles, due to his many works in that area. However, he evolved a distinctive philosophy of nature also, partly through the influence of Andreae. The presence of alchemical ideas in his *Naturall Philosophie Reformed by Divine Light* (London, 1651) was briefly noted by Kargon.[56] More recently

[51] Held, "Introduction," in Andreae, *Christianopolis*, pp. 100–25 (3, n. 44).

[52] W. H. G. Armytage, "The early Utopists and science in England," *Annals of Science* **12** (1956), 247–54.

[53] Joshua C. Gregory, "Chemistry and alchemy in the natural philosophy of Sir Francis Bacon, 1561–1626," *Ambix* **2** (1938–46), 93–111.

[54] Muriel West, "Notes on the importance of alchemy to modern science in the writings of Francis Bacon and Robert Boyle," *Ambix* **9** (1961), 102–13.

[55] Yates, *Enlightenment*, pp. 125–29 (1, n. 47).

[56] Robert Hugh Kargon, *Atomism in England from Hariot to Newton* (Oxford: Clarendon Press, 1966), pp. 154–55, hereinafter referred to as Kargon, *Atomism*.

the Czech Comenian scholar Jaromír Červenka has explored the natural philosophy of Comenius in further detail. Červenka concluded that there was considerable evolution in Comenius' thoughts on nature over the years and that at different periods he incorporated a macrocosmic–microcosmic philosophy similar to that of Robert Fludd, and a Neoplatonism, similar to that of d'Espagnet, into his natural philosophy.[57]

Also, recently, the great importance of alchemical and Hermetic thought in seventeenth-century suggestions for the general reform of education has been explicated by Debus[58] and Sadler.[59] Much of that relationship between alchemical and Hermetic thought and the reform of education is centered on the Comenian concept of pansophy. Behind the concept as Comenius developed it lay the idealist philosophic tradition of Plato which held that man does not ordinarily experience the reality or essence of things but only their shadows, as in the Platonic simile of the cave. It should however be possible to step out of the cave into the light and experience the true nature of reality to some degree. Comenius extended that concept along macrocosmic–microcosmic lines until he conceived of a light existing within the human mind as well as outside it. When man opened his eyes to both, then he would see the ultimate truth of the universe, which Comenius called pansophy. The essential nature of external reality, Comenius thought, could be conveyed by education to the simplest intelligence if all knowledge could be reduced to a basic principle.[60] The essential nature of internal reality, on the other hand, would be illuminated through religious love and/or the exercises of spiritual alchemy. Viewed as a fundamental technique for the discovery of the inner light, spiritual alchemy thus could be conceived as fundamental and even necessary to the internal part of the education process.

Yet alchemy in its more chemical form could be considered to play a crucial role in education and in the search for new knowledge too. In their rejection of the pagan accounts of natural phenomena offered by Aristotle and Galen, Renaissance Hermeticists had come to emphasize anew the importance of the first chapter of the book of Genesis. In Genesis was a

---

[57] Jaromír Červenka, *Die Naturphilosophie des Johann Amos Comenius* (Praha: Academia Verlag der Tschechoslowakischen Akademie der Wissenschaften, 1970). See also Jaromír Červenka, "J. A. Komenský, Ladislaw Velen ze Žerotína a Alchymie," *Acta Comeniana* **24** (1970), 21–44, summary p. 44 in German, and Yates, *Enlightenment*, pp. 156–70 (1, n. 47); and, on Comenius as "a late inheritor" of and thoroughly representative of "the Rudolfine mood," Evans, *Rudolf*, pp. 275–85.

[58] Allen G. Debus, *Science and Education in the Seventeenth Century: the Webster–Ward Debate* (London: MacDonald; New York: American Elsevier, 1970), hereinafter referred to as Debus, *Science and Education*.

[59] *Comenius*, ed., with introd., by John Sadler (Educational Thinkers Series; London: Collier–Macmillan, 1969), hereinafter referred to as Sadler, *Comenius*.

[60] *Ibid.*, pp. 21–22.

divine account of the creation of the world, one which could not be disputed, and one which could lend itself to interpretation as a divine chemical separation. If the act of creation itself was to be understood chemically, then all of nature was to be understood similarly. In short, chemistry was the key to all nature, the key to all the macrocosmic–microcosmic relationships sought by Robert Fludd and others. A study of chemistry was a study of God as He had Himself written out His word in the Book of Nature. Such a study could only lead one closer to God and was conceived as having moral value as well as contributing to the better grasp of the workings of nature and to the providing of better medicines for the relief of man's illnesses.[61]

The concept of chemistry as a key was one in turn which fitted well with other aspects of Comenian pansophy. For Comenius did not suppose that the true nature of external reality was readily apparent to simple intelligences, or indeed to erudite and brilliant ones. Rather, he sought in all fields the simplifying key, the ultimate reduction, which would unlock the complexities and allow the essences to shine through and be readily comprehensible. On one occasion Comenius expressed his thoughts on simplification in the following manner:

> For although things as they are in themselves may seeme to have a certaine infinitie in them, yet is it not so indeed: for the world it selfe (that admirable worke of God) is framed of a few elements, and some few kinds of formes; and all Arts whatsoever have beene invented, may easily be reduced unto a summary and generall method. Because therefore things themselves, and their Conceptions, and Words the expressions of those Conceptions are parallel one to the other, and in each of them there are certaine fuodamentalls [sic] from which the rest of them result; I thought that it is not impossible, to collect also the fundamentalls of Things themselves, and their conceptions, as well as hath been done already in Words. Also the practise of the Chymists came into my mind, who have found out a way so to cleare, and unburden the essences, and spirits of things from the surcharge of matter, that one small drop extracted out of Mineralls, or Vegetables containes more strength, and vertue in it, and is used with better successe, and efficacie, then can be hoped for from the whole, and entire lumpe.[62]

[61] Debus, *Science and Education*, pp. 15–32 (3, n. 58).

[62] John Amos Comenius, *A Reformation of Schooles, designed in two excellent Treatises: The first whereof Summarily sheweth, The great necessity of a generall Reformation of Common Learning. What grounds of hope there are for such a Reformation. How it may be brought to passe. The second answers certaine objections ordinarily made against such undertakings, and describes the severall Parts and Titles of Workes which are shortly to follow. Written many yeares agoe in Latine by that Reverend, Godly, Learned, and famous Divine Mr. John Amos Comenius, one of the Seniours of the exiled Church of Moravia: And now upon the request of many translated into English, and*

And so as Comenius tried to devise a universal language to replace multi-lingual complexities, so the pansophists who followed him came to seek a universal chemical key to unravel the incomprehensible tangle of natural phenomena.

The Hartlibians and their chemical alchemy

The honor of leading that search fell to that group in England which was heir to the ideas of Andreae, Bacon, and Comenius, the so-called Hartlib circle. Samuel Hartlib, like Andreae and Comenius, was primarily interested in educational and religious reform but was perhaps more socially and practically oriented than they. Leaving much of the effort at religious reform to his co-worker John Dury, an ecumenical Presbyterian from Scotland working for church reunion, Hartlib concentrated on building up patronage and Parliamentary support for projects on education and the gathering of knowledge. On his own initiative he undertook to be a sort of communications center for the collection and dissemination of information of all sorts. He was later to attempt to institutionalize this enormous correspondence of his into a state-supported Office of Address, but he failed to obtain funds for it and the correspondence for scientific matters passed after his death into the hands of John Dury's son-in-law, Henry Oldenburg. Oldenburg, as Secretary for the new Royal Society, personally maintained it until in 1665 he converted it into the more public format of the *Philosophical Transactions.*[63]

Hartlib's circles of acquaintance and correspondence grew ever wider from the time he first entered England as a student in 1625 and were soon to include men of every possible intellectual persuasion. Of all the different schools of thought thus brought into startling confrontation and made to cross-fertilize each other, perhaps none were more important than the alchemists and the mechanical philosophers. A number of details on their intersection has survived.

The most fruitful contacts were made when Hartlib's acquaintances established working relations with those of Marin Mersenne. In Paris Mersenne was serving as the center of a correspondence network similar

published by Samuel Hartlib, for the generall good of this Nation (London: Printed for Michael Sparke senior, at the Blew Bible in Greene Arbor, 1642), pp. 47–48.

[63] An excellent recent discussion of Hartlib and his group may be found in *Samuel Hartlib and the Advancement of Learning*, ed., with introd., by Charles Webster (Cambridge Texts and Studies in the History of Education, gen. eds., A. C. F. Beales, A. V. Judges and J. P. C. Roach; Cambridge: At the University Press, 1970). Webster provides a discussion of, and an introduction into, much of the older literature on Hartlib and his predecessors and associates. Hereinafter referred to as Webster, *Hartlib*. See also Charles Webster, "English medical reformers of the puritan revolution: a background to the 'Society of Chymical Physitians,'" *Ambix* **14** (1967), 16–41, for biographical details about the associates of Hartlib and about their interest in medical reform. Hereinafter referred to as Webster, "English medical reformers."

to Hartlib's,[64] and in the 1630's contact began. Dury had contact with Descartes by 1634/35 and Theodore Haak was in touch with both groups.[65] Furthermore, there was the Cavendish–Newcastle circle. One member of the latter group, John Pell, wrote a short pansophic mathematical tract for Hartlib in the 1630's which Haak forwarded to Mersenne.[66] In the 1640's the nucleus of the Cavendish–Newcastle group went into exile in Paris and became ever more closely associated with Mersenne, Gassendi, and Descartes. There they were joined by Sir Kenelm Digby and William (later Sir William) Petty, both of whom were already or were later to work closely with Hartlib.[67] By May, 1648, a very complex set of relationships had developed, as evidenced by a letter from Hartlib to Boyle: Hartlib has shown a letter of Boyle's to Hobbes; Hartlib reports to Boyle on an experimental demonstration of the vacuum which had been sent by Sir Charles Cavendish to "Mr. *Petty*"; Hartlib tells Boyle that "your worthy friend and mine, Mr. *Gassend*, is reasonable well, and hath printed a book... [on the life of] *Epicurus*..." and has another in the press at Lyons on "the *Philosophy of Epicurus*, in which I believe we shall have much of his own philosophy, which doubtless will be an excellent work."[68] And in the same letter Hartlib speaks of Morian and Worsley, both of which names will appear below in a thorough-going alchemical context.

Thus the situation was created whereby the reforming movement which was originally closely associated with alchemy and relied upon the most mystical and spiritual variety of that much abused study for the illumination needed for reform came under the calming and rationalizing influence of the new mechanical philosophers. The way was prepared for the rationalization of alchemy, its chemicalization and its clarification.

The role of empiricism in the new approach to alchemy which was about to be developed should not be neglected either. Although the reformers were committed philosophically to a spiritual alchemy, yet their humanitarianism prompted them to search for new chemical medicines

[64] Harcourt Brown, *Scientific Organizations in Seventeenth Century France* (1620–1680) (reissue of the 1934 edn.; New York: Russell & Russell, 1967), pp. 41–63, hereinafter referred to as Brown, *Organizations*.

[65] Webster, *Hartlib*, pp. 14–15 (3, n. 63); Pamela R. Barnett, *Theodore Haak, F.R.S. (1605–1690): The First German Translator of Paradise Lost* ('s-Gravenhage: Mouton & Co., 1962), pp. 34–50, hereinafter referred to as Barnett, *Haak*.

[66] Barnett, *Haak*, p. 38 (3, n. 65). [67] Kargon, *Atomism*, pp. 63–76 (3, n. 56).

[68] Samuel Hartlib to Robert Boyle, May 9, 1648, in Robert Boyle, *The Works of the Honourable Robert Boyle. To which is prefixed The Life of the Author* (A new edition: 6 vols.; London: Printed for J. and F. Rivington, L. Davis, W. Johnston, S. Crowder, T. Payne, G. Kearsley, J. Robson, B. White, T. Becket and P. A. De Hondt, T. Davies, T. Cadell, Robinson and Roberts, Richardson and Richardson, J. Knox, W. Woodfall, J. Johnson, and T. Evans, 1772) VI, 77–78, hereinafter referred to as Boyle, *Works*. The books of Gassendi's mentioned by Hartlib would have been *De Vita et Moribus Epicuri* (Lyons, 1647) and *Animadversiones in decimum librum Diogenis Laertii* (3 vols.; Lyons, 1649). The latter did indeed contain Gassendi's own physics as well as the *Philosophiae Epicuri syntagma*. See Kargon, *Atomism*, esp. pp. 65–68 (3, n. 56).

for the relief of man's diseases. A certain amount of laboratory work was involved in that. Also empiricism as a general approach to questions in natural philosophy was becoming acceptable. Gilbert had already published on the magnet; the barometric studies of Torricelli and Pascal were in progress. Harvey had demonstrated the motions of the heart and blood anatomically. Digby himself was profoundly empirical: he seems to have maintained a chemical laboratory wherever he stopped long enough to get one established. And of course alchemy had always been and still was a laboratory study. In the present work it has been maintained up to this point that a shift had occurred by the late sixteenth century which over-emphasized the spiritual side of alchemy. That seems to be quite true, but it is now time to reiterate the original point that alchemy had always involved *two* sides – a secret knowledge and also labor at the furnace. For the one side to be overemphasized would not mean the other was completely eliminated immediately. Even for those writers in the sixteenth and early seventeenth centuries who were of the most mystical persuasions, it can not be maintained with certainty that they never entered into laboratory work at all. Evidence is admittedly scanty, but probably there were all along numerous laboratories all over Europe in which alchemical labors were performed, in the back of pharmacy shops, in private houses, in the cellars of cottages and castles. It seems certain that alchemy was the most flourishing laboratory tradition to exist before the scientific revolution and probably the strongest laboratory tradition ever to exist anywhere in a non-institutionalized form.

But its lack of an institutional base and its individualistic character would have doomed it to stasis had not a new ideal intervened. Perhaps the most important single factor contributed by the Hartlib circle – although in this the mechanical philosophers agreed with them – was the ideal of the public communication of knowledge. It is perhaps an odd quirk of history that a group clearly descended from the flaming mysticism of late-sixteenth-century Germany should have been the one to perform the task of making alchemy clear, rational, and chemical. However, it must be realized that Andreae, Bacon, Comenius, and Hartlib had all emphasized utility and social good, and that above all they had held that all knowledge should be made public for the common benefit of mankind and that perhaps it could be reduced to easy comprehensibility by the use of a simple key. When that communal orientation of Hartlib's group is considered, it is clear that having inherited a strong undercurrent of alchemical interest, they would probably attempt to turn mystical alchemy (necessarily a rather private affair) into a more public alchemy (which, for ease of communication, had necessarily to be more rationalized and less mystical) which could then serve as a key to all natural phenomena and could be utilized as an educational device.

Many different aspects of the proto-chemistry of the seventeenth century appealed to Hartlib and his group, such as chemical medicine, new metallurgical techniques, and the possible use of chemical fertilizers, but only some of the alchemical studies of the group can be considered here. As the details are developed below, the mercurial Sir Kenelm Digby will frequently be seen flashing into view. Digby played rather a lively role in much that was going on in chemical circles and his book of alchemical and chemical *Secrets*[69] affords considerable insight into the smoky ferment bubbling just below the well-known developments in rational mechanics, astronomy, mathematics, and philosophy in the same period.

Samuel Hartlib kept a daybook, his "Ephemerides," from 1634 until the approach of his final illness in 1660. The entries from the latter part of 1643 to the spring of 1648 have been lost, but quite enough remains to document the early and long-continued interest of the Hartlib circle in alchemy-chemistry. The daybook and other papers, now deposited at Sheffield University by their owner Lord Delamere, are being studied by Wilkinson in a series of important papers, two of which have already appeared.[70] Chemical references begin as early as 1640 and continue to the end. By the mid-1640's, when Hartlib first met Robert Boyle and began to influence him, Hartlib was recording many items of chemical interest. Wandering alchemists and chemists of all sorts visited him, and Hartlib kept in touch with developments in Holland through correspondence with the chemist-divine Johann Morian.[71]

In addition to the "Ephemerides," many of the other Hartlib manuscripts contain information about the chemical activities of the group. Of the greatest interest is Bundle xvi, labelled "Chymica et Phylosophica." The bundle contains letters about alchemical processes and about potable gold, extracts from various authors, recipes for distilling, a description of

[69] Kenelm Digby, *A Choice Collection of Rare Secrets and Experiments in Philosophy. As Also Rare and unheard-of Medicines, Menstruums, and Alkahests; with the True Secret of Volatilizing the fixt Salt of Tartar. Collected And Experimented by the Honourable and truly Learned Sir Kenelm Digby, Kt. Chancellour to Her Majesty the Queen-Mother. Hitherto kept Secret since his Decease, but now Published for the good and benefit of the Publick, By George Hartman* (London: Printed for the Author, and are to be Sold by *William Cooper*, at the *Pelican* in *Little Britain;* and *Henry Faithorne*, and *John Kersey* at the *Rose* in *St. Paul's* Church-yard, 1682), hereinafter referred to as Digby, *Secrets.* For bibliographic information on this work and an analysis of selected items from it, see Dobbs, "Digby. Part III" (2, n. 59).

[70] Ronald Sterne Wilkinson, "The Hartlib Papers and seventeenth century chemistry. Part I," *Ambix* **15** (1968), 54–69, and Ronald Sterne Wilkinson, "The Hartlib Papers and seventeenth century chemistry. Part II," *Ambix* **17** (1970), 85–110, hereinafter referred to as Wilkinson, "Hartlib. Part I" and "Hartlib. Part II," respectively. Hartlib's "Ephemerides" are currently being edited for publication by Charles Webster.

[71] Wilkinson, "Hartlib. Part I" (3, n. 70). See also John J. O'Brien, "Samuel Hartlib's influence on Robert Boyle's scientific development. Part I. The Stalbridge period," *Annals of Science* **21** (1965), 1–14, and John J. O'Brien, "Samuel Hartlib's influence on Robert Boyle's scientific development. Part II. Boyle in Oxford." *Annals of Science* **21** (1965), 257–76.

the "Alkahest," and other items. One paper gives an inventory of Digby's laboratory at Gresham College, while another discusses the chemical aims of Frederick Clodius, Hartlib's son-in-law.[72]

The names of many little-known figures appear in Hartlib's papers, most of them undoubtedly involved with the most esoteric sort of alchemical thought. Of the many peripheral names, two of the better known ones may be chosen for examples. The first of these, although not among the most intimate of Hartlib's friends, was known to him and to others in the group, and his activities were discussed by Hartlib and his associates. Hence his writings form part of the background from which the Hartlibians emerged. That first one was Eugenius Philalethes, who was in reality the speculative alchemist and mystic Thomas Vaughan.

Vaughan's activities and writings were noted in Hartlib's "Ephemerides" for 1650;[73] an example of his style of writing follows. The quotation is from *Anthroposophia Theomagica* which aroused Hartlib's interest when it was published in 1650. The work was originally dedicated to the "Brethren" of the Rosy Cross.

> You see now – if you be not men of a most dense head – how man fell, and by consequence you may guess by what means he is to rise. He must be united to the Divine Light, from whence by disobedience he was separated. A flash or tincture of this must come or he can no more discern things spiritually than he can distinguish colours naturally without the light of the sun. This light descends and is united to him by the same means as his soul was at first. I speak not here of the symbolical, exterior descent from the prototypical planets to the created spheres and thence into "the night of the body"; but I speak of that most secret and silent lapse of the spirit "through the degrees of natural forms"; and this is a mystery not easily apprehended.[74]

One can only agree with Vaughan that his words do contain "a mystery not easily apprehended."

The second person to be mentioned here is the anonymous Eirenaeus Philalethes, whose true identity is still in doubt. If he should be identified as George Starkey, then he was actually a close associate of Hartlib's for a while. In any case Starkey did certainly introduce the Philalethes manuscripts to the group. The group read them eagerly and circulated them in manuscript for a considerable time before they were published.[75] The

[72] Wilkinson, "Hartlib. Part I," p. 57 (3, n. 70). The inventory of Digby's laboratory has been printed in Dobbs, "Digby. Part II," pp. 146–47 (2, n. 58).

[73] Wilkinson, "Hartlib. Part II," pp. 88–89 (3, n. 70).

[74] Thomas Vaughan, *Anthroposophia Theomagica, or A Discourse of the Nature of Man and His State after Death*, in Thomas Vaughan, *The Works of Thomas Vaughan: Eugenius Philalethes*, ed., annotated, with introd. by Arthur Edward Waite (London: The Theosophical Publishing House, 1919), pp. 1–60, quotation from p. 46.

[75] Ronald Sterne Wilkinson, "George Starkey, physician and alchemist," *Ambix* **11** (1963), 121–52, hereinafter referred to as Wilkinson, "Starkey." Also see Wilkinson,

names of several of the numerous tracts from the pen of this second Philalethes have been presented above; although in most places they are hardly
more comprehensible than Vaughan's writings, they are much more
"chemical" in the sense that Eirenaeus Philalethes evidently understood
the alchemical process to take place in matter rather than in man's soul.

Eirenaeus Philalethes was soon to become one of Isaac Newton's
favorite alchemical writers, as will appear below, and for that reason considerable interest attaches to his works. It will be recalled that Brewster
had seen among Newton's papers a copy of a book by Eirenaeus Philalethes: the *Secrets Reveal'd: or, An Open Entrance to the Shut-Palace of the King*.
It was a book Newton studied with great care and for that reason seems
especially suitable as an example of Hartlibian activity.

William Cooper, the publisher of the English version of the *Secrets
Reveal'd* of 1669, commented upon the earlier Latin version published by
John Langius in 1660, noting that his English manuscript was much fuller
than the manuscript which Langius had used. Cooper also stated that he
had been in possession of his copy for a number of years before Langius'
Latin appeared. Cooper thought his to have been made from the author's
copy or at worst to have been very little corrupted.[76] From Cooper's remarks it is apparent that a number of manuscripts in variant forms must
have been circulating in the 1650's and 1660's, one more indication of the
great interest aroused by alchemical topics during that period. Wilkinson
has discussed the difficulties of establishing a canon of the works of
Eirenaeus Philalethes even in their published forms, a difficulty which
arises in part because of the many variants.[77]

In the *Secrets Reveal'd* Philalethes first establishes the necessity of the
"Sophick Mercury" for the "Work of the Elixir." Then he turns to an
"explanation" of the principles composing the "Mercury Sophical,"
calling it "our Water" and noting that it is compounded of "Fire," "the
Liquor of the Vegetable *Saturnia*," and the "bond of ☿". It is truly "a
Chaos," and out of it he says he knows how to extract all things, even gold
and silver without the transmuting elixir. If one wishes to understand it,
he should

> learn to know, who the Companions of *Cadmus* are, and what that
> *Serpent* is which devoured them, what the hollow *Oak* is which *Cadmus*

"Identity," pp. 24–25 (3, n. 25), and Wilkinson, "Hartlib. Part II" (3, n. 70).

[76] Eirenaeus Philalethes, *Secrets Reveal'd: or, An Open Entrance to the Shut-Palace of the
King: Containing, The greatest Treasure in Chymistry, Never yet so plainly Discovered. Composed
By a most famous English-man, Styling himself Anonymvs, or Eyraeneus Philaletha Cosmopolita:
Who, by Inspiration and Reading, attained to the Philosophers Stone at his Age of Twenty three
Years, Anno Domini, 1645. Published for the Benefit of all English-men, by W. C. Esq; a true Lover
of Art and Nature* (London: Printed by *W. Godbid* for *William Cooper* in Little St. *Bartholo-
mews*, near *Little-Britain*, 1669), from "The Publishers Epistle to the English Reader,"
the larger work hereinafter referred to as Philalethes, *Secrets Reveal'd*.

[77] Wilkinson, "Identity" (3, n. 25).

fastened the *Serpent* through and through unto; Learn what *Diana's Doves* are, which do vanquish the *Lion* by asswaging him; I say the Green *Lion*, which is in very deed the *Babylonian Dragon*, killing all things with his Poyson: Then at length learn to know the *Caducean Rod* of *Mercury*, with which he worketh Wonders, and what the *Nymphs* are, which he infects by Incantation, if thou desirest to enjoy thy wish.[78]

However, the writings of the two Philalethes are far from representing the totality of the Hartlibians' approach to alchemical matters, for within the group a contrasting attitude towards alchemy began to appear during the 1640's and the 1650's. Something of the new approach which was developing may be seen in the little volume entitled *Chymical, Medicinal, and Chyrurgical Addresses: Made to Samuel Hartlib, Esquire*, published in London in 1655.[79] The second of the addresses is actually one of the tracts by Eirenaeus Philalethes, the first of them to be published. It, and the first address (an inquiry into the cabbalistic Urim and Thummin, possibly by Dury), represent the more esoteric tradition in which the Hartlib group was embedded. But in the fifth address Robert Boyle begins the call for complete openness in communication.[80] Boyle's essay was written to encourage the sharing of medical secrets but his language made his appeal quite general. Urging charity rather than covetousness, and refuting several current arguments for secrecy, Boyle goes on to reveal the piety and humanitarianism which provided the motivation for the Hartlibian reforms.

[I]f...the Elixir be a secret, that we owe wholly to our Makers Revelation, not our own industry, me thinks we should not so much grudge to impart what we did not labour to acquire, since our Saviours prescription in the like case was this: *Freely ye have received, freely give.* Should God to one of our Divines reveal some newer Truths and Secrets of his Gospel, would we not condemn him for the concealment of what was imparted but to be communicated? Those secrets that were intended for our use, are not at all profaned by being made to

[78] Philalethes, *Secrets Reveal'd*, pp. 1–6, quotation from p. 6 (3, n. 76).

[79] *Chymical, Medicinal, and Chyrurgical Addresses: Made to Samuel Hartlib, Esquire, viz.* 1. *Whether the Vrim & Thummin were given in the Mount, or perfected by Art.* 2. *Sir George Ripley's Epistle, to King Edward unfolded.* 3. *Gabriel Plats Caveat for Alchymists.* 4. *A Conference concerning the Phylosophers Stone.* 5. *An Invitation to a free and generous Communication of Secrets and Receits in Physick.* 6. *Whether or no, each Several Disease hath a Particular Remedy?* 7. *A new and easie Method of Chirurgery, for the curing of all fresh Wounds or other Hurts.* 8. *A Discourse about the Essence or Existence of Mettals.* 9. *The New Postilions, pretended Prophetical Prognostication, Of what shall happen to Physitians, Chyrurgeons, Apothecaries, Alchymists, and Miners.* (London: Printed by *G. Dawson* for *Giles Calvert* at the *Black-spread Eagle* at the west end of *Pauls*, 1655), hereinafter referred to as *CMCA*. This publication has been discussed from the viewpoint of attempts at medical reform in Webster, "English medical reformers" (3, n. 63).

[80] Boyle's essay is reprinted, with commentary, in Margaret E. Rowbottom, "The earliest published writing of Robert Boyle," *Annals of Science* **6** (1950), 376–89.

reach their end: but by being fettered from the diffusiveness of their nature. And therefore though God should address those special favours but to some single person; yet he intendeth them for the good of all Mankind, and to make that Almoner to whom he trusteth them, not the grace but the steward of his graces.[81]

Gradually the ideal of free communication which Boyle advocated was spreading to alchemical as well as medical secrets. As it did so, alchemy necessarily lost its earlier religious and psychological appeal, for, as discussed in Chapter 2, that appeal depended upon a certain amount of ignorance both of the adept's psyche and of the behavior of matter. Open discussion of the physical reactions was bound to eventuate in a loss of that necessary ignorance about the behavior of matter.

In the little volume of addresses to Hartlib, along with Boyle's appeal for open communication in medicine, there appeared two early efforts at open communication in alchemy. One was the essay contributed by Gabriel Plattes, his "Caveat for Alchymists," Essay 3 in the Hartlib volume.[82] Plattes undertook to give several practical guidelines about books and procedures to be used in alchemical studies. The *Margarita pretiosa* of Petrus Bonus was worthless, he said, as were the writings of Gaston Dulco Clavens. The latter's "many experiments" were "all false, upon my certain knowledge, and if my purse could speak it, should swear it." Sir George Ripley, Nicolas Flamel, Raymundus Lullius, and Bernardus Trevisanus were the best ones to read.[83] In his recommendations, Plattes reflected the general popularity of Ripley, Flamel, Lull, and Trevisan in the seventeenth century, noted above.

But Plattes went further and attempted a generalized statement about alchemical procedures that a thoughtful student of alchemy should be able to formulate for himself by careful reading: (1) that "he must have a mineral spirit" in order to get the first dissolvent, (2) "he must know the secret of dissolution," (3) "he must know what is meant by the hollow Oak" (which Plattes implies is the furnace in which dissolution occurs), (4) he must know how to refix his bodies after making them volatile by dissolution, (5) he must learn that there is an extra step between projection of the Philosopher's Stone onto a base metal and the testing of the metal, and (6) he must understand the nature of the fire "which is to be gentle, continual, compassing round about the matter, and not burning it."[84]

Plattes' instructions given here are not truly operational, of course, but they represent an attempt at a rational statement of exactly what the alchemical process required. Plattes also undertook to "discover" to the

---

[81] *Ibid.*, p. 384; *CMCA*, pp. 129–30 [incorrectly paginated as 147–48] (3, n. 79).

[82] D. Geoghegan, "Gabriel Plattes' caveat for alchymists," *Ambix* **10** (1962), 97–102; *CMCA*, pp. 49–87 [incorrectly paginated as 88] (3, n. 79).

[83] *CMCA*, pp. 59–62, quotation from p. 60 (3, n. 79).

[84] *Ibid.*, pp. 55–58.

public several notable "cheats" and their techniques of deception. These were the two principal presentations he wished to make in writing his tract – a statement about what the alchemical process really required and a series of practical warnings about imposters. Plattes' motivation was rooted in a Christian charity, which he made explicit: "I cannot chuse but to publish these advertisements, for that is a fundamental point in my Religion, to do good to all men, as well enemies as friends...."[85]

He went on to say that if he "could be satisfied that the publishing thereof [of the alchemical process] would do more good than hurt; then the world should have it in plain terms, and as plain as an Apothecaries receit...."[86] Since he was not convinced of that point, however, he was planning to secure the benefits of alchemy for the English nation in a roundabout way which, although public, involved the use of adequate safeguards. He had petitioned Parliament for the privilege of demonstrating to it new improvements in husbandry, "Physick," and in "the Art of the transmutation of Mettalls." The last demonstration was to be made only

> If I may have a Laboratory, like to that in the City of *Venice*, where they are sure of secrecy, by reason that no man is suffered to enter in, unless he can be contented to remain there, being surely provided for, till he be brought forth to go to the Church to be buried.[87]

Plattes' letter to the readers of his essay was dated "At Westminster, this 10. of March, 1643."[88] That period was just past the high-water mark of Hartlibian influence in Parliament. Although interest in the proposals of Hartlib, Dury, and Comenius was high in 1641, almost immediately afterward Parliament had been distracted by civil war concerns.[89] Hartlibian proposals were to be presented again in Parliament on later occasions, but Plattes probably died before he ever had an opportunity to make his demonstration.[90]

It must have been a very near miss, however, for Plattes apparently almost caught the high tide of interest in 1641. In his utopian tract entitled *A Description of the Famous Kingdome of Macaria* (London, 1641), Plattes made mention of the affair.[91] In the person of the *Traveller* in the dialogue of the book, Plattes offers to the *Schollar* a copy of his book on husbandry. Returning he asks, "Well, have you perused my book?"

---

[85] *Ibid.*, p. 52.

[86] *Ibid.* Debus has noted that Plattes actually had already published a procedure by which he believed he had transmuted a regulus of iron and copper into gold, in his *A Discovery of Subterraneall Treasure* (London, 1639). See Allen G. Debus, "Gabriel Plattes and his chemical theory of the formation of the earth's crust," *Ambix* **9** (1961), 162–65.

[87] *CMCA*, p. 86 [incorrectly paginated as 87] (3, n. 79).

[88] *Ibid.*, p. 50.   [89] Webster, *Hartlib*, pp. 25–39 (3, n. 63).   [90] *Ibid.*, p. 12.

[91] Printed in *ibid.*, pp. 79–90. The work has long been attributed to Hartlib, but Webster has recently shown that Plattes himself wrote the little book. It was dated Oct. 25, 1641, and addressed to Parliament. See Charles Webster, "The authorship and significance of *Macaria*," *Past and Present* no. 56 (August 1972), 34–48.

Sch. Yes Sir: and finde that you shew the transmutation of sub-
lunary bodies, in such manner, that any man may be rich
that will be industrious...; this booke will certainly be highly
accepted by the high Court of Parliament.

Trav. Yes, I doubt it not; for I have shewed it to divers Parliament
men, who have all promised mee faire, so soone as a seasonable
time commeth for such occasions.[92]

Plattes' "seasonable time" never came, but Hartlib perhaps hoped to
arouse Parliamentary interest in the demonstration again by publishing
Plattes' essay in the 1655 collection.

"Alchemy for the public good" one might call Plattes' remarkable plan
to demonstrate transmutation before Parliament. Along with improvements in husbandry and medicine, the benefits of alchemy were to be conferred upon the whole social fabric. Once again the humanitarianism and
the public spiritedness of the Hartlib circle are made manifest.

The other essay in the Hartlib volume that aimed at open communication in alchemy was Essay 4, "A Conference concerning the Philosopher's Stone." It derived from Hartlib's earlier counterpart in Paris,
Théophraste Renaudot. Renaudot had died in 1653, in a state of relative
disgrace after successful attacks against him by the Medical Faculty at the
Sorbonne for his advocacy of medicinal antimony and as a consequence
of a decline in his political fortunes, but his Parisian *Bureau d'Adresse* lived
on in the Hartlibian Office of Address in London. Renaudot had acted
as a clearing house for ideas and for employment and had fostered the use
of chemical medicines, as Hartlib was still doing. Renaudot had also held
weekly conferences at his house, however, which Hartlib did not undertake.
The conferences were on any subject which the attending public requested,
with the exceptions of religion and affairs of state, and their general purpose was the airing and discussing of all points of view – again the ideal of
free and open communication appearing.[93]

The essay printed by Hartlib was one of Renaudot's weekly conferences,
translated for presentation to the English public. It contained the opinions
of seven different speakers at the conference. Two of the men had argued
against the possibility of metallic transmutation, whereas five thought it
possible, probable, or certain. The whole offers a valuable panorama of
early seventeenth-century views on alchemy, which would bear detailed
analysis, but here only a few points of interest may be mentioned. The
five men who favored alchemical views presented "reasons" which were
frequently only restatements of standard Arabic or Aristotelian alchemical

[92] Webster, *Hartlib*, p. 88 (3, n. 63).
[93] Brown, *Organizations*, pp. 18–30 (3, n. 64); Howard M. Solomon, *Public Welfare,
Science, and Propaganda in Seventeenth Century France. The Innovations of Théophraste Renaudot*
(Princeton: Princeton University Press, 1972).

theory. Several analogies with organic growth or metamorphosis were offered also. Perhaps the most unusual "reason" given was that from the fourth speaker who said that "the Greek Etymologie of Mettals doth shew that they may be changed from one to another."[94] The first man, who rejected alchemical notions, did so as a sceptic and a cynic; men had always been prone to wishful thinking and their belief in the Philosopher's Stone was merely another example of that, he said. Only the sixth speaker showed much evidence of a critical judgment of concepts or of a critical use of empirical information. He rejected the alchemical maturation of metals because mines of iron, lead, etc., never become mines of gold, and art can never do what nature cannot. "And besides," he went on,

it is a mistake to say, that Mettals are all of one kind, and that they differ but in coction [state of being "done" or "cooked," roughly equivalent to maturation], for we see that Iron is more conconcocted ["cooked together," made hard] than Silver, it being harder, and not so easie to melt....[95]

The speaker was attempting to distinguish two rather divergent meanings of "coction" and to show that the "perfection" which supposedly came from full "coction" or "maturation" did not make allowance in its definition for the greater hardness which "coction" was supposed to lend metals, a good effort to break up the scalar arrangement of the metals which ideas of metallic "perfection" implied. But even he rather spoiled his effect by adding immediately that neither should the metals be ranked in a graded scale of "coction" because "their differing is needful for mans use."

The vagaries expressed at M. Renaudot's "Conference" were the opinions of individual men on the subject of alchemy. Their advocacy at an open meeting – and Hartlib's later publication of them – moved them into the area of public dialogue where assumptions underlying alchemical theory could be subjected to a critical analysis of just the sort which the sixth speaker offered. And conceptual scrutiny was being paralleled elsewhere in the group by a more open communication of empirical information on alchemical processes, which communication also tended to move alchemy into the public arena.

Even before the publication of the essays, the Hartlib group had undertaken a venture perhaps unique in the annals of alchemy up to that time: a joint commercial operation for the manufacture of gold. Benjamin Worsley, an English physician in contact with Hartlib from about 1645, and Hartlib's Dutch correspondent for chemical matters, Johann Morian, had joined together and were being financed by an unidentified "gold maker." The "Aurifaber" of Amsterdam, as Hartlib called him, had in-

[94] CMCA, pp. 101–12, quotation from p. 108 (3, n. 79).
[95] Ibid., pp. 109–10, quotation from p. 110.

vested £1200 in the scheme to transmute tin into gold, when, in 1651, Starkey developed procesess for extracting a silver from antimony and for transmuting iron into gold. Information about Starkey's discoveries was duly circulated by Dury and Hartlib, and Starkey was asked to join the gold-making venture.[96]

References in the Hartlib papers to their joint efforts fade away without comment on their success, but the affair was marked by a willingness to share alchemical secrets. "Mr. St. Luna fixa" – evidently Starkey's process for obtaining alchemical silver from antimony – circulated within the group in a fully operational form (except that the quantities are not given in Hartlib's copy).[97] The contrast between these clear directions and the arcane formulations of earlier and still current literature could hardly have been greater. Starkey's process is fully intelligible. It had to be made so, of course, in order for it to be useful to others in the group.

In the Hartlib circle all the necessary elements for this scheme of commercial alchemy were present; belief in transmutation – indeed the notion that chemical philosophy contained the key to the universe; a fairly close-knit group devoted to humanitarian projects and in constant need of money for them; an experimental tradition – partly from alchemy itself, partly from the new experimental natural philosophy of Gilbert, Harvey, Torricelli, Pascal, etc.; and the ideal of the free communication of secrets for the good of mankind.

Aside from Morian and Worsley, about whose activities little is known, the Hartlib circle had up to the mid-1650's pinned its hopes for good chemical and alchemical experimentation largely on Boyle and Starkey. Boyle's development has been studied extensively and it is beyond the scope of this chapter to enter into it again. Some of his specific ideas on transmutation and on the "opening" of metals will be explored in Chapters 5 and 6, when his influence on Newton is considered, and the interested reader may approach him more generally through Boas (Hall),[98] Fulton,[99] his own *Works*,[100] More,[101] Ihde,[102] Maddison,[103] and

[96] Wilkinson, "Hartlib. Part II," pp. 90–92 (3, n. 70).

[97] Hartlib Papers, XVI, 1, quoted in full in *ibid.*, p. 101, n. 91.

[98] Marie Boas (Hall), *Robert Boyle and Seventeenth-Century Chemistry* (Cambridge: At the University Press, 1958).

[99] John F. Fulton, *A Bibliography of the Honourable Robert Boyle, Fellow of the Royal Society* (2nd edn., Oxford: At the Clarendon Press, 1961), hereinafter referred to as Fulton, *Bibliography*.

[100] Boyle, *Works* (3, n. 68).

[101] More, "Boyle as alchemist" (2, n. 53); Louis Trenchard More, *The Life and Works of the Honourable Robert Boyle* (London, New York, Toronto: Oxford University Press, 1944).

[102] Ihde, "Alchemy in reverse" (2, n. 52).

[103] R. E. W. Maddison, *The Life of the Honourable Robert Boyle F.R.S.* (London: Taylor & Francis; New York: Barnes and Noble, 1969), hereinafter referred to as Maddison, *Boyle*.

Wilkinson.[104] Boyle was in any case generally located away from London and was spending a great deal of energy on other projects.

George Starkey was in London and was quite active experimentally for a number of years after his arrival there in 1650, but Hartlib began to have less faith in him after a while.[105] By about the end of 1653 Hartlib had given up all hope that Starkey would make a great breakthrough either in chemical medicine or in alchemy. Hartlib wrote to Boyle:

...Dr. *Stirk*...is altogether degenerated, and hath, in a manner, undone himself and his family. I know not directly how many weeks he hath lain in prison for debt; but after he hath been delivered the second time, he hath secretly abandoned his house in *London*, and is now living obscurely.... When God hath brought you over again [from Ireland], we shall leave him altogether to your test, to try whether yet any good metal be left in him or not. But the best is, that he stands more in need of us, than we of him.[106]

The reason Hartlib was so easy in his mind about giving up on Starkey was that shortly after the Amsterdam venture – about 1651–52 – there had appeared a new experimenter on the scene, Frederick Clodius. His standing with Hartlib rose as Starkey's declined. Clodius had settled in London and had become intimate with the group there, sharing opinions and research on alchemy, iatrochemistry, and husbandry with Boyle, Starkey, Hartlib, and others. He then married Hartlib's daughter and set up his laboratory in Hartlib's back-kitchen.[107] Hartlib told Boyle about the new laboratory in the same letter which carried news of Starkey's "degeneration."

As for us, poor earthworms, we are crawling in my house about our quondam back-kitchen, whereof my son hath made a goodly laboratory; yea, such a one, as men (who have had the favour and privilege to see, or to be admitted into it) affirm, they have never seen the like, for its several advantages and commodiousnesses. It hath been employed days and nights with no small success, God be praised, these many weeks together. But what particulars it hath hitherto produced, and what greater medicines he is taking in hand this week and the next, I suppose this my obedient and very chemical son will be able better to relate unto you, than myself.[108]

Not much is known about Clodius' chemical aims, but more should appear as the Hartlib papers are examined more closely. Even at the present writing it seems clear that he provided the Hartlib circle with a

[104] Wilkinson, "Hartlib. Part I" (3, n. 70).

[105] Wilkinson, "Hartlib. Part II," pp. 103–04 (3, n. 70).

[106] Samuel Hartlib to Robert Boyle, Feb. 28, 1653/54, in Boyle, *Works*, VI, 79–80 (3, n. 68).

[107] Wilkinson, "Hartlib. Part II," pp. 96–98 and 103 (3, n. 70).

[108] Samuel Hartlib to Robert Boyle, Feb. 28, 1653/54, in Boyle, *Works*, VI, 79 (3, n. 68).

renewed emphasis on alchemical experimentation. In this he had the strong backing of Sir Kenelm Digby, who shared Hartlib's enthusiasm for Clodius.

Digby had been immediately attracted to Clodius. Hartlib told Boyle that Digby

> protested seriously unto me, that in all his travels and converses with the choicest wits, both in *Italy* and *France*, he hath not met so much of theoretical solidity and practical dexterity both together, as he finds in my chemical son; and therefore is resolved to improve that prize accordingly, which providence hath brought to his hands.[109]

Digby pressed Clodius to accept from him a complete provision for himself and his family for two years which was to include the furnishing of a laboratory.[110] The laboratory was to be a "universal" one, in line with the other pansophic plans of the Hartlibians, and Digby was to contribute £600–700 to it as soon as he again received possession of his sequestered properties. In the meantime a "general chemical council" had been constituted and Digby had already contributed plans for new types of furnaces, drawn from a French correspondent of his. Hartlib noted that they were to be "erected ere long amongst ourselves, to prosecute really philosophical studies." In his exuberance Digby had also freely shared his secrets with Clodius, probably many of the very ones later to appear in Digby's *Secrets* in 1682. Hartlib said:

> He [Digby] hath many excellent secrets and experiments of all sorts, yea, some arcana of the highest nature, which he hath already freely (yet *sub fide silentii*) imparted into his [Clodius'] breast; and is purposing to send for all his papers out of *France*, that he may put them into his custody and management. Both their judgments and experiences agree mightily together, to the very amazement of each other. And there wants nothing to the perfecting of the grand design, but that you [Boyle] are not present with them, without whose interest we have no mind that it should be consummated.[111]

Digby had had numerous earlier contacts with the Hartlib group[112] and the eagerness with which he fell in with the new "chemical council" and the plans for a "universal laboratory" reflect also his own long standing interest in increasing the public stock of knowledge. By the time he met Clodius in 1654 Digby had already reached a point of development similar to that being exhibited by the Hartlib group in England. Digby had had

---

[109] Samuel Hartlib to Robert Boyle, May 8, 1654, in *ibid.*, VI, 87.

[110] Samuel Hartlib to Robert Boyle, May 15, 1654, in *ibid.*, VI, 89.

[111] Samuel Hartlib to Robert Boyle, May 8, 1654, in *ibid.*, VI, 86–87, quotation from p. 87.

[112] There are hundreds of references to Digby in the Hartlib papers. Private communication from R. S. Wilkinson, May 18, 1972.

his own alchemical and Hermetic background, but it was one which was subtly different from that of the other Hartlibians. He had been much less influenced by the reforming zeal of the Andreae–Comenius–Hartlib trio with their tinge of Rosicrucian mysticism and their interest in macrocosmic–microcosmic relationships. Rather he had been steeped in a basic Aristotelianism along with his more occult studies and by 1654 had long been strongly influenced by the mechanical philosophers.[113]

The mechanical philosophers were after all quite as eager as the Hartlibians for the reform of knowledge and the increase of it. In another context Digby had said of Mersenne,

> This Problem [in local motion] was proposed to me by that worthy religious man, Father *Mersenius*: who is not content with advancing learning by his own industry and labours; but besides, is alwayes (out of his generous affection to verity) inciting others to contribute to the publick stock of it.[114]

Digby had known Mersenne from about 1637 until Mersenne's death in 1648 and had most assuredly absorbed that ideal. So although no evidence appears to show that Digby was consciously acting upon Mersenne's program for the reform of alchemy, yet in a more general way he was certainly influenced by Mersenne's ideas. And in the "general chemical council" and the "universal laboratory" of Digby and Clodius there appears as full an embodiment of Mersenne's idea of an alchemical academy as one could ever hope to find. There alchemy was to be studied without mystery and without secrets. There mechanical philosophy, Rosicrucian reform, Aristotelianism, Neoplatonism, chemical medicine, empiricism, and Christian humanitarianism met and fused to work towards an experimental alchemy.

Evidence on the outcome of the grand plans of 1654 is scanty but the general trend of events is clear enough. Boyle did not join Digby and Clodius in London: rather he went to Oxford about 1655, as is well known. Digby returned to France, probably in 1656. Correspondence maintained contact among the three while in London building plans went forward under Clodius' sporadic guidance with at least an occasional visit from Boyle and probably some financial support from him. Clodius may have become increasingly unstable and gone into some sort of decline; in the event alchemical activities seem to have been centered on Digby's house

---

[113] Dobbs, "Digby. Part I," pp. 13–16 (2, n. 28), and Dobbs, "Digby. Part II" (2, n. 58).

[114] Kenelm Digby, *Of Bodies, and Of Mans Soul, To Discover the Immortality of Reasonable Sovls. With two Discourses Of the Powder of Sympathy, and Of the Vegetation of Plants* (London: Printed by *S. G.* and *B. G.* for *John Williams*, and are to sold in Little *Britain* over against St. *Buttolphs-Church*, 1669). In this edition of Digby's major works, *Of Bodies* has a separate pagination and *Of Mans Soul*, *Of the Powder of Sympathy*, and *Of the Vegetation of Plants* have a different but continuous pagination. Digby's reference to Mersenne occurred in his treatise *Of Bodies*, p. 91.

and laboratory after he returned to London in 1660, rather than at Clodius' establishment.

While Digby was still in London in 1654–55, both he and Clodius maintained a chemical correspondence with Boyle, as did others of Hartlib's circle such as Starkey. Evidence for this correspondence rests in the unpublished Boyle Papers at the Royal Society, especially in Boyle's "Philosophical Diary" for 1654/55 and his "Promiscuous Observations" of September–December 1655.[115] After Digby returned to France he also wrote to Clodius, sending him "observables." In May 1658, Hartlib told Boyle,

My son *Clodius* hath also received a letter from Sir *K. Digby*, with many observable things in it. He tells me that Sir *K.* hath also published of late the treatise of the sympathetical powder; and, if I be not mistaken, that he hath gotten, or is shortly to get a copy of it.[116]

Later the same year Hartlib made an extended reference to Clodius' "chemical college or laboratory" in a letter to Boyle. It is clear that plans for the laboratory were still in hand and that Boyle had considerable interest in the laboratory still, and in Clodius also.

My son *Clodius* is under new trials, his wife being brought to bed this night of a child that died soon after, the mother remaining very weak. He wrote to you last *Saturday*. His chemical college or laboratory will not be so much prostituted as you fear; but there was an absolute necessity to resolve upon such a course amongst a few confiding friends. There are near three furnaces finished, but five more must be added unto them; and having these, they shall be able to command any kind of operation whatsoever. I know not how he will be able to dispose of his business, so as to give you now a visit at *Oxford*: but I suppose that some occasion or other, in these changes of public affairs, may invite you to *London* sooner, it may be, than you had resolved.[117]

Although the published evidence gives no inkling of what Boyle's fears in the matter were, he did find occasion to visit the laboratory at some time in 1658 or thereabouts, for ten years later Agricola wrote to Boyle that he had had the pleasure of seeing him there.

It is almost ten years ago since I went out of *England*, and as I dwelt with Mr. *Frederic Clodius*, I had the happiness and honour to see your nobleness, and to wait upon you in our laboratory....[118]

[115] Boyle Papers VIII and Boyle Papers XXV, cited in Maddison, *Boyle*, pp. 85–86 (3, n. 103).
[116] Samuel Hartlib to Robert Boyle, May 25, 1658, in Boyle, *Works*, VI, 109 (3, n. 68).
[117] Samuel Hartlib to Robert Boyle, Sept. 14, 1658, in *ibid.*, VI, 114.
[118] G. [J.C.] Agricola to Robert Boyle, April 6, 1668, in *ibid.*, VI, 650.

Probably also Boyle gave money to Clodius during the period of the late 1650's, for in 1663 Clodius wrote to Boyle asking for more financial assistance.[119]

In that same letter Clodius claimed that thefts and misfortunes were all conspiring to his ruin. He had still been translating treatises on the philosopher's stone in 1659,[120] but what exactly happened to him or to his laboratory after that time remains unknown. Not even the date of his death has been recorded, but it is clear that he passed from prominence in the chemical circles of London at least by the early 1660's.

With the Restoration of 1660 Digby returned to London and Nicolas le Fèvre came too, apparently at the request of that royal experimenter who was about to receive his crown. Le Fèvre, it will be recalled, had been the chemical demonstrator at the Jardin Royale in Paris during the 1650's. While there he had instructed both Digby and John Evelyn, both of whom were shortly to become active in the Royal Society, and other Hartlibians such as Oldenburg are known to have come into contact with le Fèvrian chemical circles in Paris.[121] Le Fèvre was a thoroughgoing Neoplatonist philosophically and had much to do with the mid-century efflorescence of that philosophical position. By November 15 Charles II had made le Fèvre his "professor of chemistry" and by December 31 also apothecary in ordinary to the royal household. Le Fèvre was also put in charge of Charles' laboratory at St. James' Palace and in 1661 was elected Fellow of the Royal Society.[122] The object of Charles' and le Fèvre's experiments in the royal laboratory is not known with any certainty, but there have been some speculations on Charles' desire to "fix" mercury and on his possible death by mercury poisoning.[123] In all likelihood it was the overlap of Charles' alchemical–chemical interests with those of the Royal Society which encouraged him to extend his patronage to that body.

Probably many of the alchemical activities of the Hartlib circle centered in Digby's house after his final return to London in 1660, in continuation of the plan for a "general chemical council" or a "universal laboratory." He had taken up residence "in the last faire house westward in the north portico of Convent garden"[124] and again established his laboratory. Dropping out of court and political life, he established a salon of the French

---

[119] F. Clodius to Robert Boyle, Dec. 12, 1663, unpublished Boyle Letters at the Royal Society (B. L. II, 17), cited in Maddison, *Boyle*, p. 87 (3, n. 103).

[120] Samuel Hartlib to Robert Boyle, May 17, 1659, in Boyle, *Works*, VI, 125 (3, n. 68).

[121] Dobbs, "Digby. Part II," pp. 150–51, 156–59 (2, n. 58).

[122] *Dictionary of National Biography*...ed. by Sir Leslie Stephen and Sir Sydney Lee (22 vols.; London: Oxford University Press, 1949–50), 11, 840, hereinafter referred to as *DNB*; Partington, *History*, III, 17–24 (3, n. 20).

[123] See Dobbs, "Digby. Part III," pp. 26–27, and the references cited there (2, n. 59).

[124] John Aubrey, "*Brief Lives.*" *Chiefly of Contemporaries, set down by John Aubrey, between the Years 1669 & 1696*, ed. by Andrew Clark (2 vols.; Oxford: At the Clarendon Press, 1898), I, 227.

type which "became a *rendezvous* of mathematicians, chemists, philosophers, and authors."[125]

Balthazar de Monconys (1611–65), a French *virtuoso* who traveled widely and gathered up all he could of the "New Learning" wherever he went, was in London in 1663. "Cheualier d'Igby" is much in evidence in Monconys' journal of that visit, advising Monconys on a good microscope maker and discussing the "Touching for the King's Evil" with him. Monconys visited Digby's laboratory and talked with his operator, whom he thought to be unusually honest.

> De là ie fus au Laboratoire de M. d'Igby, où son Artiste me sembla
> vn des plus honestes hommes de cette profession, en me confessant
> qu'il n'auoit rien appris dans tout le temps de sa vie qu'il auoit
> employée à cette sçience, si non qu'il n'y sçauoit rien.[126]

This honest artist may have been George Hartman who was steward and laboratory assistant to Digby at the time of his death and who was later to gather together Digby's collection of alchemical recipies and publish them as Digby's *Secrets* in 1682.

Many of the materials upon which Hartman drew for that book seem to have been collected by Digby in the last ten years of his life, as those items which are dated are all from the late 1650's and early 1660's. Some of the recipes derive from Digby's last trip into the Germanies in 1659; several are le Fèvrian in nature; one came from Frederick Clodius. Several came from French correspondents of Digby's about whom nothing is known; others are clearly Digby's own or were developed in his laboratories. A few items are not recipes at all but are notes comparing terms from older alchemists and interpreting them. Recipes for cosmetics (another of Digby's many interests) and for medicines (among them Starkey's pills and Digby's own sympathetic powder) are included. Almost every selection in the work is given in fully operational terms, and the book as a whole is undoubtedly a product of that strange confluence of

---

[125] T. L. [Thomas Longueville], *The Life of Sir Kenelm Digby by One of his Descendants* (London, New York and Bombay: Longmans, Green and Co., 1896), pp. 290–91. Longueville added a footnote to "See Appendix F," apparently intending to add details on Sir Kenelm's final establishment. One wishes that he had done so; since he had access to family documents for his biography, some significant new information might have emerged. But Appendix F seems never to have been printed; the book ends with Appendix E in the middle of the last page.

[126] Balthazar de Monconys, *Iovrnal des voyages de Monsievr de Monconys, Conseiller du Roy en ses Conseils d'Estat & Priué, & Lieutenant Criminel au Siege Presidial de Lyon. Ou les Sçauants trouueront vn nombre infini de nouueautez, en Machines de Mathematique, Experiences Physiques, Raisonnemens de la belle Philosophie, curiositez de Chymie, & conuersations des Illustres de ce Siecle; Outre la description de diuers Animaux & Plantes rares, plusieurs Secrets inconnus pour le Plaisir & la Santé, les Ouurages des Peintres fameux, les Coûtumes & Moeurs des Nations, & ce qu'il y a de plus digne de la connoissance d'vn honeste Homme dans les trois Parties du Monde. Enrichi de quantité de Figures en Taille-douce des lieux & des choses principales. Auec des Indices tres-exacts & tres-commodes pour l'vsage. Publié par le Sieur de Liergves don Fils.* (3 vols.; Lyon: Chez Horace Boissat, & George Remevs, 1665–66), II, 57.

the Hartlibian school of thought with that of the mechanical philosophers which called forth both the experimental examination of alchemical procedures and also their rational description. As such it may be considered as representative of much of the later activity of the whole Hartlib circle, at least until a fuller exploration of the Hartlib papers proves otherwise.

Such organization as was present in the group diminished in the later 1660's. Hartlib died in 1662, Digby in 1665, le Fèvre in 1669, and the Royal Society furnished a new focus for many chemical activities. But the Hartlib circle cannot be said to have died intestate, for it left to its most famous member, Robert Boyle, an incurable interest in matters alchemical and an aptitude for expressing alchemical ideas in the idiom of orthodox mechanical philosophy. Boyle published a work in 1666 which did just that – offer a mechanical explication of alchemical transmutation – and when Isaac Newton read that book he became interested in alchemy too and began to collect the manuscripts the Hartlibians had circulated. The details to substantiate this notion of the direct influence of the Hartlib circle upon Newton will be developed in Chapters 4, 5, and 6, but first it will be well to pause to relate historical events in alchemy in the later sixteenth and in the first two-thirds of the seventeenth century to the ideas drawn from Jung's treatment of alchemy and to treat very briefly the historical developments which seem to have occurred in the last half of the century.

Mid-century summary and preview

Following Jung's analysis, the present study has held to the view that the older alchemy served a real though largely unconscious religious function for the adepts and that that spiritual aspect of alchemy received emphasis during a time of religious unrest and dissatisfaction after the Reformation. At that time the Church was no longer able to offer a clear, single, and undisputed path to salvation, and out of religious frustration and disgust with theological squabbling many men turned to alchemical studies in which the mystical and spiritual side of alchemy was exaggerated to the neglect of the side which called for physical work in the laboratory. In reaction to that shift in emphasis, more practically and materially oriented men attacked the mystical movement and evoked a counter-movement which overemphasized the physical side of alchemy. A complicated series of events ensued, and many conflicting philosophical positions interacted in ways which require a great deal of further study to be well understood. But it seems clear that both the mechanical philosophers and the reformers who were descended intellectually from the mystical Rosicrucians con-tributed to the new alchemy which insisted upon full communication of alchemical secrets, experimental study of alchemical processes, and full description of experimental results in common chemical terminology.

That was excellent for chemistry, which was thereby enabled to incorporate into itself a rational alchemical paradigm, but it was deadly for the older alchemy. It had been too thoroughly chemicalized to carry out its older functions of a religious and psychological nature, for those functions required a considerable ignorance about the substances with which the alchemist worked. From that time on the intertwined halves of the older alchemy were irrevocably separated.

It is far beyond the scope of this study to consider the subsequent development of spiritual alchemy or theosophy, which has never really quite faded away, but a few of the developments in chemical alchemy in the last half of the seventeenth century and on through the work of Boerhaave will be treated. It will be seen that during that period the decline of alchemy entered the second stage suggested in Chapter 2. Digby and his contemporaries had with their chemicalization of alchemy enacted the first step. After them came the experimenters who slowly chipped away at the processes of chemical alchemy, proving them unverifiable, until finally nothing was left but nostalgia and a wistful hope. Digby's book of *Secrets* offers examples of the type of material with which the later experimenters began their work.

Two somewhat contradictory developments seem to have taken place in alchemy in the last half of the century. One was the re-writing of alchemical theory in mechanical terminology; the other was the second stage in the decline of alchemy suggested earlier, in which new experimenters found alchemical procedures inherited from Digby's generation to be unverifiable. No attempt will be made here to be at all exhaustive; rather the intent is to suggest by a few examples the lines of development which were contemporary with Newton's work in alchemy.

The first of these developments, the theoretical, may be dealt with here in a summary fashion with a single example in the person of John Webster, famous for his part in the Webster–Ward debate on education.[127] Webster's theory differed very little in its basic concepts from the Aristotelian theory of Digby's, which probably dates from the 1630's,[128] or from the more mechanical theory of Boerhaave, which dates from early in the eighteenth century and which has already been presented in part in Chapter 2.

Webster's theory is of interest primarily in that it does mark a chronological mid-point between Digby's early work and the time of Boerhaave. Also because Webster's corpuscularianism seems a little fuzzy still – he uses "parts," "atoms," and "Minima" interchangeably – he marks a

[127] *DNB* **20**, 1036–37 (3, n. 122); Debus, *Science and Education*, pp. 37–43 (3, n. 58).
[128] Dobbs, "Digby. Part III," pp. 9–17 (2, n. 59).

half-way house in the mechanical re-thinking of alchemy. Digby, although he used mechanical concepts for many years, seems never to have gotten around to re-thinking much of his alchemy in terms of mechanical particles. There was no example of mechanical alchemy in Digby's book of *Secrets*. The time of most influence of the mechanical philosophies on alchemical thought seems to have come during or a little after Digby's active collection period, and perhaps derived from some of Boyle's writings which will be considered in Chapter 6. Webster's theory, published in 1671, does show such an influence. That fact is of special interest because Webster and Digby had both done alchemical work with the same alchemical operator, one Hans Hunneades, in the 1630's.[129]

Webster says that "the Vulgar" and also most of the "Mine-men" that he has talked to deny that metals grow in the earth, but he insists that most learned men believe that they do.[130] He himself argues at length for transmutation on the grounds of experience, theory, and authority. Unusual is his argument based on petrifaction, which he takes to be a form of transmutation. "From all this we may plainly gather," he says, "what Transmutation of Metals is, and how it is wrought:"

> So that if Metals be in their root all one Mercurial and Homogeneous nature, and that there be perfect Sulphur and Mercury equally as well in the imperfect as perfect Metals, then must their Transmutation be easie; for then the Heterogeneous matter, or combustible Sulphur, Scoria, or Dross, being removed, and some of the Tincture added, the parts are most closely joined, and so united *per minima*, and tinged, by which means they are maturated in a short time by the help of Art, that Nature could not perform in many years. So that all metallick Mercury wants nothing of the degrees and nature of Gold; but removing of its Heterogeneous parts, and the adding something more of the fire of Nature, and then it becomes most dense, and to have all the requisites that are necessary to Gold."[131]

In Webster's formulation the Aristotelian "forms" are missing, but his "one Mercurial and Homogeneous nature" at the root of all metals is surely equivalent to an Aristotelian "substantial form." Likewise Webster's "Heterogeneous parts" are similar to Aristotelian "accidents" which prevent the baser metals from being gold. The chief difference,

[129] Dobbs, "Digby. Part II," p. 148 (2, n. 58).

[130] John Webster, *Metallographia: or, An History of Metals. Wherein is declared the signs of Ores and Minerals both before and after digging, the causes and manner of their generations, their kinds, sorts, and differences; with the description of sundry new Metals, or Semi Metals, and many other things pertaining to Mineral knowledge. As also, The handling and shewing of their Vegetability, and the discussion of the most difficult Questions belonging to Mystical Chymistry, as of the Philosophers Gold, their Mercury, the Liquor Alkahest, Aurum potabile, and such like. Gathered forth of the most approved Authors that have written in Greek, Latine, or High-Dutch; With some Oberservations and Discoveries of the Author himself* (London: Printed by A. C. for *Walter Kettilby* at the *Bishops-head* in St. *Pauls Church-yard*, 1671), p. 40.

[131] *Ibid.*, pp. 356–88, quotation from p. 378.

rather a subtle one but perhaps important in the long run, was that the more mechanical formulation of "parts" suggested in Webster's theory that the heterogeneous ones might be removed mechanically and something more appropriate added. The mechanical approach seems to lend itself more readily to experimental verification or disproof than the Aristotelian approach did, where the process of maturation depended on an internal "cementation" to change the "accidental form" but not remove it.

Of greater interest, during the last part of the seventeenth century, than the theoretical formulations of alchemy are the failed experiments. Usually, although perhaps not in every case, the operator knew the experiments had failed, and in most cases he undertook to notify the public of that fact.

The first example is drawn from the Englishman George Wilson. Wilson was a London "chymist" and had evidently been subjected to a number of influences from the mid-century Hartlibians, as his book carried recipes from Digby and from Starkey, and in addition he had read Boyle and le Fèvre, as will become apparent in the discussion to follow. His book, *A Compleat Course of Chymistry*, first appeared in 1691. It had five editions in all: 1691, 1699 or 1700, 1709, 1721, and 1736.[132] In 1746 Wilson's work was incorporated verbatim into *A Course of Practical Chemistry* published by William Lewis, whose work has been discussed elsewhere by the present writer.[133]

Beginning with the third edition of 1709 Wilson added an appendix to his book entitled "Of the Transmutation of Metals," which appendix was also incorporated into Lewis' work of 1746. In this appendix Wilson offered the results of his own alchemical experiments, a long series of them, some begun as early as 1661, some as late as 1704. He recognized that the doctrine of the transmutation of metals was meeting with very little approval at the time he wrote but said nevertheless that it had been

> positively asserted by many authors, both antient and modern; by men of great learning and experience, of solid virtue and piety; of which (amongst others) the late honourable Mr. *Boyle* is, I think, a complete instance.[134]

Wilson went on to say that although he himself lacked the "great blessings of academical education" and had no more philosophy than he had "fetched out of the fire" and would not undertake to be an advocate

---

[132] Thorndike, *Magic*, VIII, 166–69 (3, n. 5).

[133] Dobbs, "Digby. Part I," p. 8 (2, n. 28).

[134] William Lewis, *A Course of Practical Chemistry. In which are contained All the Operations Described in Wilson's Complete Course of Chemistry. With Many new, and several uncommon Processes. To each Article is given, The Chemical History, and to most, an Account of the Quantities of Oils, Salts, Spirits, yielded in Distillation, &c. From Lemery, Hoffman, the French Memoirs, Philosophical Transactions, &c, and from the Author's own Experience. With Copper Plates.* (London; Printed for J. Nourse, at the *Lamb*, against *Katherine-street*, in the *Strand*, 1746), p. 406, hereinafter referred to as Lewis, *Chemistry*.

for the doctrine of transmutation, yet "the expensive and tedious experiments I have made, abundantly convince me, that metals may be very much meliorated, if not entirely transmuted."[135] But what Wilson reported was a series of failures, and his first set of experiments, begun in August of 1661, is peculiarly well adapted to illustrate the point of the present argument.

The experiments dealt with the "mercurial water," a pure liquor extracted from common mercury and supposed to have large alchemical properties. In the older alchemy, it was called the "philosophical mercury," the "$\female$ of $\female$," the "water that does not wet one's hands," or simply "'our' mercury." There it was not physically described at all, but in many writers it was the vital mystery at the heart of the alchemical process which the adept must discover at all costs. Jung has pointed out that in the older alchemy it was the bearer of many contradictory natures, and, symbolizing both the "self" and the process of individuation which produced the "self," it was beginning, means, and end to the adept.[136]

For Wilson it was a physical liquid which might be extracted from common mercury if the mercury was "philosophically opened" and would in that case yield "the true metalline menstruum."[137] He set about obtaining it by a methodical fractional distillation of common mercury, using a heated cast-iron body to which a glass head, seven aludels, and a receiver were attached. In the space of sixteen hours Wilson and his friend "Mr. T. T." treated one pound of mercury in this apparatus, a few drops at a time, and obtained about one pint of water which weighed thirteen ounces and six drams.[138] "We pleased ourselves wonderfully with our supposed treasure," Wilson said, "which we concluded could be no less than the universal menstruum..."[139]

But then came the anticlimax, for Wilson and "Mr. T. T." were not adepts of the old school but were chemical alchemists. Where an older adept would have been content to discuss vaguely the wonderful properties of his water, Wilson and "Mr. T. T." went on to experiment with it, trying it on various definite substances.

[B]ut, after many trials, we found it of no more virtue than common water. We spent above half of it upon leaf gold, gold calcined with mercury, and upon a lunar calx, &c. but the more experiments we made, the more we were convinced that our mighty expectations were vainly founded....[140]

In a note added in the last edition of his book by way of explanation

[135] *Ibid.*, pp. 406–07.
[136] Jung, *Works*, vol. 13: *Alchemical Studies*, pp. 193–250 (2, n. 3).
[137] Lewis, *Chemistry*, p. 75 (3, n. 134).
[138] *Ibid.*, pp. 407–08.
[139] *Ibid.*, p. 408.
[140] *Ibid.*

and postscript to his experiments on the extraction of a water from mercury, Wilson indicated statements on the subject by Boyle and le Fèvre which showed that both of those illustrious gentlemen had believed it might be done and that Boyle thought he had even done it once.[141] And although he (Wilson) had himself believed it in 1661, now he thinks that Boyle and le Fèvre must have been mistaken, for neither "Mr. *Hales*" nor himself had been able to obtain a water in later experiments when proper precautions were taken.

But it is to be suspected, as Mr. *Hales* very well observes, (in his *Statical Experiments*, p. 200) that Mr. *Boyle*, and others, were deceived by some unheeded circumstance, when they thought they obtained a water from mercury, which should seem rather to have arisen from the lute, and earthen vessels, made use of in the distillation; for Mr. *Hales* could not find the least sign of any moisture, upon distilling mercury in a retort made of an iron gun-barrel, with an intense degree of heat, although he frequently cohobated the mercury which came over into the recipient.[142]

Wilson proceeded to repeat Hales' process himself after he read the *Statical Experiments*. Wilson took the added precaution of first heating the mercury to be used in a crucible "in order to evaporate any moisture that might have been accidentally mixed with it," and, like Hales, found and reported not "the least perceptible appearance of any aqueous humidity" but only "the common form of running mercury."[143] And so in the application of the most prosaic laboratory techniques one avenue which had been supposed to lead to the "philosophicall mercury" was marked "closed." No one reading Hales or Wilson needed to try that route again.

Wilson had likewise closed off one path utilized by the Neoplatonic alchemists, for Wilson said that in 1661

The same gentleman [Mr. *T. T.*] and I tried many experiments

---

[141] *Ibid.*, pp. 429–30. Boyle's statement was in his "Experiments and Notes About the Producibleness of Chymical Principles; Being Parts of an Appendix, designed to be added to the Sceptical Chymist," and may be seen in Boyle, *Works*, I, 653 (3, n. 68). It was first published with the second edition of the *Sceptical Chymist* in 1680. See Fulton, *Bibliography* (2nd edn.), pp. 25–29 (3, n. 99). Wilson gave no reference for le Fèvre's statement but indicated that he had recommended a process similar to the one he (Wilson) had used.

[142] Lewis, *Chemistry*, p. 430 (3, n. 134). Wilson referred to Stephen Hales, *Vegetable Staticks: Or, An Account of some Statical Experiments on the Sap in Vegetables: Being an Essay towards a Natural History of Vegetation. Also, a Specimen of an Attempt to Analyse the Air, By a great Variety of Chymico-Statical Experiments; Which were read at several Meetings before the Royal Society* (London: Printed for W. and J. Innys, at the West End of St. *Paul's*; and T. Woodward, over-against St. *Dunstan's* Church in *Fleetstreet*, 1727), pp. 194–96. On p. 195 Hales referred to Wilson's first experiment and said that he "with several others" had repeated it "about 20 years since...at the elaboratory in Trinity College Cambridge ...." Hales and his co-workers had obtained some water at that time, but had suspected the lute and an earthen vessel as its source.

[143] Lewis, *Chemistry*, pp. 430–31 (3, n. 134).

with air attracted by several magnets; with *May*-dew, and other insipid menstruums, &c. and we never found them more useful than
distilled rainwater.[144]

The alchemical experiments of the great Herman Boerhaave, physician
to Europe and at one time holder of the three chairs of medicine, chemistry,
and botany at the University of Leyden, were begun in 1718. Boerhaave
continued to experiment, primarily with mercury, until 1734. At that
time he published his results in three papers, all of which appeared in the
Philosophical Transactions. The second of these also appeared in the
Histoire de l'Acadèmie Royale des Sciences. One series of experiments involved
heating mercury continuously at a temperature above 100°F. for fifteen
years and six months. Another series studied the influence of distillation
on amalgams of gold, tin, and lead. A specimen of gold amalgam was distilled 877 times; a sample of lead amalgam had already been heated
continuously for twenty years when a clumsy laboratory assistant knocked
it over. Boerhaave also studied the effects of movement on quicksilver by
attaching a flask of it to the jumping wood-block of a fulling mill operated by wind power, so that it was agitated for several months whenever
the wind blew.[145]

In addition to publication in the journals of two learned societies, the
material received a separate English edition in 1734, and it is that on
which the present writer has relied for a study of the experiments and the
rationale behind them. Boerhaave had syncretized the writings of the alchemists to the point where he thought the chief ones had all said essentially the same thing; that metals are generated in their veins in the earth;
that there "Metallic Seeds" are nurtured by suitable warmth and aliment
until they mature; that the aliment is probably quicksilver, "the common
Matter of all Metals," which requires to be "concocted" by a "Metal-
lific Power" called "Sulphur." Quicksilver has inherent in it "an original
Flaw or Blemish," but if that could be removed,

> then it would become liquid, metallic, most weighty, and most
> simple; neither by any Art or Nature divisible into different Things;
> and in which the vivified Seed of every dissolved Metal wou'd most
> perfectly multiply itself; in which the Gold itself dissolving, being
> cherish'd and maturated, wou'd be the last so much sought for, and
> so much celebrated Reward of the Labour.[146]

When he had established these general principles, Boerhaave said, he
then "endeavour'd to learn by Experience, by what Artifice a pure un-

[144] *Ibid.*, pp. 408–09.

[145] G. A. Lindeboom, *Herman Boerhaave. The Man and his Work*, foreword by E.
Ashworth Underwood (London: Methuen & Co., 1968), pp. 195–98.

[146] Herman Boerhaave, *Some Experiments Concerning Mercury. Translated from the Latin,
communcaited by the Author to the Royal Society* (London: Printed for J. Roberts, near the
Oxford-Arms, in *Warwick-Lane*, 1734), pp. 9–14, quotation from p. 14.

mix'd Mercury might be obtained...."[147] Some of his conclusions, stated in the form of corollaries to the experimental results, are as follows:

1. Quicksilver persists in the Fire, retaining its Nature unalterable.
2. Simple, and not separable, into different Parts by Distillation.
3. It is fixed by Fire, and seems changed in its outward Form.
4. Appearing so, in various Parts, it acquires different degrees of Fixedness.
5. Yet none of these Parts acquired, by so strong and lasting a Fire, the Fixedness of Gold or Silver.[148]

Thus not only have his experiments demonstrated to him that he cannot purge mercury to a purer form by fire but also they have shown him that he cannot "fix" it permanently by fire into anything approaching gold or silver. For, he continues, "from the Mercury so fixed by Fire, a greater Fire restores true Mercury...." And "therefore Fire, by these Experiments, is not demonstrated to be the Sulphur of the Philosophers that fixes Mercury into Metals." However, not to leave Boerhaave sounding too modern, it must be noted that he immediately added, even after almost twenty years of failed experiments, that "it seems probable that the Sulphur of the Philosophers is something else very near it."[149]

Indeed it was never Boerhaave's intent in reporting his failures in the purgation and fixation of the common mercury to attempt to dissuade others from alchemical pursuits. On the contrary, he offered his results to the public so that others could use them for a starting point and pass on to other attempts.

[These] laborious Experiments...are so very certain, that they may justly pass for true. Others will not need to repeat them, but may safely make use of these as true upon Occasion. And a diligent Artist, by assuming (or supposing) these Experiments, may apply his Mind farther to others, in order to promote the Study of Chemistry the more.[150]

Thus all Boerhaave did with his almost twenty years of laborious experimentation in alchemy was to show that a few more paths in the labyrinth led nowhere, and he was fully conscious that that was all he had done.

The early alchemical studies and experiments of Sir Isaac Newton, which are to form the subject of Chapter 5 of this study, began precisely at the time when the Hartlib circle was breaking up, i.e., in the 1660's. With the perspective of the mid-century Hartlibian activities in mind, it begins to seem that Newton's alchemical interests might have been very much a product of his own time. And indeed he was a chemical alchemist, and he did draw on the ideas and literature of the Hartlib circle. Boyle

[147] *Ibid.*, p. 14.   [148] *Ibid.*, pp. 43–44.
[149] *Ibid.*, pp. 44–45.   [150] *Ibid.*, p. 15.

was a strong early influence on him;[151] he read Starkey[152] and Eirenaeus Philalethes.[153] He had John Webster's *Metallographia*[154] and some of his surviving alchemical manuscripts contain recipes quite similar to those Digby collected.[155]

But although apparently starting his alchemical work with much the same sort of approach that the Hartlibians, and especially Boyle, engaged in, Newton went on beyond their endeavors to probe the whole vast literature of the older alchemy as it has never been probed before or since. Isaac Newton was no mere collector of recipes; he sought a depth of understanding which was never dreamed of by Digby and the others associated with Hartlib.

Newton looked for no less than the structure of the world in alchemy – a system of the small world to match with his system of the greater. Having found out the force which held the planets in their orbits, he remained unsatisfied. "I wish," he said in the preface to the first edition of the

[151] University Library, Cambridge, Portsmouth Collection MS Add. 3975, ff. 12v–41v (Newton's pp. 22–80), contains many notes from Boyle's early works, interspersed with Newton's own experiments and observations, which probably date in the later 1660's. More evidence of Boyle's large influence on Newton appears in subsequent chapters. In this notebook, in which the chemical laboratory notes appear later, Newton himself numbered the pages. But he mis-numbered once and also he made notes on the flyleaf which did not have pagination, so it has been thought useful to give to the manuscript folio numbers beginning with the flyleaf. In subsequent references the folio numbers so assigned will be given, but Newton's own pagination will be noted also where applicable. In the only other published study of MS Add. 3975 (as the notebook will be referred to hereafter), the authors utilized Newton's pagination for their folio numbers and future investigators should note the resulting discrepancy in citation. Cf. Boas and Hall, "Experiments" (1, n. 41).

[152] MS Add. 3975, ff. 80r–112r (Newton's pp. 159–223), contains notes on Boyle and Starkey, probably dating from the 1670's (3, n. 151).

[153] Newton's heavily annotated copy of Philalethes' *Secrets Reveal'd* (3, n. 76) was a part of the Duveen Collection and is now at the University of Wisconsin. Newton's copy of Philalethes' *Enarratio methodica trium gebri medicinarum* (London, 1678) is in the Newton Collection, Trinity College, Cambridge, NQ. 16. 86. It also shows signs of considerable use by Newton. In addition many of Newton's alchemical manuscripts at King's College, Cambridge, contain notes on Philalethes' tracts or are transcripts of them: Keynes MSS 25, 34, 36, 51, 52, and 221. Keynes MS 52 is especially intriguing. It is in Newton's very early handwriting (i.e., that of the 1660's) and is an autograph transcript of Philalethes' *Epistle to K. Edward unfolded*. The text does not correspond to either of the versions of the tract printed in 1655 and in 1677–78. In all probability it is Newton's copy of a manuscript version of the tract which had originally been put into circulation by the Hartlibians, as will be considered in more detail in Chapter 4. Keynes MSS 25, 36, 51 and 52 all belong to the early period of Newton's alchemical work and consequently are to be discussed in Chapter 5; also the interested reader is reminded that the descriptive titles for all of Newton's alchemical manuscripts are to be found in Appendix A.

[154] Trinity College NQ. 16. 150.

[155] Keynes MSS 31 and 58. Keynes MS 67, most of which is not in Newton's hand, contains some recipes of an earlier type from Paracelsus, Croll, Scala, Sennert, etc.; it seems also to have been the source from which Newton drew the recipes in Keynes MS 62. All four of these manuscripts, now at King's College, Cambridge, derive from the early period of Newton's alchemical career and so are to be discussed in Chapter 5; their descriptive titles may be found in Appendix A.

Principia, "that we could derive the rest of the phenomena of Nature by the same kind of reasoning from mechanical principles...."[156] And a partial draft of that preface provides a fuller expression of his desire:

> For if Nature be simple and pretty conformable to herself, causes will operate in the same kind of way in all phenomena, so that the motions of smaller bodies depend upon certain smaller forces just as the motions of larger bodies are ruled by the greater force of gravity. It remains therefore that we inquire by means of fitting experiments whether there are forces of this kind in nature, then what are their properties, quantities and effects.[157]

Newton never found out the forces which governed the action of small bodies, but it was not for want of trying, for he experimented with all his unique skill and patience from about 1668 or 1669 until he finally left Cambridge for London in 1696. Ideas drawn from Basilius Valentinus, Sendivogius, Eirenaeus Philalethes, and many more, came under his expert laboratory attack. And although he does not seem to have maintained a laboratory in London, he continued to work on his alchemical papers even after he moved.[158]

Against the awesome quantity of thought and charcoal which Newton lavished upon alchemy, the activities of all the Hartlibians appear trivial. They were trivial to a certain extent, yet the energies of Hartlib, Clodius, Digby, Starkey, Boyle, le Fèvre, and the rest were not mis-directed. Their work created an intellectual climate in which alchemy was not mysterious and was no longer treated as closely guarded esoteria but as an honorable branch of natural philosophy which would someday reveal its secrets to rational and chemical investigations. This was the historical milieu into which Newton stepped, and it helps to explain not only some of the ideas he drew from alchemical literature and some of the results he tried to attain in the laboratory but also something of why he undertook the study of alchemy in the first place. He undertook it, at least in part, because during his formative years it was the common assumption of an important group of English natural philosophers that alchemy could be made to yield up its secrets by experimental study. And he saw that the man who obtained those secrets would know a great deal more about the actions of small bodies.

---

[156] Newton, *Principia,* I, XVIII (1, n. 4); Koyré and Cohen, *Newton's Principia,* I, 16 (1, n. 4).

[157] Isaac Newton, draft "Praefatio," in Isaac Newton, *Unpublished Scientific Papers of Isaac Newton. A Selection from the Portsmouth Collection in the University Library, Cambridge. Chosen, edited, and translated by A. Rupert Hall and Marie Boas Hall* (Cambridge: At the University Press, 1962), pp. 304 and 307, hereinafter referred to as Newton, *Unpublished Papers.*

[158] King's College, Cambridge, Keynes MSS 13 and 56 both contain notes about affairs at the Mint in handwriting identical with that of the remainder of the manuscript, the bulk of the material in each case being alchemical. For their descriptive titles, see Appendix A.

Newton's inheritance from the Hartlibians was strongly reinforced by his belief in a general *prisca sapientia*, but at the same time his alchemical studies were soon given a considerably different cast by that belief. As McGuire and Rattansi have shown, Newton thought that in earliest times God had imparted the secrets of natural philosophy and of true religion to a select few. The knowledge was subsequently lost but partially recovered later, at which time it was incorporated in fables and mythic formulations where it would remain hidden from the vulgar. In modern days it could be more fully recovered from experience – prophecies may be understood by historical experience, Newton thought, and natural philosophy by experimentation. The results of experience enable mankind to understand correctly the true meanings hidden in the antique literature.[159]

In the case of alchemy Newton came to hold a similar belief in the *prisca sapientia*. Keynes MS 29 contains an early indication of Newton's interest in treating and interpreting fables alchemically.

The Dragon kild by Cadmus is ye subject of our work, & his teeth are the matter purified.

Democritus (a Graecian Adeptist) said there were certain birds (volatile substances) from whose blood mixt together a certain kind of Serpent ♐ (☿) ♐ was generated wch being eaten (by digestion) would make a man understand ye voyce of birds (ye nature of volatiles how they may bee fixed)

St John ye Apostle & Homer were Adeptists.

Sacra Bacchi (vel Dionysiaca) instituted by Orpheus were of a Chymicall meaning.[160]

There is no reason to believe that Newton ever gave up his belief in the *prisca sapientia*, and it must be assumed that he continued to think that the ancients knew the secrets of alchemy. Did he also think that he had discovered those secrets experimentally? An attempt to answer that question will be reserved for Chapters 5 and 6, but clearly he must have known all along that some of his processes did not work. Had he been willing to publish all the procedures which he had found to be inadequate, alchemical experimentation might have come to an end much sooner than it did. But that second stage in the decline of alchemy – the publication of failed experiments – was entered into only by other men. Newton's secret failures died with him and are only now, three hundred years later, being partially reconstructed.

[159] J. E. McGuire and P. M. Rattansi, "Newton and the 'Pipes of Pan,'" *Notes and Records of the Royal Society of London* **21** (1966), 108–43, hereinafter referred to as McGuire and Rattansi, "Newton and the 'Pipes of Pan.'"

[160] King's College, Cambridge, Keynes MS 29, f. 1v. This manuscript consists of notes on a work by Michael Maier, who was one of the greatest exponents of the *prisca sapientia* doctrine in alchemy. The manuscript will be considered in more detail in Chapter 5, and its descriptive title may be found in Appendix A.

Conclusion

The semanticist may rest content with the demonstration that "Alchemy$_1$ is not Alchemy$_2$," but the historian must ask certain other questions about an intellectual movement. How did it arise? What was its function during its period of maturity? What caused it to decay? Tentative answers to these questions have been offered in the last two chapters.

Scientific alchemy arose partly because the original religious aspects of the older alchemy made it appealing to men interested in reform. It became associated with efforts at general reform – reform of man, reform of human knowledge, of society itself. As those efforts became more widespread and public-spirited, alchemy shared in the general change. As Joseph Ben-David has recently put it, men discovered a new pattern for social cooperation in these efforts for reform, especially in the endeavors connected with the collection and dissemination of knowledge. The new pattern for cooperation slowly supplanted that of the religious community. And, as the intellectual and spiritual poverty stemming from theological controversy and hardening orthodoxies grew greater, the new paradigm provided a new social base from which the role of the scientist in society might grow and become prominent. Ben-David, in his sociological analysis, fixes the crucial area of change in the middle of the seventeenth century, and even more precisely in the Hartlibians of mid-century England.[161] A similar conclusion has been reached in the present study by a more detailed investigation of one facet of the Hartlibians' activities.

Cooperative intellectual endeavors presuppose communication. Language has to achieve precision: words in common use have to be given common referents. As alchemy participated in the general movement towards open communication, its language was divested of vagueness and its terminology made to refer to precisely described chemical events. It was a big step for alchemy to take – this clarification, rationalization, and chemicalization of its language – and the alchemy of Digby's *Secrets* is much closer in spirit to the alchemy of Boerhaave and even to the chemistry of Lavoisier than to the alchemy of Sendivogius. The function of the movement towards the rationalization of alchemy was to join alchemy to the mainstream of scientific revolution, destroy its quasi-religious aspect, and set it on a path of gradual evolution into objective chemistry.

It has been suggested here that the decay of alchemical theory did not come about because of the rise of mechanical philosophies. A number of the giants of the movement – Boyle, Newton, and Boerhaave – never disavowed the possibility of transmutation. Rather they seem to have kept

---

[161] Joseph Ben-David, *The Scientist's Role in Society. A Comparative Study* (Foundations of Modern Sociology Series, ed. by Alex Inkeles; Englewood Cliffs, N.J.: Prentice-Hall, 1971), esp. pp. 69–74.

the general theory and translated it into mechanical terms. And the other major philosophical positions of Aristotelianism and Neoplatonism had long been accommodated to alchemy with ease also.

Once one realizes that philosophy was inadequate to the task of refuting alchemical claims, it becomes apparent that empiricism was left with the job. Experiments like those in Digby's *Secrets* were the starting point. Experiments like those of George Wilson and Herman Boerhaave marked a second stage. The second stage seems to have been closely related to the slow rise during the eighteenth century of analytical techniques and standards of chemical purity and a growing sophistication in the use of empirical techniques. Irrelevant factors were slowly eliminated while relevant ones were seized upon and elaborated until the complexity of experimental situations was reduced to manageable proportions.[162]

Of course, too, there were always the natural sceptics, as well as those who thought chemistry should be pursued in the interest of medicine, or agriculture, or industry, etc., so chemical research was much less focused on metallic transmutation after the time of Boerhaave.

Nevertheless chemical alchemy could never quite be killed outright by point-by-point empirical refutations, and its most basic assumption remained subtly appealing: matter should have a unity behind all its apparent diversity. After the Lavoisierian revolution, when an operational definition of chemical elements had been generally accepted and the list of elements was growing rapidly towards its modern length, there was considerable unease among many chemists.[163] Nature ought somehow to be simpler than they were finding her to be, and it was with an almost audible sigh of relief that some of the chemical thinkers of the present century accepted the implications of radioactivity, of nuclear fission, and of nuclear fusion. The "Philosophers by Fire" were finally receiving some justification.

Yet those latter-day developments were far in the future when the young Isaac Newton began his alchemical studies in the 1660's. Then the "Philosophers by Fire" needed no justification in many circles, but on the contrary were thought to possess – or to be about to grasp – the key to the universe.

[162] See Kopp, *Die Alchemie*, II, *passim* (3, n. 2); Schmieder, *Geschichte*, pp. 438–597 (3, n. 2).

[163] This point is discussed and documented in Robert Siegfried and Betty Jo Dobbs, "Composition, a neglected aspect of the chemical revolution," *Annals of Science* **24** (1968), 275–93.

4 Chemistry and Alchemy at Cambridge

Historical problems

Newton's earliest studies in chemistry and alchemy confront the historian with several problems, which the present chapter makes an effort to solve in a tentative way. Many inferences have had to be drawn in the attempt, inferences which are based on scattered and often inconclusive bits of evidence. But the problems are such that their full solution should enable one to see Newton's alchemy as growing naturally out of his own time and place, and it is hoped that the proposals offered here contribute to that end.

The problems, which will be discussed in more detail below, may be briefly stated. In the first place, Newton does not seem to have entered into his alchemical work as a chemical tyro but rather to have had extensive knowledge of chemicals and chemical manipulations before that time. A number of facts mesh together to set the date of Newton's first experiments on transmutation at about 1668 or 1669; thus he must have learned his chemistry while an undergraduate at Cambridge, or perhaps even before. The question then arises as to when and where, to which the tentative answer may be given: both in Grantham in the late 1650's and in Trinity College, Cambridge, in the 1660's.

In the second place, Newton apparently had access, from the very first period of his alchemical studies, to a variety of unpublished materials which could only have come to him in the form of circulating manuscripts. A number of items among his surviving alchemical papers are treatises, not by Newton, which have never been printed, and at least one unpublished version of an alchemical tract which he copied out verbatim in his fine "very early" hand must have originated with the Hartlib circle in London.

It is not necessary to assume that he had some hitherto unnoticed alchemical friend in London in the late 1660's, which person was also in contact with the Hartlibians, but it does not seem very likely on the other hand that Newton simply purchased alchemical manuscripts through a bookseller or other tradesman. Were there, then, any of Newton's acquaintances in Cambridge who might have been in touch with alchemical circles in London during the 1660's, and who were themselves interested enough in alchemy to have passed such manuscripts to Newton? Isaac Barrow and Henry More have been recognized for a long time to have been Newton's friends. It now appears also that both Barrow and More, although their primary concerns lay elsewhere, did examine alchemical literature fairly carefully and may even have engaged in alchemical experimentation upon occasion. They both had extensive relationships with the Hartlib circle, both directly and through mutual friends, which antedated their friendships with Newton. Therefore it is suggested here that it was through Barrow and More and their friends, as his relationships with them developed in the 1660's and 1670's, that Newton was put in touch with the larger scene of English alchemy.

In the third place, there are certain peculiarities which characterize Newton's alchemy and set it off from the very beginning from lesser endeavors at transmutation. A rigorous and quantitative experimentation is apparent in the laboratory notes, and yet the alchemical manuscripts force one to the conclusion that Newton usually turned to the most esoteric sections of the alchemical literature when he took notes, as if he believed real secrets were hidden there. Thus there seems to be a gaping chasm in Newton's attitude towards alchemy. That chasm is in part what has given rise to the wildly divergent rationalistic and mystic interpretations of Newton's alchemy discussed in Chapter 1. Yet the two approaches utilized by Newton in his study of alchemy are not really so starkly different if one takes the doctrine of *prisca sapientia* into account. To Newton, the search for hidden secrets in the occult literature was quite as rationalistic as his experimental search, and it will be a part of the present chapter to suggest that his alchemical experimentalism and his studies of strange and apparently meaningless passages both grew naturally out of the milieu in which he matured.

Finally, there is the hint given early on by William Stukeley, that Newton's intent in his "chymistry" was to carry his inquiries down into "the ultimate component parts of matter," as he had carried them up by other means into "the boundless regions of space." In a word, it was not transmutation *per se* that Newton sought. He had a larger philosophical goal, and although this cannot be pursued in this chapter and will be taken up later, there is evidence that Newton at one time intended to include the results of his alchemical studies in the *Principia*. Surely such giant ambitions were not customary among the alchemists and the question is, whatever gave Newton the notion that alchemy could be used in that way?

Probably many influences worked together to turn his thoughts in that direction and no strict determination of intellectual genesis can be made. There was, of course, that general position prevalent in many circles at the time, that chemistry was the key to the universe, and there were undoubtedly many more specific influences. In the next chapters it will be seen, for example, that shortly before he made his first recorded alchemical experiments Newton had been reading a work of Boyle's in which Boyle spoke of transmutation experiments as being especially "luciferous." And about the same time, or perhaps even before that, Newton may have gotten the idea from Barrow that alchemy was extraordinarily useful – or might be – in dealing with a certain critical problem in natural philosophy. It is the last mentioned possibility that will be raised in the present chapter: that Isaac Barrow encouraged Newton to undertake the study of alchemy, to do it in an experimental way, and to do it for a particular purpose. Henry More also, whose Neoplatonism in many ways ran parallel to Barrow's position and who believed in a *prisca sapientia*, might well have exerted a formative influence on Newton's thought.

Although it may be argued that genius charts it own course and that these influences were unimportant, still it seems to set Newton's alchemy in better historical perspective to realize that two of his closest friends and associates held the ideas to be discussed below, were themselves interested to some extent in alchemy, and were in contact with several more active students of alchemy. Since both the ideas and the personal contacts seem to have been important in establishing the immediate context in which Newton began his alchemical work, an attempt will be made here to discuss both, even though the evidence is not always conclusive.

Isaac Barrow (1630–77)

Isaac Barrow was Newton's mathematics instructor and friend and one of the first persons to suspect the extent of his genius. According to his seventeenth-century biographer, Abraham Hill, Barrow had entered Cambridge in 1645 and had been chosen Fellow of Trinity in 1649. Finding his opinions on church and state out of favor at that point, Barrow had undertaken the study of medicine and for some years had trained himself in anatomy, botany, and chemistry, after which he took up mathematics in a serious way. But missing an appointment as Professor of Greek, he went abroad in 1654 and remained traveling until the Restoration in 1660. At that time he did receive the Greek professorship, but in 1662 he was also appointed Professor of Geometry at Gresham College, London. For about two years he held both posts (and filled in as well for the Astronomy Professor at Gresham, who was away), traveling between London and Cambridge weekly. In 1663 he resigned his other positions to accept appointment as the first Lucasian Professor of Mathematics at Cambridge and within a year or two he encountered his most famous pupil.[1]

Very little is known about Barrow's activities other than those mentioned by Hill, but some elaboration of the bare facts may be made from

[1] A. H. [Abraham Hill], "Some Account of the Life of Dr. Isaac Barrow: To the Reverend Dr. Tillotson, Dean of Canterbury," in Isaac Barrow, *The Works of the Learned Isaac Barrow, D.D. Late Master of Trinity-College in Cambridge. (Being All his English Works.) Published by His Grace, Dr. John Tillotson, Late Archbishop of Canterbury* (3 vols. in 2; London:

Barrow's incidental Latin compositions, which were frequently topical in nature. In the nineteenth century, as an accompaniment to the Napier edition of Barrow's *Works*, the historian and philosopher of science William Whewell, then himself Master of Trinity, wrote a piece entitled "Barrow and His Academical Times, As Illustrated in his Latin Works."[2] From Whewell's discussion and from Barrow's own statements, it would seem that Barrow was involved before he went abroad in 1654 with a circle of active experimenters who might have been interested to some extent in chemistry or alchemy. It was in 1654, about ten years before he seems to have met Newton, that Barrow provided his valuable sketch of the scientific community at Cambridge.

In that year Barrow was appointed to give one of the customary academic exercises, an *Oratio ad academicos in comitiis*, and his oration was a defense of the current state of university learning at Cambridge. Probably his choice of subject matter was occasioned by the recent controversies over university teaching in which John Webster, John Wilkins, Seth Ward, and Thomas Hall were involved.[3]

Barrow praised the extent and quality of language instruction at Cambridge and the recent revival of mathematics. Then he began to discuss the studies in natural philosophy which were current as he spoke. He mentioned the "innocent cruelty" being practised in the anatomizing of dogs, fishes, and birds, and also the botanical knowledge of plants in the neighborhood, "better than Dioscorides" would have had, had he been alive. And he then went on – unexpectedly to the modern reader – to refer in that same natural-philosophical context to the arcane knowledge of Hermetic philosophy, saying that the Hermetic philosophy was comparable to mathematics in its great productions and that it was the only art which might well be able "to complete and to bring to light not only Medicine but also a universal Philosophy."[4]

It will be recalled that according to his seventeenth-century biographer Barrow had studied anatomy, botany, and chemistry in the 1640's, and

Printed for Brabazon Aylmer, at the *Three Pigeons* against the *Royal-Exchange* in *Cornhill*, 1700), sig. a1r–b3v. Hill's "Life of Barrow" has also been published in Isaac Barrow, *The Theological Works of Isaac Barrow, D.D., Master of Trinity College, Cambridge*, ed. by Alexander Napier (9 vols.; Cambridge: At the University Press, 1859), I, xxxvii–liv. The edition by Napier is hereinafter referred to as Barrow, *Works*. See also Percy H. Osmond, *Isaac Barrow. His Life and Times* (London: Society for Promoting Christian Knowledge, 1944), hereinafter referred to as Osmond, *Barrow*.

[2] Barrow, *Works*, IX, i–lv (4, n.1).

[3] See Debus, *Science and Education* (3, n. 58), and Osmond, *Barrow*, pp. 33–36 (4, n. 1).

[4] Osmond, *Barrow*, pp. 33–41 (4, n. 1); William Whewell, "Barrow and His Academical Times, As Illustrated in his Latin Works," in Barrow, *Works*, IX, i–v (4, n. 1). The *Oratio ad academicos in comitiis* may be found in Barrow, *Works*, IX, 35–47, quotation from p. 46: "...quaeque sola non Medicinam tantum, sed et universam Philosophiam valde perficere et illustrare possit." The present writer is indebted to Professor Michael McVaugh for the references to Barrow's Latin works.

1650's. At that time any study of "chemistry" might very well involve one in a study of alchemy and Hermetic philosophy, and that is what seems to have happened to Barrow. But there were evidently others besides himself engaged in the studies to which he referred, for he continued:

> Truly I know [some] whose minds are enkindled by a desire for these studies hotter than a Chymic flame: others who do not hesitate to take hold of and try to comprehend the extant writings of Lully, Villanova, and other Philosophers of the same sort, and even the most obscure writings of Paracelsus himself; not to mention other distinguished men who are so bold that they fear not to espouse the cause of the noble Goldmaking Stone with lavish faith, whether it be fable or history.[5]

Barrow probably had reference in his oration to his own circle of friends at Trinity which included especially John Ray, the great naturalist, who had entered Trinity exactly when Barrow had and who was Barrow's *socius studiorum* in his mathematical studies. An older Fellow of Trinity, also a member of the group, John Nidd, probably maintained for a while a communal laboratory used by Barrow, Ray, and others. Nidd kept a vivarium for the study of the breeding of frogs, and the dissections mentioned by Barrow seem to have been done in Nidd's rooms. Nidd furthermore owned a copy of Johann Rudolph Glauber's *A Description of New Philosophical Furnaces* (Amsterdam, 1648, and London, 1651), and at one time the group of friends set out to procure "an iron retort like to Glauber's in the second part of his Philosophical Furnaces" so that their experiments could go forward.[6]

Nidd died in 1659 but John Ray was at Trinity until 1662, overlapping both with Barrow's return and with Newton's arrival. One can only speculate about what may have happened to the "iron retort" and any other pieces of equipment the group may have acquired, but it is not at all impossible that they remained in more or less continuous use and eventually passed to Newton himself. If, for instance, Ray took charge of the laboratory arrangements at Nidd's death, or if the group had had built a more extensive laboratory facility at any time, then it is easy to imagine that a continuing laboratory tradition was maintained.

The speculation centers around the location of Ray's rooms. Ray had "a little spot of ground belonging to his chamber" in which he maintained about seven hundred plants for his botanical studies. According to the

[5] Barrow, *Works*, IX, 46 (4, n. 1): "Equidem novi quorum animos ad haec studia igni chymico ferventius desiderium inflammavit: alios qui se Lullii, Villanovae, et quae ejusdem farinae Philosophorum extant monumenta, imo et ipsius Paracelsi obscurissima scripta se capere et comprehendere non dubitarent; ne memorem alios egregios viros, quorum magnanima audacia de Chrysopoeo Lapide nobilem sive fabulam sive historiam generosâ fide amplecti non pertimesceret."

[6] Charles E. Raven, *John Ray, Naturalist. His Life and Works* (2nd edn.; Cambridge: At the University Press, 1950), pp. 43–59.

information that Ray's biographer, Charles Raven, was given, the only chambers which had grounds attached to them during Ray's tenure were those on either side of the Great Gate of Trinity: in the seventeenth century there were small gardens between those rooms and the street on either side of the Gate, and the chambers to which the gardens were attached were the ground floor to the south of the Gate and the first floor to the north.[7] Now it has never been determined precisely which of those sets of rooms Ray actually had, but a seventeenth-century print of Trinity College by Loggan shows those two gardens clearly, and in fact shows additional ones.[8] But the gardens in question have quite distinctive appearances: the one to the north of the Gate is laid out in numerous rectangular plots rather than in the walks and avenues of the others and is altogether more what one would expect of a "working" garden which a botanist such as Ray might have maintained. And in the garden to the north there was a laboratory.

Unfortunately no date for the building of the laboratory has been recorded, and since Loggan's print dates from about 1688, it is of no use in deciding the present question. But the laboratory shown in the print was undoubtedly the one Newton used and no one has ever suggested that he himself had it built. It was a small wooden structure of two stories at the Chapel end of the garden, which conforms to Humphrey Newton's later description of Sir Isaac's laboratory, and the garden and the laboratory were reached from the first floor by means of a stairway attached to a veranda raised on wooden pillars and projecting into the garden from the range of buildings.[9] Probably it would have been beyond Newton's financial means to have built those elaborate arrangements in the days of his Trinity Fellowship, and it seems rather more likely that they had been built by the earlier experimenters of whom Barrow spoke in 1654 and that Newton inherited the use of the laboratory and probably its equipment along with John Ray's rooms and garden. Newton's laboratory was well-furnished, it will be recalled, with a large number of items that Sir Isaac made very little use of, according to Humphrey Newton's description of it, and one wonders if those items did not in fact represent a residue of equipment from the Barrow–Ray–Nidd group.

According to C. D. Broad, who himself occupied Newton's old rooms for many years in the present century, Newton settled into the staircase

---

[7] *Ibid.*, p. 38.

[8] A portion of the print is reproduced in Keynes, "Newton, the Man," facing p. 30 (1, n. 33). The entire print, reduced in size, appears in G. M. Trevelyan, *Trinity College. An Historical Sketch* (Cambridge: At the University Press, 1943), facing p. 120. The present writer is indebted to Professor R. S. Westfall for calling attention to the additional gardens, which appear both in the print and in the Trinity College records: private communication, Nov. 18, 1973.

[9] Keynes, "Newton, the Man," p. 30 (1, n. 33).

called E. Great Court, between the Great Gate and the Chapel (i.e., north of the Gate) in 1668. At first he had rooms on the left hand side of the ground floor but soon moved into those on the right hand side of the first floor, those which Broad had at the time of writing. Those were the rooms Newton remained in until he removed to London in 1696.[10] The clue to Newton's interest in moving to the first floor may lie in the fact that it was those rooms to which the garden was attached, and also the laboratory if, as seems likely, it had already been built. But one manuscript source indicates that Newton had access to that garden, and even some sort of proprietary interest in it, before he moved to the first floor. The item is a list of expenses in Newton's hand, thought to refer to the fitting up of his rooms on the ground floor. One expense listed was for the making of a new cellar behind "ye Chappel," another was for a "stone roller in ye Garden besides ye frame."[11]

Some items gleaned from Newton's notebooks also may be relevant, for it seems that Newton had some access to laboratory facilities even before 1668. In the student notebook reported on in separate studies by Hall[12] and Westfall,[13] there is given, for example, a procedure for making a speculum metal,[14] and in the midst of the optical experiments recorded in MS Add. 3975 (the notebook in which the chemical laboratory notes appear later) Newton described a dissection of the optic apparatus, the nerves connecting eye and brain.[15] It does not seem at all unlikely that Newton had been encouraged by Barrow to use whatever equipment the group had in the 1660's, if Barrow realized that the young Newton's interests turned in the direction of natural philosophy as well as towards mathematics. The evidence is hardly definitive, but the cumulative effect of all these bits is to make it appear that there was an on-going community of experimentalists in Trinity College from the early 1650's, some of whom were interested in chemistry or alchemy, and that Newton was early incorporated into that group.

Whether Barrow himself was an active experimenter during the 1660's has not been recorded. But he certainly approved of it in principle, saying in his oration of 1654, as he introduced his descriptions of anatomical and

[10] C. D. Broad, "Annual lecture on a master mind, Henriette Hertz Trust; Sir Isaac Newton," *Proceedings of the British Academy* **13** (1927), 173–202, esp. p. 176.

[11] Jewish National and University Library, Yahuda MS Var. 1, Newton MS 34 (Sotheby Lot 201), f. 2r.

[12] A. Rupert Hall, "Sir Isaac Newton's Note-Book, 1661–65," *Cambridge Historical Journal* **9** (1948), 239–50, hereinafter referred to as Hall, "Note-Book."

[13] Richard S. Westfall, "The foundations of Newton's philosophy of nature," *The British Journal for the History of Science* **1** (1962–63), 171–82, hereinafter referred to as Westfall, "Foundations."

[14] University Library, Cambridge, Portsmouth Collection MS Add. 3996, f. 111, as quoted in Hall, "Note-Book," pp. 244–45 (4. n. 12).

[15] MS Add. 3975, ff. 9r–11v (Newton's pp. 15–20) (3, n. 151).

botanical research, "You get your eyes to help your ears: you make experiment the companion of reason."[16] He would hardly have failed to convey his belief in the efficacy of experimentation to his favorite pupil, and Newton's mature scientific methodology, a compound of experimentation and mathematics, is thought to be derived directly from Barrow.[17] And from Barrow's own words, taken in context and as quoted above, he quite clearly conceived alchemical experimentation to have scientific status equal to that of experimental anatomy and botany, if not actually superior to them.

Another of Barrow's attitudes was probably even more significant for Newton's study of alchemy, however, and that was Barrow's sympathy for the Neoplatonic philosophical position as opposed to strict Cartesian mechanism. It was a position quite similar to that taken by the Cambridge Platonists against Descartes about the same time and involved on the part of Barrow and of the Cambridge Platonist Henry More a considerable interest in matter–spirit relationships. In their effort to revitalize the dead mechanism of Descartes, they turned especially to what the Neoplatonists had said. More perhaps gave greater emphasis in his studies to the philosophical sources of Neoplatonism and Barrow, at least to a greater extent than More, relied on alchemical writers. Since Newton was reading Neoplatonic alchemy before the decade of the 1660's was out, Barrow's early adoption of that position is of some importance.

Barrow had been impressed with Descartes' new philosophy – in many ways quite impressed – seeing it as the production of a great mathematical mind able to strip away popular error and prejudice and formulate first principles clearly and straightforwardly. But Barrow thought perhaps Descartes had stripped away too much when he eliminated the spiritual and immaterial from the physical world and left only matter and motion.

> He thinks unworthily of the Supreme Maker of things who supposes that he created just one homogeneous Matter, and extended it, blockish and inanimate, through the countless acres of immense space, and moreover, by the sole means of Motion directs those solemn games and the whole mundane comedy, like some carpenter or mechanic repeating and displaying *ad nauseam* his one marionettish feat...[18]

[16] Barrow, *Works*, ix, iv (Whewell's translation), and 45–46 (4, n. 1): "Quin et oculos auriculis succenturiatis, ac duci rationi comitem adjungitis experientiam."

[17] Kargon, *Atomism*, pp. 120–21 (3, n. 56). But cf. E. W. Strong, "Newton's 'mathematical way,'" *Journal of the History of Ideas* 12 (1951), 95–110, where it is suggested that it was in *outgrowing* Barrow's geometrical influence that Newton's mathematics achieved scientific maturity.

[18] Barrow's critique of Descartes was in a Latin oration of 1652: *In comitiis 1652. Cartesiana hypothesis de materia et motu haud satisfacit praecipuis naturae phaenomenis*, Barrow, *Works*, ix, 79–104 (4, n. 1). It has been discussed by Whewell in Barrow, *Works*, ix, vi–viii, and by Osmond in Osmond, *Barrow*, pp. 27–33 (4, n. 1). Quotation (Osmond's translation) from Osmond, *Barrow*, p. 31; Barrow, *Works*, ix, 89–90: "Nam imprimis,

What was missing from Descartes, it seemed to Barrow, was "soul" or "vital spirit" – something immaterial to direct the motions of matter. Barrow marshalled a number of arguments bearing on his point, such as magnetism and the fact that living organisms vary in what seemed to him a non-mechanical way. But his final argument is based on the Hermetic philosophers, who find a subtle spirit as well as gross matter in all things.

[T]hese Philosophers observe that every natural mixed body, such as are animated beings, vegetables, minerals, stones, and the like, are composed from two parts entirely diverse and distinct; just as men [are composed] of soul and body: undoubtedly from a spirit subtle, pure, most potent, and a body dark, foul, impure, and feeble: and [they observe] that these two are separated by the benefit of fire, and that they display themselves separately.[19]

Barrow was probably a little ambivalent about the findings of the Hermetic philosophers – he never quite says that he himself believes their claims – but he recognizes that their evidence at the very least renders the Cartesian hypothesis of one single species of matter suspect. And because they claim to have found out their ideas on body and spirit by questioning nature directly – i.e., by experimentation – they deserve careful consideration. In Barrow's eyes the Hermetic philosophers were going about solving the question in the right way with their experimentation, whereas Descartes had not. Descartes indeed was completely askew in his methodology, Barrow thought.

For it appears, out of his [Descartes'] own mouth, that he has inverted the order of philosophizing which seems the most expedient and soundest, inasmuch as it seemed good to him, not to learn from things, but to impose his own laws on things. He proceeded in such a way that, first he collected and set up metaphysical truths which he considered suitable... from notions implanted in his own mind...; next he descended to general principles of Nature; and then gradually advanced to particulars explicable from principles which, forsooth, he had framed without consulting Nature....[20]

videtur de supremo rerum Conditore indignè sentire, quisquis autumat, eum solummodo unam materiam homogeneam, eamque hebetem et inanimem, per tot jugera immanis spatii extensam creasse, ac cum unico praeterea motus instrumento ludos hosce solemnes, omnemque mundanam comaediam dispensare; quasi fabrum, aut artificem Χειρώυακτα technam hanc suam singularem neurospasticam ad nauseam usque repetentem et ostentantem...."

[19] Barrow, *Works*, IX, 101 (4, n. 1): "...observant hi Philosophi omne corpus naturale mixtum, qualia sunt animantia, vegetabilia, mineralia, lapides et similia, ex duabus partibus omnino diversis atque distinctis constare; non secus ac hominem ex anima et corpore: nimirum ex spiritu subtili, puro, potentissimo, et corpore opaco, foeculento, impuro atque imbelli: atque haec duo ignis beneficio separari, et seipsa seorsim ostentare."

[20] Osmond, *Barrow*, pp. 30–31 (4, n. 1) (Osmond's translation); Barrow, *Works*, IX, 87 (4, n. 1): "Nam ex proprio ore constat, eum philosophandi ordinem, qui et utilissimus videtur et solidissimus, plane invertisse, dum non a rebus discere, sed rebus leges suas

After 1669, when he resigned the Lucasian Chair of Mathematics in favor of Newton, Barrow turned his attention towards theology and preaching – in both of which he seems to have been quite as successful as he was with languages and mathematics. In 1672 Charles II advanced him to the Mastership of Trinity, remarking that he was giving that dignity to the best scholar in England. While Master for the next five years, Barrow initiated the building of the magnificent Wren Library of Trinity and spent much time in planning it and soliciting funds for it. But in 1677 he took a cold or pneumonia while preaching in London and died, perhaps from an overdose of the laudanum with which he had learned to doctor himself while on his eastern travels. All of this seems to point to those few years of the 1660's as being the time of Barrow's greatest influence upon Newton, for afterwards Barrow moved on to other things. But sometime between 1664 and 1669 Barrow had given Newton's mind a firm set towards experimentation, induction, and mathematicizing in "philosophizing." That influence has been recognized for some time, but it also appears now that Newton might well have been directed into alchemical studies by Barrow, which has not been noticed heretofore. For Barrow thought that the Hermetic philosophy, because of its experimental nature, offered a better methodology than Descartes' deduction. Furthermore, Barrow thought perhaps the subtle spirits found by the Hermetic philosophers in their experiments might be used to modify Cartesian mechanism in exactly the right direction, lending to dead matter just what it required to direct its motion.

Henry More (1614–87)

Isaac Newton's intellectual relationship with Henry More began in a way before Newton ever arrived at Cambridge, for he had been More's pupil's pupil at Grantham. More was himself from Grantham and had been tutor at Christ's College, Cambridge, to the Dr. Clark who was usher at the King's School when Newton first attended it and brother to Mr. Clark, apothecary, with whom Newton lodged.[21] It seems that Dr. Clark's "great parcel" of books came to rest sometime in the garret of the apothecary's house, where Isaac a little later consumed them in lieu of his dinners, as will be discussed below. One can well imagine that More's philosophical position had influenced Dr. Clark's original choice of those books.

imponere ei visum est. Ita enim processit, ut primum quas proposito suo idoneas censuit metaphysicas veritates colligeret et stabiliret, nempe ex notionibus naturâ menti suae insitis; deinde ab iis ad generalia principia Naturae descendit; et inde gradatim progressus est ad particularia quaeque explicanda, ex principiis scilicet quae ipse inconsulta natura struxerat...."

[21] William Stukeley to Dr. Mead, Grantham, June 26, 1727, in Turnor, *Collections*, pp. 174–80, esp. p. 176 (1, n. 7).

By the time Newton got to Cambridge, More had long been involved with a study of Descartes' philosophy and with attempted modifications of it somewhat similar to Barrow's. Pleased with much of Descartes' work – for More was ever interested in natural studies and understood that natural philosophy could bolster the rational theology in which he placed great store – at the same time More could see defects in it and feared that Descartes' complete severance of matter and spirit would lead to no good end. In 1648 he began a personal correspondence with Descartes in which he attempted to get Descartes to qualify some of his positions,[22] and failing in that he continued to argue his points in an expanded way in many of the works he published after Descartes' death. One of the most important of these, *The Immortality of the Soul* of 1659, came into Newton's hands while he was still an undergraduate and was referred to by him in the student notebook dated 1661–65.[23]

In that work of More's, Newton would have found numerous criticisms of Descartes' impact physics, criticisms of details of supposed events within Descartes' system and also of Descartes' basic assumption that spirit was absent in the operations of the universe. The two things were linked together in More's mind and after he demonstrated that several of Descartes' mechanisms were inadequate to account for the natural effects attributed to them, he offered a universal spirit, or soul of the world, to compensate for the inadequacies of mechanism by the *directing* of natural mechanical motions.

It was not that More objected to "Mechanick" motions as such. By that term "Mechanick" More always meant impact operations, and he was quite ready to accept their general prevalence, although he did insist that the motion so exhibited could not have arisen from matter itself but had to have been supplied by a "*Substance Incorporeall*," which in the beginning was God Himself.[24] More's framework was thoroughly Cartesian in many respects: he did not at all attempt a refutation of Descartes, only certain modifications. The principal ones in the realm of motion were for

---

[22] This correspondence has been published in Latin and French in René Descartes, *Correspondence avec Arnauld et Morus*, introd. and notes by Geneviève Lewis (Bibliotheque des textes philosophiques, Henry Gouhier, Directeur; Paris: Librairie philosophique J. Vrin, 1953), pp. 94–187. See also Serge Hutin, *Henry More. Essai sur les doctrines théosophiques chez les Platoniciens de Cambridge* (Studien und Materialien zur Geschichte der Philosophie, Herausgegeben von Heinz Heimsoeth, Dieter Henrich, und Giorgio Tonelli, Band 2; Hildesheim: Georg Olms Verlagsbuchhandlung, 1966), pp. 97–108; also Alexandre Koyré, *From the Closed World to the Infinite Universe* (Johns Hopkins Paperbacks edn.; Baltimore: The Johns Hopkins Press, 1968), pp. 110–54.

[23] Hall, "Note-Book," p. 243 (4, n, 12); Westfall, "Foundations," pp. 172–73, n. 5 (4, n. 13).

[24] Henry More, *The Immortality of the Soul, So farre forth as it is demonstrable from the Knowledge of Nature and the Light of Reason* (London: Printed by *J. Flesher*, for *William Morden* Bookseller in Cambridge, 1659), pp. 75–78, hereinafter referred to as More, *Immortality*.

those motions which More thought could not be fully explained by "Mechanick" operations.

The supra-mechanical motions in nature, More asserted, were such things as the sympathetic vibrations of strings tuned in unison, certain sympathetic cures and pains and fetal malformations and "the Wines working when the Vines are in the flower," magnetism, and gravity. In several of his examples More followed Sir Kenelm Digby's *Discourse on the Powder of Sympathy* but rejected Digby's mechanical explanations of the supposed sympathetic events.[25] More also mentioned the effects supposedly passed from Paracelsian "*Astral* bodies" back to the earthly bodies from which they had ascended, but does express some degree of incredulity about those reports. Had More's incredulity waxed a little stronger in his discussion of "sympathetic" relationships, his discourse would have had a more modern ring, of course, but when he turned to questions on the actions of magnetism and gravity, what he had to say did have considerable validity, for Descartes had offered fantastic explications of those phenomena.

More's discussion of magnetism makes his position clear.

The Attraction of the Load-stone seems to have some affinity with these instances of *Sympathy*. This mystery *Des-Cartes* has explained with admirable artifice as to the immediate corporeal causes thereof, to wit, those wreathed particles which he makes to pass certain screw-pores in the Load-stone and Iron. But how the efformation of these particles is above the reach of meer mechanical powers in Matter, as also the exquisite direction of their motion, whereby they make their peculiar *Vortex* he describes about the Earth from Pole to Pole, and thread an incrustated Star, passing in a right line in so long a journey as the Diameter thereof without being swung to the sides; how these things, I say, are beyond the powers of Matter, I have fully enough declared [elsewhere]....[26]

Granting Descartes his own "immediate corporeal causes" of magnetism, and of gravity also, More thought – and rightly so – that mere impact of one particle on another could never produce such a succession of events. A directing or guiding agent was required, the "*Spirit of Nature*," which he called in one place "the great *Quarter-master-General* of divine Providence."[27] It was defined as

A Substance incorporeal... pervading the whole Matter of the Universe, and exercising a plastical power therein according to the sundry predispositions and occasions in the parts it works upon, raising such Phaenomena in the World, by directing the parts of the Matter and their Motion, as cannot be resolved into meer Mechanical powers.[28]

[25] Cf. Dobbs, "Digby. Part i," pp. 5–13 (2, n. 28).
[26] More, *Immortality*, pp. 457–58 (4, n. 24).
[27] *Ibid.*, p. 469.   [28] *Ibid.*, p. 450.

Now More's "Spirit of Nature" had substantial differences from the "Universal Spirit" of d'Espagnet and also from the chemical form of that idea being taught by le Fèvre in Paris at the same time More was writing *The Immortality of the Soul*. Both d'Espagnet and le Fèvre conceived of interchanges between matter and spirit, and imagined that their "Universal Spirit" could be materialized into specific forms of matter. More held no such idea. He kept his spirit as rigidly separated from matter as Descartes did, as far as their interchangeability was concerned. But More allowed for the *action* of his spirit on all kinds of matter – insisted upon it even – and so it could as readily be applied as the active agent in chemical phenomena as anywhere else. In the following passage More speaks to the generality of his concept.

> Out of what has been said may be easily conceived why I gave it this name [the Spirit of Nature], it being a Principle that is of so great influence and activity in the *Nascency*, as I may so call it, & *Coalescency* of things: And this not onely in the Production of Plants, with all other *Concretions* of an inferior nature, and yet above the meer *Mechanical* lawes of Matter; but also in respect of the *birth* of *Animals*, whereunto it is preparatory and assistent.[29]

One had only to judge that chemical productions were "*Concretions* of an inferior nature, and yet above the meer *Mechanical* lawes of Matter" to find More's concept useful in chemistry or in alchemy.

More does not appear to have done any chemical or alchemical experimentation himself until after 1670 and although he read some alchemy, he was particular about what he accepted from it. These facets of his work and thought will be discussed below, when the larger circle of Newton's alchemical contacts is outlined. For the moment the discussion must continue to concentrate on what Newton might have drawn from More's *Immortality of the Soul* with which he is known to have become familiar between 1661 and 1665. For there was an additional dimension to that work of More's into which Newton entered whole-heartedly: the doctrine of the *prisca sapientia*.

The idea of *prisca sapientia* or ancient wisdom appeared in More's work in the form of an argument in favor of another concept which More wished to establish, the pre-existence of souls. More designed his presentation in the following manner: arguments based on reason, arguments from the Wisdom and Goodness of God and "the face of Providence in the World," and then finally indications that many of the wise men of antiquity had held the same opinion. In other words, he based his reasoning on three great authorities: the authority of reason, divine authority, and the authority of antiquity. The range of his antique authorities is enormous, which may be indicated by excerpts from his discussion.

[29] *Ibid.*, pp. 467–68.

In Egypt, that ancient Muse of all hidden Sciences, that this Opinion [of the *Praeexistency of the Soule*] was in vogue amongst the wise men there, those fragments of *Tresmegist* doe sufficiently witness. ...[O]f which Opinion not onely the *Gymnosophists* and other wise men of *Egypt* were, but also the *Brachmans* of *India*, and the *Magi* of *Babylon* and *Persia*; as you may plainly see by those *Oracles* that are called either *Magicall* or *Chaldaicall*.... To which you may adde the abstruse Philosophy of the *Jewes*, which they call their *Cabbala*....

...And in the first place, if we can believe the *Cabbala* of the Jewes, we must assign it to *Moses*, the greatest Philosopher certainly that ever was in the world; to whom you may adde *Zoroaster*, *Pythagoras*, *Epicharmus*, *Empedocles*, *Cebes*, *Euripides*, *Plato*, *Euclide*, *Philo*, *Virgil*, *Marcus Cicero*, *Plotinus*, *Iamblicus*, *Proclus*, *Boethius*, *Psellus*, and several others....[30]

More's use of ancient wisdom in this manner was far from being idiosyncratic. Newton would have seen similar passages in other books he was reading about the same time, for example, in Charleton's *Physiologia*.[31] There Charleton presented arguments against the concept of a plurality of worlds in exactly the same format which More used: first arguments based on reason, then on divine and ancient authorities.

If *Humane Authority*; we may soon perceive, that those Ancient Philosophers, who have declared on our side, for the Unity of the World, do very much exceed those *Pluralists*..., both in Number and *Dignity*. For, *Thales*, *Milesius*, *Pythagoras*, *Empedocles*, *Ecphantus*, *Permenides*, *Melissus*, *Heraclitus*, *Anaxagoras*, *Plato*, *Aristotle*, *Zeno* the Stoick, attended on by all their sober Disciples, have unanimously rejected and derided the Conceit of many Worlds....[32]

Neither was such use of antique authority localized in England or limited to the mid-seventeenth century. On the contrary, it had a history which stretched back to the late fifteenth-century Academy at Florence and behind it lay an extremely interesting argument and a rationale which had been devised to deal with the fundamental problems of acculturization faced by the Latin West in the Renaissance.

The problem was the accommodation of the non-Christian learning, then being revived, to the Christian framework of western Europe.[33] A similar problem had followed on the heels of the flood of translations

---

[30] *Ibid.*, pp. 246–47.   [31] Westfall, "Foundations," p. 172 (4. n. 13).

[32] Walter Charleton, *Physiologia-Epicuro-Gassendo-Charltoniana: Or a Fabrick of Science Natural, Upon the Hypothesis of Atoms, Founded by Epicurus, Repaired by Petrus Gassendus, Augmented by Walter Charleton, Dr. in Medicine, and Physician to the late Charles, Monarch of Great Britain*, indexes and introd. by Robert Hugh Kargon (Reprint of the London edn. of 1654; The Sources of Science, no. 31; New York and London: Johnson Reprint Corporation, 1966), p. 14.

[33] The following discussion adheres in a general way to D. P. Walker, "The *Prisca Theologia* in France," *Journal of the Warburg and Courtauld Institutes* **17** (1954), 204–59, and

from the Arabic in the twelfth and thirteenth centuries, but Aristotle and even Averroës had proved easier to Christianize than were the esoteric materials of the Hermetic Corpus, of Eastern mysticism, and of the Jewish Cabbala which the Renaissance rediscovered. Since, however, the Christian revelation had to be the touchstone of truth, all of those fascinating new–old ideas in the esoteric writings of antiquity had to be reconciled with Christianity or else rejected out of hand.

Orthodox thinkers were prone to hold strictly to the Christian revelation and discredit all other formulations as false and useless, or worse, because they came from damned pagans. But even the most orthodox had to admit some validity to the Jewish revelation because it was plainly from God, Who had chosen the Hebrew people as the receptacle of His prophecies and wisdom, until in the fullness of time the revelation of Christ should supercede the earlier partial revelations. One must think in terms of a literal and full belief in Holy Scripture as the revealed Word of God to comprehend the argument at all.

There were other thinkers in the Renaissance, however, who moved from a position of strict orthodoxy to a syncretic use of Christian and non-Christian revelations, by extrapolation from the accepted idea that the Jews had had at least a partial truth. Some held that Gentiles as well as Jews were prepared by God for the ultimate Christian revelation by partial revelations; others held that Moses had had that great gift from God and that other peoples had learned their wisdom from the Jews who were bearers of the pure Mosaic tradition. Occasionally the discovery of partial truths by the use of natural reason was allowed to the pagans, but the emphasis throughout all the arguments was on revelation.

Following the syncretic thinkers a little further along their difficult journey, it may be seen that if all true knowledge is to be considered as stemming from an original divine revelation, whether one or many, then all true traditions must be reconcilable with each other. Such reconciliations presented complex problems in interpretation for the syncretists, but they were assisted in their efforts by two attitudes which had long been adopted in Biblical exegesis.

One was the realization that parts of the Bible had to be interpreted allegorically if they were ever to be reconciled with orthodox Christian doctrine. For example, if the Christian doctrine of the Trinity be true, then the Old Testament must also bear witness to that truth. Thus the ancient Hebrew invocation, "Hear, O Israel, the Lord our God is one Lord," received a tortuous Trinitarian interpretation. In like manner it was said that Solomon's Song of Songs must be about Divine Love rather than about a very human and sensual love.

D. P. Walker, *The Ancient Theology. Studies in Christian Platonism from the Fifteenth to the Eighteenth Century* (London: Duckworth, 1972).

There was in addition a belief anciently and widely held that many mysteries, religious and other, had been deliberately disguised or hidden by initiates so that the secrets could be guarded from minds not fit to receive them. One needed only to point to Christ's use of parables to find Christian justification for that belief, and it could then be extended to cover all sorts of veiled esoteric literature and even to account for certain sects of supposed wise men who had left no writings but had relied on oral traditions to convey their knowledge to initiates.

Armed with those two approaches, the Renaissance syncretists might readily transform the Greek pantheon into one god with many names, claim that Hermes and Zoroaster had acquired the original revelation of Moses by oral tradition, or find every new scientific discovery hidden in the fables and myths of the ancients. For More to claim that the Gymnosophists of ancient Egypt – who left no writings – had held to a belief in the pre-existence of souls, or for Charleton to find arguments for the unity of the world in so many and varied philosophers, show how far the arguments had been carried by the middle of the seventeenth century.

More evidently gave great weight to the theory that the original revelation had been given to Moses, as he called Moses "the greatest Philosopher certainly that ever was in the world." Similarly Newton later tended to emphasize the importance of the Hebraic transmission of God's Word, for he thought that the Brachmans of India had learned their religion, albeit in a corrupted form, from the "Abrahamans," or sons of Abraham, from which he thought the name "Brachman" derived.[34]

For the purposes of the present study, however, Newton's exact belief in possible lines of transmission of ancient secrets is less important than his ready acceptance of that other aspect of the doctrine of *prisca sapientia*, the belief that the ancients had veiled their deepest knowledge in myths and fables or in deliberately obscure language. Studies on widely separated aspects of Newton's total work have demonstrated his lasting commitment to that view.

Newton understood original true religion to have been vested in certain of the patriarchs and in Christ, as has already been noted in Chapter 1. To Newton it seemed that the religious revelation granted to those figures had been clearly conveyed and required no particular interpretation, only care that it not be diminished by later corruptions. Not so with the Biblical prophecies, however, for there the language was "mystical" and required careful treatment. Newton set out his technique for interpreting prophecies in detail in one of his theological manu-

[34] Isaac Newton, *The Chronology of Ancient Kingdoms Amended. To which is Prefix'd, A Short Chronicle from the First Memory of Things in Europe, to the Conquest of Persia by Alexander the Great* (London: Printed for J. Tonson in the *Strand*, and J. Osborn and T. Longman in *Pater-noster Row*, 1728), p. 351.

scripts, published in part by McLachlan, and since it was a technique he used across the whole range of his studies, with only minor variations, an extensive excerpt from that manuscript on "The Language of the Prophets" will be given here. The rational matter-of-factness of Newton's approach should be noted, as well as his willingness to draw upon the several techniques of textual comparisons, cross-comparison to other systems of mystical interpretation, and comparisons of the prophetic language with the natural world in which it was founded by "analogy."

He that would understand a book written in a strange language must first learn the language.... Such a language was that wherein the Prophets wrote, and the want of sufficient skill in that language is the main reason why they are so little understood. John did not write in one language, Daniel in another, Isaiah in a third and the rest in others peculiar to themselves, but they all write in one and the same mystical language.... The Rule [for fixing the signification of the Prophets' types and phrases] I have followed has been to compare the several mystical places of scripture where the same prophetic phrase or type is used, and to fix such a signification to that phrase as agrees best with all the places: and, if more significations than one be necessary, to note the circumstances by which it may be known in what signification the phrase is taken in any place: and, when I had found the necessary significations, to reject all others as the offspring of luxuriant fancy, for no more significations are to be admitted for true ones than can be proved. And as Critics for understanding the Hebrew consult also other oriental languages of the same root; so I have not feared sometimes to call in to my assistance the Eastern expositors of their mystical writers (I mean the Chaldee Paraphrast and the Interpreters of dreams).... For the language of the Prophets, being Hieroglyphical, had affinity with that of the Egyptian priests and Eastern wise men, and therefore was anciently much better understood in the East than it is now in the West. I received also much light in this search by the analogy between the world natural and the world politic. For the mystical language was founded in this analogy, and will be best understood by considering its original.[35]

Following his rules of operation, Newton gave several examples of his work in which the "mystical" language of the Prophets is reduced to historical or political terminology, as for example a new moon signifying a people's return from dispersal. In his historical work also Newton moved from the esoteric to the commonsensical, and in his hands complex mythologies were transformed into prosaic prehistory, for he understood

[35] Isaac Newton, "The Language of the Prophets," in Newton, *Theological MSS*, pp. 119–26, quotation from pp. 119–20 (1, n. 12).

the gods to represent divinized kings and the myths themselves to encapsulate real datable events. Eventually he was to work out a fixed date for the expedition of the Argonauts and to offer a revised chronology of the ancient kingdoms based on it. His method involved a complex euhemeristic interpretation of the mythological figures of the constellations to arrive at a catalogue of the fixed stars as they had appeared to the ancients, and then calculations of the subsequent precession of the equinoxes to fix the date.[36]

Newton considered his work in these areas to be fully as scientific as his work in optics and astronomy. Indeed his methodology was quite as rigorous and rational in his studies of the esoteric language systems of prophecy and myth as it was in his studies of the natural world, and one need only question his basic assumption, i.e., that real truths were embodied in the myths, fables, and prophecies. But for Newton that assumption was not questionable because it stemmed from his belief in the *prisca sapientia*, the ancient wisdom granted by God to mankind through revelation. That wisdom was hidden in the esoteric language of the ancients and could be recaptured by rational methods, and any knowledge discovered by other methods – as by the experimentation, induction, and mathematizing he applied in natural philosophy – could always be reconciled with the old knowledge occultly preserved. Furthermore, Newton certainly thought there could be interaction between the two approaches to knowledge, that the one validated the other, and that the one approach might give clues for interpretation in the other.

The paper by McGuire and Rattansi noted in Chapter 3 has shown how Newton utilized this double approach in certain draft Scholia and draft Queries which he at one time intended to include in revised editions of the *Principia* and the *Opticks*.[37] There it appears that Newton had decided that some discoveries of Pythagoras on musical harmonies had been applied by that famous ancient to celestial relationships, and that Pythagoras had as a consequence of that application recognized the inverse square law of gravity, the "true harmony of the heavens." Pythagoras had hidden his knowledge in parables to keep it from the vulgar, but the knowledge was nevertheless kept alive in the myths which dealt with the musical instruments of the gods – the Pipes of Pan and Apollo's Harp. But – and this is a crucial point – Newton also took his reasoning back in the other direction, and from the myths which he had interpreted in the light of his own scientific discoveries, he ventured to suggest that the ancients had thought God was the direct cause of gravity. The conclusion Newton drew from his interpretations of the myths then undoubtedly influenced his scientific thinking in its turn.

[36] Manuel, *Historian* (1, n. 13).
[37] McGuire and Rattansi, "Newton and the 'Pipes of Pan'" (3, n. 159).

By what means do bodies act on one another at a distance? The ancient Philosophers who held Atoms and Vacuum attributed gravity to atoms without telling us the means unless in figures: as by calling God Harmony representing him & matter by the God Pan and his Pipe. . . . Whence it seems to have been an ancient opinion that matter depends upon a Deity for its laws of motion as well as for its existence.[38]

It has seemed well to dwell at such length at this point upon Newton's early introduction to the *prisca sapientia* tradition through his reading of More and others because that tradition played such an enormous role in Newton's study of alchemy that any real understanding of Newton's alchemy is precluded if his adherence to the *prisca sapientia* doctrine is ignored. As will be explained in the next chapter, Newton applied his rules for understanding strange languages to the language of alchemy just as he did to the language of prophecy. He took certain myths to be the bearers of alchemical secrets and on occasion allowed the story line of the myth to determine his experimental procedure. Moreover, the fact that Newton believed so strongly in the notion that the ancients had deliberately hidden their secrets in their esoteric languages explains why he so often seems to have chosen some of the most obscure alchemical literature and terminology for study – no doubt he thought the choicest secrets were concealed there.

To complete the immediate setting for Newton's alchemy at Cambridge, however, before following him into the laboratory and the library, some note must be taken of the larger circle of alchemists with whom he came in contact in the 1660's and the 1670's.

"Mr F." (1633–75)

Towards the end of an alchemical tract included among Newton's papers, a tract entitled "Manna," the body of which is not in Newton's hand, there is this note in Newton's handwriting: "Here follow several notes & different readings collected out of M. S. communicated to Mr F. by W. S. in 1670, & by Mr F. to me 1675."[39] Newton's handwriting seems to be what has been designated in the present study as "late early" (see Appendix D) and the "several notes" at the end probably do date from about 1675. "Mr F." then was someone with whom Newton had personal contact on alchemical matters about 1675, and in all likelihood for some time preceeding that also. But who was "Mr F."?

He may be identified, at least tentatively, through a much later manuscript of Newton's which also refers to "Mr F." The later manuscript,

---

[38] Draft variant to Query 23 of the 1706 Latin edition of the *Opticks*, University Library, Cambridge, Portsmouth Collection MS Add. 3970, f. 619r, quoted in *ibid.*, p. 118.

[39] King's College, Cambridge, Keynes MS 33, Sotheby Lot 44, f. 5r, hereinafter referred to as Keynes MS 33.

"De Scriptoribus Chemicis," seems to date in the "late" period by the handwriting. There, at the end of a long list of alchemical works, is mentioned "The Chemical weddin ⟨sic⟩ translated by Mr F."[40] The work referred to was the first English translation of that Rosicrucian allegory written by Andreae, the *Chymische Hochzeit: Christiani Rosencrantz*, and it appeared in 1690 as *The Hermetick Romance: or the Chymical Wedding. Written in high Dutch By Christian Rosencreutz. Translated by E. Foxcroft, late Fellow of King's Colledge in Cambridge.*[41] It is entirely possible, of course, that Newton called one man "Mr F." in 1675 and another man "Mr F." in the 1690's or after. But it seems more likely that the two are identical, especially as E. Foxcroft was in Cambridge in the 1660's and part of the 1670's, was friendly with Henry More, and was interested in "Chymistry."

Ezekiel Foxcroft was the son of a London merchant, George Foxcroft, had attended Eton, and had been admitted to King's College, Cambridge, as a scholar in 1649. There he took his B.A. in 1652/53 and his M.A. in 1656. He was a Fellow of King's from 1652 until 1675.[42] Also, according to Nicolson, he was a Mathematical Lecturer in the University for some period. His mother was Elizabeth, sister of Dr. Benjamin Whichcote, one of the Cambridge Platonists with whom Henry More was intimate. Ezekiel's first cousin Mary Whichcote married John Worthington, Master of Jesus College and another friend of More's. Both Ezekiel and his mother, Elizabeth Whichcote Foxcroft, are mentioned frequently in More's correspondence.[43]

The larger circle

The identification of Ezekiel Foxcroft as "Mr F." and the realization that both Barrow and More had alchemical interests yield a fairly secure vantage point from which the larger circle of Newton's alchemical contacts may be surveyed.

That Newton had such contacts is unquestionable. Not only does the "Manna" manuscript, Keynes MS 33, point directly to that fact, but

[40] Stanford University Library, "De Scriptoribus Chemicis," Sotheby Lot 6, f. 3r.

[41] [John Valentin Andreae,] *The Hermetick Romance: or the Chymical Wedding. Written in high Dutch by Christian Rosencreutz. Translated by E. Foxcroft, late Fellow of Kings Colledge in Cambridge* (n.p. [London]: Printed, by *A. Sowle*, at the *Crooked-Billet* in *Holloway-Lane Shoreditch*: And sold at the *Three-Kyes* in *Nags-Head-Court Grace-Church-street*, 1690).

[42] *Alumni cantabrigienses. A Biographical List of all Known Students, Graduates and Holders of Office at the University of Cambridge, From the Earliest Times to 1900*, compiled by John Venn and J. A. Venn (10 vols. in 2 pts.; Cambridge: At the University Press, 1922–54), pt. I, vol. II, p. 170.

[43] *Conway Letters. The Correspondence of Anne, Viscountess Conway, Henry More, and their Friends, 1642–1684*, collected and ed., with biographical account, by Marjorie Hope Nicolson (New Haven, Connecticut: Published by *Yale University Press* and to be sold in London by Humphrey Milford, *Oxford University Press*, 1930), p. 217, n. 2, hereinafter referred to as *Conway Letters*.

some of the other manuscripts do also, for some are copies of tracts which were never published. Keynes MS 52, the "Epistle to K Edward Unfolded" by Eirenaeus Philalethes, for example, mostly in Newton's "very early" hand (see Appendix D), does not correspond to either of the published versions of that work. The tract had first been published in 1655 in the *Addresses to Hartlib* discussed in Chapter 3, and it also appeared in an expanded form separately in 1677 and in *Ripley Reviv'd* in 1678. Newton's versions are intermediate between those two. Wilkinson, who has recently studied the variants of the tract, has noted that he has collated Newton's completed copy (the bulk of the manuscript) with another manuscript, British Museum Sloane 633, and that Newton had excerpted yet another variant at the end of Keynes MS 52 which is the same as a version found in British Museum Sloane 3633.[44] The versions Newton copied could only have come to him as circulating manuscripts, and that presupposes a group of friends or acquaintances within which alchemical manuscripts circulated. Also, since the Philalethes tracts were originally introduced into the Hartlib circle in London by George Starkey, it presupposes some contact between alchemical circles in Cambridge and in London.

The alchemical tracts utilized by Newton which were not available in published form and which he must have obtained as circulating manuscripts do not, for the most part, bear convenient internal evidence concerning their origins, excepting Keynes MS 33, of course, the tract on "Manna" received at least in part from "Mr F." Nevertheless, it may safely be assumed that Barrow or More or both were among the likeliest transmitters of alchemical manuscripts to Newton because of their close friendships with Newton. The date when personal friendship began has been recorded in neither case, but it seems on both counts to have been early, intimate, and long continued. Furthermore, Barrow and More had more than a nodding acquaintance with each other. For example, the theory of absolute space which Newton incorporated in the *Principia* in 1687 and which is thought to derive from Henry More[45] appears in a similar form in Barrow's *Mathematical Lectures* in the 1660's,[46] indicating More's influence on Barrow during that period (or a converse influence). Barrow knew and respected Newton well enough by 1669 to acquire, circulate, and recommend his mathematical work,[47] and also well

[44] Ronald Sterne Wilkinson, "Some Bibliographical Puzzles Concerning George Starkey," *Ambix* **20** (1973), 235–44.

[45] Edwin Arthur Burtt, *The Metaphysical Foundations of Modern Physical Science* (revised edn.; Doubleday Anchor Books; Garden City, N.Y.: Doubleday & Co., 1954), pp. 143–48 and 244–64.

[46] *Ibid.*, pp. 155–61; Osmond, *Barrow*, pp. 122–24 (4, n. 1).

[47] Isaac Barrow to John Collins, Cambridge, July 20, 1669, in Newton, *Correspondence*, I, 13–14 (1, n. 52). D. T. Whiteside has argued conclusively that Newton was essentially self-trained in mathematics and for that reason has also argued against the notion of an

enough to resign the Lucasian Chair of Mathematics in his favor.[48] If Newton and More were not already personally acquainted by that time, it seems likely that Barrow would have introduced Newton to More. Certainly, in 1680 More was arguing problems in the interpretation of prophecy with Newton in terms which imply a friendship of considerable previous duration: "We have a free converse and friendship, which these differences will not disturb."[49] And in 1687, when he died, More willed Newton a funeral ring, Newton being the only member of the University of Cambridge so honored except the Master and Fellows of Christ's College among whom More had lived and worked for fifty years.[50] The problem raised by the fact of Newton's acquisition of unpublished alchemical materials then may be given a reasonably concrete historical solution if Barrow and More (and, likewise, Ezekiel Foxcroft) can be shown to have had contact with alchemical circles in London either before or during the period of Newton's collecting the manuscripts.

Barrow, as noted earlier, had held simultaneous appointments at Cambridge and at Gresham College in London for about two years in 1662 and 1663, during which time he traveled weekly between the two places. On October 29, 1662, he had also been made a Fellow of the Royal Society, then meeting at Gresham College.[51] He seems to have attended Royal Society meetings with considerable regularity, and one need only note the presence at those meetings of Boyle, Digby, le Fèvre, and John Winthrop, Jr., to demonstrate Barrow's ample opportunities for the acquisition of alchemical manuscripts. Boyle had been one of the original founders of the Royal Society at the historic meeting of November 28, 1660, and Digby's name had been proposed for membership that same day.[52] On December 19, 1660, Digby was voted in[53] and before the

early relationship between Barrow and Newton, in Newton, *Mathematical Papers*, "Introduction," I, 10–11, n. 26 (1, n. 30). But on the face of the evidence cited by Whiteside, Newton had occasion to be examined by Barrow in 1664 and attended Barrow's mathematical lectures in 1665. Cf. *ibid.*, I, 11, n. 26, and *ibid.*, I, 18. To the present writer there seems nothing inherent in Newton's early private mathematical studies, mostly done in 1664, to preclude friendship with Barrow shortly thereafter. On the contrary, if Barrow became aware of Newton's competency, as he probably did when Newton attended his lectures in 1665, it would seem more likely that Barrow's interest in that remarkable undergraduate would have been thoroughly aroused.

[48] John Collins to James Gregory, London, Nov. 25, 1669, in Newton, *Correspondence*, I, 15 (1, n. 52).

[49] Henry More to Dr. John Sharp, Cambridge, Aug. 16, 1680, in *Conway Letters*, pp. 478–79, quotation from p. 479 (4, n. 43).

[50] "The Will of Henry More," in *ibid.*, pp. 481–83, esp. p. 482.

[51] Thomas Birch, *The History of the Royal Society of London for Improving of Natural Knowledge, From its First Rise, In which The most considerable of those Papers communicated to the Society, which have hitherto not been published, are inserted in their proper order, As a Supplement to The Philosophical Transactions* (facsimile reprint of the London edn. of 1756–57; 4 vols.; Bruxelles: Culture et Civilisation, 1968), I, 119.

[52] *Ibid.*, I, 3–4. [53] *Ibid.*, I, 7.

first month of the new year was out he had read his paper on *The Vegetation of Plants* to the Society.[54] The King's physician and alchemist, "Monsieur Le Febure," was admitted December 4, 1661,[55] and Winthrop a month later, January 1, 1661/62.[56] One might considerably enlarge this alchemical roll call of the Royal Society in its early years, but the present list suffices to place Barrow in frequent contact with the most important known students of alchemy in London in the 1660's. All of them had been at one time more or less closely associated with the Hartlib circle of alchemists, as shown in Chapter 3, and any or all of them might have shared alchemical manuscripts with Barrow.

Henry More's contacts with the Hartlibians came even earlier than Barrow's, although their origin is difficult to pinpoint, since few facts appear in the eighteenth-century "Life" of More,[57] and few have appeared since then in biographical form.[58]

Part of the difficulty lies in the fact that More led such a retiring life. He entered Christ's College about the age of seventeen and hardly ever left it against except for brief visits to his home in Grantham, to London, and to the home of Lord and Lady Conway at Ragley. But his friends were many and various, his correspondence ranged widely, and through the lives of his students and of the acquaintances encountered at Ragley, he maintained full contact with the world outside the University. Much about his interests and friends may be deduced from his own writings and even more from that admirable collection of correspondence, the *Conway Letters*. For More – the "Angel of Christ's," as he came to be called – found in Anne, Viscountess Conway, his own "dearest dear," and in his long Platonic love affair with her a remarkably full communication was grounded.

Not long after taking his degree in the 1630's, More began to read "the Platonick Writers, Marsilius Ficinus, Plotinus himself, Mercurius Trismegistus; and the Mystical Divines," according to his own account.[59] Presumably his interest in mystical, Platonic, and Neoplatonic writings did not flag in the 1640's (although he added Descartes, Hobbes, and Digby to his intellectual diet about then), for in 1650 his attention became focused on the new mystical works of Eugenius Philalethes (Thomas Vaughan).

[54] *Ibid.*, I, 41.   [55] *Ibid.*, I, 66.   [56] *Ibid.*, I, 68.

[57] Richard Ward, *The Life of the Learned and Pious Dr Henry More Late Fellow of Christ's College in Cambridge By Richard Ward, A. M. Rector of Ingoldsby in Lincolnshire, 1710. To which are annexed Divers Philosophical Poems and Hymns*, ed., with introd. and notes, by M. F. Howard (London: *Published and Sold by* The Theosophical Publishing Society at 161 New Bond St., 1911), hereinafter referred to as Ward, *More*.

[58] For example, see Henry More, *Philosophical Writings of Henry More*, ed., with introd. and notes, by Flora Isabel Mackinnon (The Wellesley Semi-Centennial Series; New York, London, Melbourne, & Bombay: Oxford University Press, American Branch, 1925).

[59] Ward, *More*, p. 64 (4, n. 57).

One of these works was *Anthroposophia Theomagica*, which it will be recalled had also caught Hartlib's attention, and the other was *Anima Magica Abscondita*. Vaughan's tracts prompted More to write and publish some *Observations* on them, signing himself "Alazonomastix Philalethes." Vaughan was stung by More's criticisms and replied with *The Man Mouse Taken in a Trap*. The pamphlet war continued in 1651 with More's reply, *The Second Lash of Alazonomastix*, and Vaughan's rebuttal, *The Second Wash: or the Moor scour'd again*. More declined at that point to enter into further controversy, but he still thought enough of his original arguments to have his pamphlets republished in 1656 and to write an introductory treatise, *Enthusiasmus Triumphatus*, to go with them.[60] In that introductory tract written in 1656 and revised slightly for the 1662 edition of his collected philosophical writings, one may find More's clearest expression of his views on alchemy and alchemists of different sorts.

More did not wish to speak contemptuously, he said, of the good intentions of "the *Theosophist*" nor disallow the industry of "the *Chymist*," but he wished to excuse himself "from giving any credit to either, any further than some lusty Miracle, transcendent Medicine, or solid Reason shall extort from me."[61] The fact was, he thought, that many of those engaged in such pursuits spoke and wrote as "*Enthusiasts*." Since they did so, More thought they should be ranked with sectarian religious enthusiasts who claimed to receive illumination directly from God and so denied man the use of his reason in religious matters. To More, as to all the Cambridge Platonists, reason was "the candle of the Lord" in man, and its proper use should enable man to arrive at religious truths upon which all sober and sane men could agree. Reliance upon private illumination, not subject to the reasoning powers of the general community, could only lead to further religious division and strife and even laid the door open for the growth of atheism.

The three chief cures of enthusiasm More thought to be temperance, humility, and reason, and he went on to say:

By *Reason* I understand so settled and cautious a Composure of Mind as will suspect every high flown & forward Fancy that endeavours to carry away the assent before deliberate examination; she not enduring to be gulled by the vigour or garishnesse of the representation, nor at all to be born down by the weight or strength of it; but patiently to trie it by the known Faculties of the Soul, which are either the *Common notions* that all men in their wits agree upon, or the *Evidence of Outward Sense*, or else a *clear and distinct Deduction from these*.

[60] Henry More, *Enthusiasmus Triumphatus* (*1662*), introd. by M. V. DePorte (The Augustan Reprint Society, Publication 118; Los Angeles: William Andrews Clark Memorial Library, 1966), pp. ii–iii.
[61] *Ibid.*, p. 36.

Whatever is not agreeable to these three, is *Fancy*, which testifies nothing of the *Truth* or *Existence* of any thing, and therefore ought not, nor cannot be assented to by any but mad men or fools.[62]

Many of the "*Chymists*" and several "*Theosophists*" failed to achieve the right use of reason in More's judgment. They

dictate their own Conceits and Fancies so magisterially and imperiously, as if they were indeed Authentick messengers from God Almighty. But that they are but Counterfeits, that is, *Enthusiasts*, no infallible illuminated men, the gross fopperies they let drop in their writings will sufficiently demonstrate to all that are not smitten in some measure with the like Lunacy with themselves.[63]

More followed the above statement with a number of selections from various authors which justified his point by their patent irrationality. Passages from Sendivogius and from Paracelsus may be identified among his selections, and some came also from Thomas Vaughan, who had originally inspired More's criticisms along these lines. But perhaps the important point is not what More chose to criticize but the fact that he had read so many authors of that ilk in the first place. Also significant is the fact that the main thrust of his criticism is directed against astrological ideas and against a general "*Enthusiastical*" spirit which "is more likely to fill [men's] Brains full of odde fancies, then with any true notions of Philosophy." He does not criticize the general idea of transmutation or the search for the Philosopher's Stone. Neither is chemical or alchemical experimentation criticized as such, but only when it flies "above its sphere" and leads the experimenter into system-making. When it becomes "*Architectonical*" then More criticizes it in terms reminiscent of Bacon's comments about Gilbert's building a world-system out of his few experiments with the magnet.

This is that that commonly makes the *Chymist* so pitiful a *Philosopher*, who from the narrow inspection of some few toys in his own art, conceives himself able to give a reason of all things in *Divinity* and *Nature*; as ridiculous a project, in my judgment, as that of his, that finding a piece of a broken Oar on the sand, busied his brains above all measure to contrive it into an entire Ship.[64]

More thus appears as a person who had intimate knowledge of a wide selection of occult literature. He probably held the notion that theosophy and chemistry might be as useful in building a general natural philosophy as the mechanical ideas of Descartes, Hobbes, and Digby were. Indeed his philosophical writings may fruitfully be treated as attempts to synthesize the mechanical philosophy with the Neoplatonism which underlay so much seventeenth-century theosophy, chemistry, and alchemy. Per-

[62] *Ibid.*, p. 38. [63] *Ibid.*, p. 29.
[64] *Ibid.*, pp. 29–36, quotations from p. 36.

haps he even inclined to the principle that chemistry held the key to the universe, as so many others of the period did, but of one thing he was sure: that whatever natural philosophers suggested, be they mechanical or chemical, those ideas had to be inspected in the clear light of reason.

More's pamphlet war with Vaughan in 1650–51 would certainly have attracted the attention of the Hartlibians to him even if they had not known of him beforehand, and, as the years of the 1650's went by, a few peripheral contacts led to fuller relationships.

Frederick Clodius seems to have been known to the Conways and to Lady Conway's brother, John Finch, who had been More's pupil. In 1653 Clodius was conveying a portrait of Anne, Viscountess Conway, to her brother John Finch, then in Padova,[65] and the next year Clodius undertook to treat Anne's severe recurrent headaches. What he did to her is not clear – perhaps his medicine made her worse, which does not seem at all unlikely if one considers what he probably gave her – but whatever it was, the result upset Henry More mightily.

> This Clodius has moved my indignation above all measure. . . . For if thinges holds as ill as they represent themselves for the present, he is as accurs'd a Raskall as ever trod on English ground, and I am sorry Mr Hartlib should have such a wretch to his son-in-law. But vengeance will find him out and his end will be miserable if he hold on. I profess Madame till I heare better of him, the very remembrance of him will make me half sick, and as vomituriant at this knave Physicion as Alexander is reported to have been at the sight of a corrupt judge. If Sir Kenhelm Digby can informe you any thing better, of this fals man, I shall be heartily glad if it be true.[66]

This episode took place just about the same time that Clodius and Digby were laying plans for their "universal laboratory," and one wonders if the invocation of Digby's advice on Clodius indicates some knowledge on the part of More or of Anne Conway of that grandiose project. The *Conway Letters* show that Sir Kenelm was an old friend of the Conway family, but are silent on that specific point.

By that time when More was learning something about Clodius, More's great friend John Worthington, Master of Jesus College, had become Samuel Hartlib's chief correspondent in Cambridge. Through their correspondence, it appears that More visited Hartlib in London in 1655,[67] and that in 1659 More met other Hartlibians – Brereton and

[65] John Finch to Anne Conway, Padova, Nov. 30/Dec. 10, 1653, in *Conway Letters*, pp. 89–90 (4, n. 43).

[66] Henry More to Anne Conway, Cambridge, April 24, [1654], in *ibid.*, pp. 94–95.

[67] Samuel Hartlib to John Worthington, London, Nov. 20, 1655, in John Worthington, *The Diary and Correspondence of Dr. John Worthington, Master of Jesus College, Cambridge, Vice-Chancellor of the University of Cambridge, etc., etc. From the Baker MSS. in the British Museum and the Cambridge University Library and Other Sources*, ed., by James Crossley and

Pell – at Ragley, the Conway home.[68] In the meantime Worthington and Hartlib had begun to send each other books and manuscripts regularly. Worthington, for example, had sent Barrow's 1655 edition of Euclid to Hartlib and also More's 1659 *Immortality of the Soul*. A little later he was to send John Ray's 1660 *Catalogus Plantarum circa Cantabrigiam nascentium* and a manuscript on Turkish politics which Barrow had evidently brought back from his eastern travels. Many a packet passed from London to Cambridge also. Their contents were not usually described in the correspondence: they might well have contained some of the alchemical materials Newton later copied, passing from Worthington to More or to Foxcroft and thence into Newton's hands.[69]

Another possible avenue of transmission lay in Foxcroft's relationship with Henry More. Foxcroft's father was living in London or was abroad with the East India Company during the period in question, whereas at least part of the time his mother was residing at the Conway home in Ragley, serving as a companion to the ailing Viscountess. The *Conway Letters* show "Mrs Foxcrofts son" traveling to London upon occasion and Henry More gave news of his journeys and his health to the Viscountess to pass on to Mrs. Foxcroft. Sometimes Ezekiel seems to have visited with Robert Boyle in London or else to have been in correspondence with him, for in 1671 or thereabouts, after More's *Enchiridion metaphysicum* was published, Foxcroft went by More's rooms to tell him exactly what it was that Boyle had "taken offence at" in that book.[70] By that time More himself had been in correspondence with Boyle for several years,[71] but in any case, any alchemical manuscripts which Boyle wanted to share could easily have come to More directly or to More through Foxcroft or directly to Foxcroft or to Foxcroft through More.

Although there is no solid evidence that Boyle did pass alchemical manuscripts to More, or to Barrow, or to Newton through either of the others, it appears that by 1673 some sort of customary and comfortable four way relationship existed among Boyle, More, Barrow, and Newton. In September of that year Boyle requested Oldenburg to convey his most recent book to Newton and to enclose copies for Barrow and More in the same packet. Oldenburg said:

> I herewth send you Mr Boyle's new Book of Effluviums, wch he desired me to present to you in his name, wth his very affectionat

Richard Copley Christie (Remains Historical and Literary Connected with the Palatine Counties of Lancaster and Chester, vols. XIII, XXXVI, and CXIV; 2 vols. in 3; Manchester: The Chetham Society, 1847–86), I, 55–56.

[68] Samuel Hartlib to John Worthington, London, June 26, 1659, in *ibid.*, I, 140–42.

[69] *Ibid., passim.*

[70] Henry More to Robert Boyle, Cambridge, Dec. 4, [1671], in Boyle, *Works*, VI, 513–15 (3, n. 68).

[71] Henry More to Robert Boyle, Cambridge, June 5, 1665, and Henry More to Robert Boyle, Cambridge, Nov. 27, 1665, in *ibid.*, VI, 512–13.

service, and assurance of ye esteem he hath of your vertue and knowledge. I take ye liberty to Joyne in the same pacquet two copies more of ye same Book, one for Dr Barrow, and ye other for Dr More, wch he intreats you to send to ym from him, if they be now at Cambridge; if not, to keep ym in your hands, till they shall returne thither.[72]

Undoubtedly those four illustrious men had many attitudes in common – a devout belief in protestant Christianity and an abiding delight in natural philosophy – and there is no reason to think that they would not have been friends and would not have exchanged ideas even if none of them had ever heard of alchemy. But – since they all expressed at some time considerable interest in alchemy or Hermetic philosophy or in the twin off-shoots of the older alchemy which had developed by then, i.e., chemistry and theosophy – there is equally no reason to deny that there probably were alchemical interchanges among them, as well as ones in the natural philosophy which the twentieth century finds more orthodox.

It might be appropriate to end this survey of the larger circle of Newton's alchemical contacts with a brief account of More's association with Francis Mercury van Helmont, scholar, gypsy, alchemist, philosopher, and son of the famous Jean Baptiste van Helmont.[73] It was apparently through the good offices of Hartlib that More came into contact with the younger van Helmont,[74] who arrived in England in the fall of 1670 and came to Cambridge, bringing letters for More out of Germany. It is thought that the letters he brought probably concerned cabbalistic studies then being made in Germany, but More was aware of van Helmont's interest in chemical matters too, and when he came, More had him to dine and invited Ezekiel Foxcroft to be present also.

> For Van Helmont has been with me and dined with me yesterday at my chamber, with Mr Doyly and Mrs Foxcrofts son. Whose curiosity I thought it would gratify to converse with Van Helmon [sic], they both haveing a genius to Chymistry.[75]

More immediately undertook to get van Helmont to treat Lady Conway's headaches, which he soon did. In fact, van Helmont settled himself comfortably at Ragley until Lady Conway's death in 1679. He fitted up a laboratory there, and, when More came to visit and the Viscountess was too ill to see him, the two men evidently experimented together.[76] Probably that was Henry More's only excursion into experimental science.

[72] Henry Oldenburg to Isaac Newton, Sept. 14, 1673, in Newton, *Correspondence*, I, 305 (1, n. 52).

[73] On F. M. van Helmont, see *Conway Letters*, pp. 309–22 (4, n. 43), and Partington, *History*, II, 242–43 (3, n. 20).

[74] Rosalie L. Colie, *Light and Enlightenment. A Study of the Cambridge Platonists and the Dutch Arminians* (Cambridge: At the University Press, 1957), p. 7.

[75] Henry More to Anne Conway, Cambridge, Oct. 13, 1670, in *Conway Letters*, p. 323 (4, n. 43).

[76] *Ibid.*, p. 317.

But it is clear that that close association with van Helmont would have made it easy for him to acquire alchemical manuscripts, not only from English circles but from continental ones also. For Francis Mercury van Helmont held fame as an alchemist everywhere, thanks to his father's name, his own work, and his wide travels.

The young Newton's chemistry

About the year 1667 or 1668 – when he already knew Barrow but probably had not yet met More and almost certainly did not yet know Boyle – Isaac Newton wrote out a dictionary of chemical terminology.[77] Because it was separated from the other papers in 1936, when it was purchased by the American R. V. Sowers, it has never before been studied in relationship to the Keynes Collection, or indeed been studied at all. It demonstrates that Newton was an accomplished chemist before his interest in alchemy was aroused.

The only book mentioned in this Newtonian manuscript is Boyle's *Of Formes*, which had appeared in 1666.[78] It is possible that Newton contrived to obtain a copy of the book and use it while he was in the country during that plague period, but it seems much more likely that he read it after he returned to Cambridge in 1667. Yet the writing in the manuscript is so small – it should be recalled that Newton's earliest script was the smallest of all – that it does not seem possible for it to have been written much later than that. The writing is so small in fact as to be literally microscopic in the view of the present investigator, who worked from photographic copies, in which the original was enlarged about $1\frac{1}{2}$ times, and still had to apply a hand lens to much of it. It is smaller than the writing of any of the alchemical manuscripts in the Keynes Collection but is comparable to the first writing in the calf-bound notebook which contains the later laboratory notes.[79]

Some of the specific information in Newton's chemical dictionary derived from his reading of Boyle's *Of Formes*, as his references indicate, but much of it did not. In many places the mass of empirical detail provided makes it appear that Newton had himself handled the apparatus

[77] Bodleian Library, Oxford, MS Don. b. 15, Sotheby Lot 16, hereinafter referred to as MS Don. b. 15.

[78] Robert Boyle, *The Origine of Formes and Qualities, (According to the Corpuscular Philosophy,) Illustrated by Considerations and Experiments, (Written formerly by way of Notes upon an Essay about Nitre)* (Oxford: Printed by H. Hall Printer to the University, for Ric: Davis, 1666), hereinafter referred to as Boyle, *Of Formes*. A second edition "Augmented by a Discourse of Subordinate Formes" appeared in 1667 and several subsequent Latin editions followed, beginning in 1669. See Fulton, *Bibliography*, pp. 55–68 (3, n. 99). Newton's references have been collated by the present writer with the first edition of 1666.

[79] MS Add. 3975 (3, n. 151).

121

and worked through the processes. His definition of "ffurnace," for example, which follows, in a model of concise, lucid, yet full description, the like of which could probably not be found in any contemporary literature. Newton knew all there was to know about furnaces, and his definition makes Humphrey Newton's later statement, that Sir Isaac made and altered his own furnaces as occasion required, without troubling a bricklayer, seem reasonable indeed.

> ffurnace. As 1 ye Wind furnace ↗ (for calcination, fusion, cementation &c ↙ wch blows it selfe by attracting ye aire through a narrow passage 2 ye distilling furnace by naked fire, for things yt require a strong fire for distillation. & it differs not much from ye Wind furnace only ye glasse rests on a crosse barr of iron under wch bar is a hole to put in the fire, wch in ye wind furnace is put in at ye top. 3 The Reverbatatory ⟨sic⟩ furnace where ye flame only circulating under an arched roof acts upon ye body. 4 ye Sand furnace when ye vessel is set in Sand or ↗ ⟨sifted⟩ ↙ Ashes heated by a fire made underneath. 5 Balneum ↗ or Balneum Mariae ↙ when ye body is set to distill or digest in hot water. ⟨6⟩ Balneum Roris or Vaporosum ye glasse hanging in the steame of boyling water Jnstead of this may be used ye heat of hors dung (cald venter Equinus) i:e: brewsters grains wheat bran, Saw dust, chopt hay or straw, a little moistened close pressed & covered. Or it may in an egg shell bee set under a hen. 7 Athanor, Piger Henricus, or ffurnus Acediae for long digestions ⟨the vessel⟩ being set in sand heated wth a Turret full of Charcoale wch is contrived to burne only at the ⟨bottom⟩ the upper coales continually sinking downe for a supply. Or the sand may bee heated ⟨by a⟩ Lamp. & it is called the Lamp ffurnace. These are made of fire stones, or bricks[80]

The Sotheby Catalogue described this manuscript as an alchemical dictionary, but alchemical content is largely lacking in it. Newton does define "*Sanguis draconis*" – the blood of the dragon – and "magistery," both definitions being eminently rational, but a number of words which are alchemically sensitive are listed by him and left undefined, such as "Alcahest," "Anima," "Elixar ⟨*sic*⟩," "Minera Work," and "Projection." And the definitions Newton did write – of chemicals, of processes, and of laboratory equipment – are so precise and operational that one could draw from this single manuscript a fairly full description of the state of straightforward chemical knowledge and practice as it stood in general in 1667, excepting medicinal preparations.

The question immediately arises as to when and where Newton had acquired his thorough knowledge of chemistry and some speculations may be offered in answer. When attending the King's School at Gran-

---

[80] MS Don. b. 15, f. 3r (4, n. 77).

tham, Newton had lodged with the Mr. Clark who was apothecary there. Seventeenth-century apothecaries, of course, of necessity performed their own operations, there being no established drug companies to supply them with neat pre-packaged pills and syrups. In all likelihood Isaac was introduced to the intriguing potentials of pyrotechny there: an early notebook of his contains recipes for some remedies and preventive medicines and directions for some simple chemical tricks, and is thought to date from about 1655–58.[81] Those dates fall about the period when the young Newton was out of school, supposed to be learning to manage the manor but in reality grasping at every other sort of learning available to him. As Stukeley described Newton's trips to Grantham with a servant on "market business" during that period,

> ...no sooner were they come...but he left man and horses and ran up to the garret at Mr Clark's where he had formerly lodg'd. The room was then fill'd with a great parcel of books which belong'd to Dr Clark desceased, consisting of physic, botany, anatomy, philosophy, mathematics, astronomy and the like. This was a feast to him, exclusive of any thoughts of dinner, any regard to market business. There he staid till the servant calld upon him to goe home.[82]

Perhaps there were chemical books in that "great parcel" also; perhaps Isaac tore himself away from the books long enough to visit the apothecary's laboratory; perhaps both.

But it is certain that he read Boyle with diligence almost from the time he entered Cambridge in 1661. And a little later, while he was yet an undergraduate, it may be that those other opportunities suggested above in connection with Barrow's experimenting friends gave him a chance to enlarge his chemical interests. In fact, the rather mature chemical knowledge evidenced in the chemical dictionary of 1667–68 serves to strengthen the notion that he had had substantial laboratory experience during the years immediately preceeding the compiling of that dictionary.

With respect to Boyle, there is first the evidence of Newton's student notebook of 1661–65, discussed by Hall and Westfall. Hall noted that the two writers with the greatest influence upon Newton during that period appeared to be Descartes and Boyle,[83] and although Hall does not pinpoint Newton's early selections from Boyle, there were certainly a number of works from the prolix and prolific Boylian pen available for Newton's perusal. The scientific works – all of which contained thoughts of interest on chemistry or matter theory – which were published by Boyle during

---

[81] David Eugene Smith, "Two Unpublished Documents of Sir Isaac Newton," in *Isaac Newton 1642–1727. A Memorial Volume Edited for the Mathematical Association by W. J. Greenstreet* (London: G. Bell and Sons, 1927), pp. 16–34.

[82] Stukeley, *Memoirs*, p. 50 (1, n. 16).

[83] Hall, "Note-Book," p. 243 (4, n. 12).

or just before the period in question were: *The Spring and Weight of the Air* (1660), *Certain Physiological Essays* (1661), *The Sceptical Chymist* (1661), *The Usefulness of Experimental Philosophy* (1663), *The Experimental History of Colours* (1663), and *The Experimental History of Cold* (1665).[84] Westfall noticed notes from the first, fifth, and sixth of these.[85] Towards the end of the notebook, perhaps stimulated by Boyle's *Of Colours*, are records of the first known optical experiments of Newton, entered under the title "Of Colours."[86]

The topic "Of Colours" is continued in MS Add. 3975,[87] the notebook which contains the later chemical laboratory notes, and it seems probable that Newton began to write in MS Add. 3975 shortly after he completed entries in the notebook studied by Hall and Westfall. "Of Colours" is followed in MS Add. 3975 by notes and experiments principally on cold.[88] In part the material on cold consists of notes from Boyle, corresponding to that publication of his which followed *Of Colours*, and in part of Newton's own experiments. It is beyond the scope of the present study to discuss any of the experiments which preceded Newton's interest in transmutation, but he had entered rather fully into studies of several aspects of natural philosophy during this early period.

Following "Of Cold, & Heate," the next major topic to appear is "Of fforms and transmutations wrought in them."[89] Newton's notes are partly from Boyle and partly from other sources. The Boylian material in this particular section of Newton's notebook comes, as might be expected, from the first edition of Boyle's *Of Formes*, and some of it corresponds to some of the items entered by Newton in his chemical dictionary.

Presumably the dictionary and the notes on forms in MS Add. 3975 were written about the same time, about 1667–68, since the two sets of materials contain so much that is comparable. And although student-type notes on Boyle and also on Starkey appear further on in MS Add. 3975, the date 1667–68 is crucial, for it marks the threshold between Newton's straightforward chemistry and his alchemy. The experimental notes which follow "Of fforms and transmutations wrought in them" are in fact Newton's independent experiments on transmutation. They soon overflowed the few pages in the notebook which he had left blank for them and eventually filled most of the rest of the notebook and much of the rest of his life. For these experimental records following the notes on forms *are* Newton's "chemical experiments" described by Boas and Hall and called the "laboratory notes" in Chapter 1 of this study. No others

[84] Fulton, *Bibliography*, pp. 9–52 (3, n. 99).
[85] Westfall, "Foundations," pp. 172–73, n. 5 (4, n. 13).
[86] Hall, "Note-Book," pp. 246–49 (4. n. 12).
[87] MS Add. 3975, ff. 2r–12v (Newton's pp. 1–22) (3, n. 151).
[88] *Ibid.*, ff. 14r–19v (Newton's pp. 25–36).
[89] *Ibid.*, ff. 32r–41v (Newton's pp. 61–80).

are known to exist except the loose sheets MS Add. 3973 which in part duplicate MS Add. 3975 and undoubtedly form part of the same experimental series. The first recorded experiments probably date from 1669 and were perhaps preceeded by other unrecorded ones. That Newton seems to have begun his chemical experimentation in 1668 or 1669 serves to confirm the speculation entered into above concerning the laboratory in that little garden to the north of the Great Gate of Trinity. Did it not stand ready for his use when he first moved into E. Great Court?

The long series of experiments begun then was apparently originally stimulated by Boyle's *Of Formes* and amplified by circulating Hartlibian manuscripts. In some of the first experiments Newton followed Boyle's lead. But he seems rather quickly to have decided that Boylian procedure was not totally adequate, and he turned to research the vast literature of alchemy and to explore experimentally ideas drawn from it, though not for all that ever completely abandoning Boyle's ideas on transmutation. These themes will be examined more fully in the next chapters when Newton's earliest attempts at transmutation are assessed.

But first it seems desirable to recapitulate briefly his immediate situation. The scientific and scholarly community in which Newton lived and worked at Cambridge was in some important respects the direct heir of those who found in chemistry or alchemy a key to the universe, and from the very first Newton was subjected to certain large philosophical considerations concerning alchemy, thanks to Barrow and More. Alchemy might be used to modify Descartes, whose mechanical system had much to recommend it but which system needed correction in several crucial respects.

Newton was soon to inherit directly the alchemical manuscripts and secrets the Hartlibians had collected and shared. Passed from hand to hand, the papers accumulated in Newton's study, where he copied them with care, passed them on, and began his laborious process for learning the "significations" of a strange language. Perhaps it was from Henry More that he learned his respect for antique and esoteric literature and acquired his belief that real secrets were hidden there.

Already in 1668 Newton knew considerable chemistry and was a thorough-going experimenter. His native genius for physical comprehension of the world had no doubt been encouraged by Barrow, and perhaps by a larger group of experimentalists, and he had already carried out fundamental experimental studies in optics and in other areas of natural philosophy. Perhaps it was from Barrow then that Newton learned to approach alchemy experimentally, as from More he may have learned to subject it to rational scrutiny.

And in 1668 he began.

5 Newton's Earliest Alchemy: 1668-75

Introduction: the manuscript materials

The problem inherent in Newton's alchemy which demands the highest "degree of fire," as an alchemist might say, and the question which has certainly drawn from historians the greatest expenditure of ink over the centuries, is that which involves the relationship of Newton's alchemy to his recognized scientific work. Were Newton's studies in alchemy mystical, recreational, "lucriferous," or otherwise aberrant from the main thrust of his efforts in natural philosophy and mathematics?

It has been the contention of the present study that they were not, that on the contrary Newton sought through them to elucidate the actions of small bodies, in order to fill out and complete his universal system. In the event that that is the correct position, two things should appear. One is that his alchemical studies should demonstrate the same care and the same rigorous methodology that his other work evinces. The other is that there should appear signs in his scientific writings of attempts to integrate alchemical ideas into them, especially in his discussions on the sub-structure of matter and the events taking place there.

Nothing would be more suitable, it seems, than to demonstrate both of those evidential supports for the present writer's position in this single chapter and perhaps settle the question. But historical facts do not always lend themselves to organizational convenience. Newton undoubtedly studied alchemy with great care and rigor during the period 1668–75 – and it is hoped that this chapter will demonstrate that – but no documents have appeared to show that he was actively engaged in integrating his alchemy with his other science until later. Probably he was, for he had read Boyle's corpuscularian ideas on the universal matter and on transmutation by mechanical rearrangements of parts before he began his alchemical experiments. Since he later wrote several items which show that he accepted and elaborated on certain of Boyle's ideas, it is extremely likely that Newton did think in corpuscularian terms as he did his alchemical experiments throughout this early period and that he was actively engaged in working out his thoughts about small bodies the whole time. Furthermore, the scientific writings which began to appear in 1675 indicate that, although his thoughts on these matters were not yet settled, he certainly had already been working at an integration of alchemy and the mechanical philosophy at that time.

But to tell the story in its historical sequence, questions dealing with Newton's use of alchemy in his general system of the world will be reserved for the next chapter. Here Newton will be treated as an alchemist, albeit a scientific one, as it is that characterization which seems most accurately to reflect the surviving documents of the early period which have chemical or alchemical content. With the single exception of Newton's chemical dictionary, introduced in the last chapter, the terminology of the manuscripts is radically alchemical in nature. Only the barest occasional mention of an atom or a corpuscle crept into these manuscripts: in them Newton dealt with substances and essences, with spirits, souls, and fermentive virtues, rather than with bodies in motion. And as will be seen, his laboratory work likewise was predicated on radically alchemical foundations.

Nevertheless, even though nothing shows directly that Newton was integrating his alchemy with the rest of his science during these early years, the period 1668–75 seems peculiarly appropriate for a study of the foundations of Newton's alchemy. Not only can it be demonstrated that it was during this period that he first turned to alchemy, but also, as will be seen in Chapter 6, it was at the end of this period, in 1675, that he first made public certain thoughts which must have derived from his alchemical studies. That latter fact indicates that by the year 1675 at least some of his alchemical ideas had become securely rooted. And, as also will be seen, there is one unique manuscript datable perhaps to this early period, and certainly before 1680, that indicates enough empirical alchemical success on his part to explain how Newton the Great Experimenter came to believe in alchemy so strongly.

A disclaimer concerning the exact end of this so-called early period of Newton's alchemical studies must be entered, however. As noted in Chapter 1, the dating of the manuscript material depends heavily upon the relative sizes of Newton's handwriting during the different periods. While that technique is adequate to separate the materials into decades with considerable accuracy, it does not allow one to establish the exact year of writing with much certainty.

Newton clearly did not cross his symbols for lead in the 1660's, however, and for some time into the 1670's he continued that unconventional custom. When exactly he began to use the more conventional crossed symbol with regularity has not been precisely established, but in the 1680's he was definitely doing so. Therefore the nature of his symbolic representation for lead has been used as a supplementary dating technique: when that symbol is uncrossed, the manuscript materials may be assigned to the period before 1675 with considerable certainty. But in other cases – in materials which contain no symbols for lead or which contain crossed symbols but have handwriting that seems to belong to the 1670's – then the dating is less certain. In those instances it is entirely possible that the manuscript materials to be discussed below as though they belong to the period before 1675 should actually be dated between 1675 and 1680. But although the final demarcation line of this early period has a hazy quality about it, the dating techniques applied do seem to warrant the assertion that all of the Keynes Collection alchemical manuscripts that belong to the period before 1675 are being considerd.

It is intended in the present chapter to reach for an understanding of Newton's earliest alchemical experiments, both the chemistry of them and the alchemical rationale or rationales behind them. The problem is complicated in the extreme since the experimental record is almost entirely lacking in theoretical content, as noted in Chapter 1. Often Newton used symbolic representations for his chemicals, some of which representations were individual with him, and furthermore most of the experiments make no sense in terms of modern chemistry even if one is able to establish the nature of the chemicals. Some suggestions of Newton's intentions do occur along with the experimental notes, however, and a variety of other materials shed some light on the questions involved. All of the earliest recorded experiments will be considered in the chronological sequence which MS Add. 3975, the laboratory notebook, seems to give them. The first two sets perhaps date from 1669, not long after the chemical dictionary was compiled. Probably the next set was done in 1669 or 1670, and the remaining ones to be considered here in the early 1670's. Although, as in the case of the alchemical manuscripts, it does not seem to be possible to say with certainty that all of these experiments were actually done before 1675, it does seem certain that all of the recorded experiments which were done before 1675 will be considered here. Then the exploration of the laboratory notebook will be terminated and the experimental results recorded in that unique alchemical manuscript mentioned above will be considered. Because the next series of experiments in the laboratory notebook may be dated with considerable certainty to the late 1670's, it will not be discussed here. The present writer hopes to consider those experiments in a future study.

The primary concept with which Newton seems to have operated in the earliest period of his alchemical studies was that of the "mercury" of metals. It was an idea common at the time. In all likelihood he was familiar with it from his earlier chemical studies, and he certainly very early copied manuscript material concerned with that concept. But he probably drew his first two recorded experimental approaches – in which he attempted to isolate the "mercury" of various metals – directly from the writings of Robert Boyle. These first two sets of experiments will be explored in turn.

Neither of the first two experimental procedures having yielded him the "mercury" he sought, Newton went to the alchemical literature to find new techniques. Over the next few years he tried out several which he drew from alchemical writers.

It is neither possible nor profitable to examine all of Newton's early alchemical readings in detail: both potentialities are precluded by the sheer volume of material. On the one hand, the mass of alchemical papers, even for the limited early period of 1668–75 is somewhat overwhelming, and many would-be students of Newton's alchemy seem to have retreated hastily as soon as they comprehended the quantity of effort Newton put into alchemy. On the other hand, as for the *meaning* of the material for Newton, one might as well try to draw intelligible ideas directly from the voluminous alchemical literature itself as to try to draw them directly from most of the alchemical manuscripts which Newton wrote out during this period or any period.

It is at that point – when one attempts to understand the meaning Newton found in his earliest alchemical readings – that the dual approach of the present study becomes of value, an approach in which the laboratory notes and the alchemical manuscripts are compared with each other. If one works back and forth between the "meaningless" laboratory notes and the "meaningless" alchemical manuscripts, it becomes possible to draw the two sets of materials into a correlation which lays bare the meaning for Newton in both, at least in some instances. The comparisons which have been found most useful, as would be expected, are those between laboratory notes and alchemical manuscripts which were written about the same time, but occasionally a gloss on the earlier material has come from the "middle" or "late" manuscripts.

Most of what Newton wrote out in the study may be safely ignored, for the laboratory notes show him to have been working with a few key substances and ideas. Nevertheless, since it is proposed to attempt as full an unraveling of Newton's alchemical work for the period 1668–75 as possible, a survey of all the Keynes Collection alchemical manuscripts which seem to be datable to that period is offered here. Some of them will be examined in considerable detail later in the chapter. For their full titles, and for information on the other scattered or later alchemical papers used in this study, the reader is referred to Appendix A.

The Keynes manuscripts which seem to have been written before 1676, and may fairly certainly be dated before 1680, fall into four basic groups. The first two groups contain no Newtonian material or virtually none. The first set is comprised of notes or transcripts from published works, the second of notes or transcripts from unpublished manuscripts.

In the first group, Keynes MSS 17 and 64 are, respectively, notes on Sir George Ripley's *Opera omnia* of 1649[1] and Basilius Valentinus' *Currus*

---

[1] Newton's copy of this work (3, n. 17) is extant as Trinity College, NQ. 10. 149; the

triumphalis antimonii of 1646.[2] Keynes MS 64 appears to be datable to the late 1660's because of the early, small handwriting. Because it contains no cross-references to other alchemical materials, it perhaps represents one of Newton's earliest ventures into alchemical literature. It was often Newton's custom in his note-taking to enter references to works other than the one being noted, presumably when it seemed to him that the other works expressed similar alchemical ideas. As he read more and more widely, his alchemical manuscripts were frequently more and more copiously annotated. Since Keynes MS 64 is not annotated at all, it probably was written very early, perhaps even in 1668. Keynes MS 17 is quite similar but does contain a reference to the *Theatrum chemicum* and so probably dates in 1669 or a little after, since Newton bought the *Theatrum chemicum* in 1669, as will be seen below.

In addition to Keynes MSS 17 and 64, Keynes MS 14 falls in the first group. Keynes MS 14 consists of autograph transcripts, with the exceptions noted below, of "Artephius His Secret Book" and "The Epistle of John Pontanus, wherein he beareth witness of ye book of Artephius." The manuscript has been collated by the present writer with English versions of those works published in 1624.[3] Newton undoubtedly used this book although he did not include page references to it. But he emended it (with reference to the *Theatrum chemicum* probably for the Pontanus), modernized the spelling somewhat, and occasionally added paragraphing. Sometimes he supplied emphasis or removed the emphases of the originals. Occasionally he omitted Artephius' explanatory parentheses. Otherwise Newton's versions are absolutely verbatim and complete.

The emendation for the Pontanus occurs where a line of text was evidently omitted from the English version and the sense is wanting. Newton supplied the line, in English translation, to complete the meaning. Probably Newton used the Latin versions found in the *Theatrum chemicum* to supply what was lacking, because "Eirenaevs Orandvs," the translator of the 1624 English edition, had evidently used an early version of the *Theatrum chemicum* for his "Latine" copy, as indicated by the heading of

present writer has collated Keynes MS 17 with it. The book also contains some manuscript notes by Newton.

   [2] The present writer has collated Keynes MS 64 with that work: Basilius Valentinus, *Cvrrvs trivmphalis antimonii. Fratris Basillii Valentini Monachi Benedictini. Opvs Antiquioris Medicinae & Philosophiae Hermeticae studiosis dicutum. È Germanico in Latinum versum operâ, studio & sumptibus Petri Ioannis Fabri Doctoris Medici Monspeliensis. Et notis perpetuis ad Marginem appositis ab eodem illustratum* (Tolosae: Apud Petrvm Bosc, 1646).

   [3] Nicolas Flamel, *Nicholas Flammel, His Exposition of the Hieroglyphicall Figures which he caused to be painted vpon an Arch in St. Innocents Church-yard, in Paris. Together with The secret Booke of Artephivs, And The Epistle of Iohn Pontanus: Concerning both the Theoricke and the Practicke of the Philosophers Stone. Faithfully, and (as the Maiesty of the thing requireth) religiously done into English out of the French and Latine Copies. By Eirenaevs Orandvs, qui est, Vera veris enodans.* (London: Imprinted at *London* by *T. S.* for *Thomas Walkley,* and are to bee solde at his Shop, at the Eagle and Childe in *Britans Bursse,* 1624.)

the Pontanus "Epistle" in the 1624 edition of it with Flamel. Newton had also emended the reference given by "Orandvs," presumably to bring it into line with his own edition of the *Theatrum chemicum* of 1659–61. Newton is known to have purchased that six volume edition in April 1669.[4] Therefore it seems likely that Keynes MS 14 was written after that time, and the size of the handwriting makes the early 1670's seem to be the most probable date.

The emendations for the Artephius consist of occasional words or short phrases which are not in the published English version, or of variant readings of certain phrases. These seem to have all been added after the original transcript was made, as they are all interlineated, the regular lines all following the published version of 1624. Newton's handwriting in the Artephius transcript is identical with that in the Pontanus, so both parts of the manuscript should be datable to the same period, i.e., early 1670's.[5]

Keynes MS 29 probably also falls in the first group of manuscripts. There are references in it to a work by Michael Maier: *Symbola aureae mensae duodecim nationem...* (Frankfurt, 1617). And although the present writer has not collated the manuscript with the book, the manuscript material seems all to be notes on or paraphrases of Maier's book as the book is summarized by Craven.[6] Keynes MS 29 probably dates before May 1669, as it contains material used by Newton in a letter to Francis Aston at that time.[7] The size of the handwriting – very small – indicates that it might even be earlier than that.

Likewise Keynes MS 36 seems to belong to the first group. Although it bears no page references, and it is not a verbatim transcript, it may be and has been collated by the present writer with a book published in 1668 by Martinus Birrius in which three of the tracts by Eirenaeus Philalethes appeared anonymously.[8] Newton owned this book and it is still extant in the Newton Collection, Trinity College.[9] It shows signs of considerable use by Newton: some manuscript notes in his hand and many pages turned

[4] Brewster, *Memoirs*, I, 32–33, n. 1 (1, n. 2).

[5] It may be noted in passing that Newton probably made an autograph transcript of the Flamel of the 1624 English edition by "Orandvs" also and that it may be extant still. Cf. Sotheby Lot 25: "The Book of Nicholas Flamel conteining the explication of the Hieroglyphical Figures wch he caused to be put in the Church of the SS. Innocents at Paris," about 15,000 words, 61 pp., autograph transcript (not seen).

[6] Craven, *Maier*, pp. 68–84 (3, n. 37).

[7] Isaac Newton to Francis Aston, May 18, 1669, in Newton, *Correspondence*, I, 9–13, esp. 11 and 12–13, n. 5 (1, n. 52).

[8] Martinus Birrius, *Tres tractatus De metallorum transmutatione. Quid singulis contineatur, sequens pagina indicat. Incognito auctore. Adjuncta est Appendix Medicamentorum Antipodagricorum & Calculifragi. Quae omnia ad bonum publicum promovendum nunc primum in lucem edi curavit Martinus Birrius, Philosophiae & Medicinae Doctor, Practicus Amstelodamensis, Apud quem Medicamenta ista reperiuntur.* (Amstelodami: Apud Johannem Janssonium à Waesberge, & Viduam Elizei Weyerstraet, 1668).

[9] Trinity College NQ. 10. 144.

down. The language of the notes in Keynes MS 36 follows the text published by Birrius almost exactly, except in places where Newton summarizes a few pages in a line or two, although much of the published material is omitted. Only one passage, an explanatory remark in square brackets on f. 1r, seems to be Newtonian. From the size of the handwriting and from Newton's uncrossed symbols for lead, Keynes MS 36 seems to date from 1668 or 1669, and since there are no cross-references in it, probably 1668 is more likely.

Keynes MS 25 is an anthology of extracts from Flamel, Eirenaeus Philalethes, Trevisan, Artephius, and the *Turba philosophorum*. As such, it falls between the first two groups of the present survey, for some of the material in it came from published items and some probably from unpublished material. The Flamel apparently depends on the 1624 English version cited above and the Philalethes seems to depend in part on the Birrius publication which Newton had abstracted in Keynes MS 36, and in part on the manuscript version of the " Epistle to K. Edward unfolded " of Keynes MS 52, discussed in Chapter 4. The other items may all have come from the *Theatrum chemicum* and there is in fact one reference to that work on f. 4v of the manuscript, in the extract from the *Turba*, although a different edition of the *Turba* seems to be indicated also. The reference to the *Theatrum chemicum* probably dates the manuscript after April 1669 and, since some of the symbols for lead are crossed and some are uncrossed, a date between 1670 and 1675 seems to be indicated, which date is consistent with the size of the handwriting. There is very little Newtonian material in Keynes MS 25.

Nine of the manuscripts in the Keynes Collection datable to the period 1668–75 seem to depend primarily on unpublished sources. Keynes MS 62, a collection of recipes, is in the handwriting of the very early period, and, since the symbols for lead are not crossed, it probably does date from 1668 or 1669. Newton evidently extracted it from another manuscript which is still extant in the Keynes Collection, Keynes MS 67.[10] Keynes MS 31 is similar to Keynes MS 62, consisting of recipes apparently drawn from an earlier compilation, although the source for it is no longer with Newton's papers. It may also be dated to the late 1660's for the same reasons which apply to Keynes MS 62.

Keynes MS 22 is an autograph transcript of "The Epitome of the Treasure of Health" by Edwardus Generosus, a sixteenth-century English alchemist. It is probably datable to about 1675 by the handwriting and the fact that, although most of the symbols for lead are crossed, one is not. Likewise Keynes MS 39 seems to date from the same period. It consists

---

[10] Although Keynes MS 67 was described in the Sotheby Catalogue as containing nothing in Newton's hand, R. S. Westfall has found that several minor items were added to the earlier compilation by Newton himself. Private communication, Dec. 11, 1973.

of transcripts of two anonymous alchemical tracts: "Observations of ye Matter in ye Glass" and "Emanuel." Keynes MS 58, which is in three distinct parts, is datable to the same period for similar reasons. The first and last sections of Keynes MS 58 seem to derive from unpublished manuscripts, but the middle section contains some of Newton's own ideas. All parts will be treated in some detail later in the chapter.

Keynes MS 33, the tract on "Manna" which contains Newton's note about "Mr F.," was considered in Chapter 4. It contains the internal date of 1675, and the last portion of it, the only part in Newton's hand, probably does date from that year if Ezekiel Foxcroft passed the material to Newton, for that was the year of Foxcroft's death.

The remaining two manuscripts in this second group, Keynes MSS 51 and 52 date in the late 1660's by the handwriting, and they derive from unpublished versions of tracts by Eirenaeus Philalethes. The reasons for assigning that status to Keynes MS 52, the "Epistle to K. Edward unfolded," were given in Chapter 4. Much the same sort of reasoning applies to Keynes MS 51. It contains material not in the published version and although page numbers are given, they do not correspond to the published pagination. The various individual tracts comprising Keynes MS 51 are taken note of below.

The third classification of manuscripts from the early period consists of notes from books with substantial Newtonian material accompanying them. Only one manuscript, Keynes MS 19, may be so classified for the early period as far as the Keynes Collection is concerned. It probably dates from before April 1669, as is explained below where some of its contents are examined.

The fourth and last division to which Newton's earliest alchemical manuscripts may be assigned is that of papers which contain primarily his own material. Keynes MS 12, Part A, may be said to fall in this category. It consists of a series of alchemical propositions which appear to be Newton's own formulations, at least in part, drawn from his readings. By handwriting and the uncrossed symbols for lead, that particular part of the manuscript would seem to date from the 1660's, but, since it contains references to items in the *Theatrum chemicum*, it should probably be dated late in 1669. The ideas in Keynes MS 12 do not seem to be reflected in the laboratory notes of the early period to any great extent and they will not be pursued here, although they may have come into play later. Professor R. S. Westfall, with whom the present writer has conferred on the dating of the manuscripts, leans towards a much later dating for Keynes MS 12, Part B, probably in the 1690's;[11] consequently it is here omitted from the survey of the earliest manuscripts.

The middle section of Keynes MS 58 and all of Keynes MS 18 both

[11] Private communication, July 28, 1973.

seem to be Newton's own in one way or another. They complete the last classification for this survey and are of extraordinary importance for the understanding of Newton's early alchemy. All sections of Keynes MS 58 seem to date from the same period, i.e., 1670–75, as already noted. Keynes MS 18 was written probably a little later, as indicated by the handwriting and the fact that the symbols for lead are crossed. But Newton was still using Latin accent marks in Keynes MS 18, a fairly good indicator of an early composition. So although perhaps it dates after 1675, it was almost certainly not written later than 1680. Both of these manuscripts, Keynes MSS 58 and 18, will be analyzed in detail in the present chapter.

What Newton was looking for in the alchemical material at this time was a way of extracting the "mercury" of metals, if one may judge by his first two sets of experiments. The other experiments of the period may be seen as a continuation of that same general idea, but differing in that the techniques he used derived directly from alchemy. Apparently, Newton thought for some while that his new procedures were working and would lead on to greater things, for in the 1670's he wrote out his "Clavis," a key explaining – more or less – what he had been doing with his experiments and expressing satisfaction with them. The "Clavis," in Keynes MS 18, is considered here to be the summit and crown of Newton's earliest alchemical efforts and will be taken to mark their end.

The "mercury" of metals: the early recipes

The old chemistry and alchemy flourished for a long time before particulate theories of matter became popular in the seventeenth century, and one of the most characteristic ideas in the older chemical studies was that of substances. It was not the corpuscles or the *minima naturalia* of matter which were anciently considered important by "chymists," for such minute parts cannot be seen or felt or tasted or smelled, and as far as common sense is concerned, they are not useful abstractions. But the ordinary substances of the world taken in massy form have secondary characteristics which the human sensory apparatus readily comprehends, and over the centuries substances came to be classified in various ways, according as they were heavy or light, moist or dry, had tastes which were bitter, salt, or sharp, and so forth.

In a very useful two-part article in which some of the basic concepts of chemistry are examined from a philosophical viewpoint, Paneth has argued that that was and is quite a proper approach for chemistry to take even though it is naively realistic; for chemistry, in contrast to physics, is directed towards the secondary qualities of bodies and not to the property-less world of the atom where only size, shape, and motion are real.

Chemistry *is* "bangs and stinks," Paneth points out, and in it sodium chloride *is* salty and hydrogen sulfide *is* unpleasant to smell.[12]

Newton's older contemporary Nicolas le Fèvre certainly would have agreed with Paneth's formulation, as would chemists of all periods. After deriding Scholastic arguments on points and parts and on the divisibility of the parts of bodies, le Fèvre says:

> You see then, that Chymistry doth reject such airy and notional Arguments, to stick close to visible and palpable things, as it will appear by the practice of this Art: For if we affirm, that such a body is compounded of an acid spirit, a bitter or pontick salt, and a sweet earth; we can make manifest by the touch, smell, taste, those parts which we extract, with all those conditions we do attribute to them.[13]

Classifications on the basis of secondary characteristics having been made, it was perhaps inevitable that "chymists" would conceive that certain chemical "principles" underlay the classes, principles which were themselves abstract substances and were the bearers of the important distinguishing properties of the classes. It may be argued, as Paneth has done, that the Empedoclean–Aristotelian elements (Earth, Air, Fire, and Water) were all just this sort of abstract substance, devised to account for the secondary characteristics of bodies. The elements were "metaphysical" or "transcendental" and were themselves unobtainable, but they served as the "basic substances" which were the carriers of physical and chemical properties.[14]

While the Arabs nursed the "chymic" torch during the Dark Ages, they developed an entirely different set of chemical principles to explain one of the customary classifications, that of "metals." Seven metals were known: gold, silver, iron, copper, tin, lead, and mercury. They were all lustrous and dense and the solid ones were malleable in some degree. When they were melted, the six solid ones showed many of the characteristics of the normally liquid one, mercury or quicksilver, and so it came to be said that all seven were characterized by a mercurial principle. The abstract "mercury" was not to be identified with ordinary quicksilver although it was similar to it. To account for the differences within the metallic class, another principle was created, an equally abstract "sulfur," which by interacting with the "mercury" in various ways contributed the specific properties of each metal.[15]

[12] F. A. Paneth, "The epistemological status of the chemical concept of element (I)," *The British Journal for the Philosophy of Science* **13** (1962), 1–14.

[13] le Fèvre, *Chymistry*, I, 9–10, quotation from p. 10 (2, n. 34).

[14] F. A. Paneth, "The epistemological status of the chemical concept of element (II)," *The British Journal for the Philosophy of Science* **13** (1962), 144–60.

[15] Holmyard, *Alchemy*, p. 75 (2, n. 2); Redgrove, *Alchemy* (2nd edn.), pp. 18–23 (2, n. 2); Taylor, *The Alchemists*, pp. 12–13, 80–81 (2, n. 2). Taylor suggests that the sulfur–mercury theory of metallic composition may also have historical roots in Aristotle's

The general cultural fluidity of the Renaissance period fostered inter-action between Aristotelian and Arabic ideas, and new sets of chemical principles were constantly being devised. By the seventeenth century most of them contained five or six members, one of which was usually a liquid one combining in some manner Aristotelian Water and Arabic "mer-cury."[16] When "chymists" considered the class of metals, however, the mercurial aspect of the liquid principle unquestionably predominated, and many, many attempts were made to isolate a "mercury" from various individual metals in the laboratories of the period. Such attempts were especially important in alchemical circles.[17]

Alchemy, with its aspirations to transmutation and its psychological and religious functions, had long since extended the original concept of a mercurial principle. The Arabic idea, although an abstraction certainly, had had roots in empirical considerations. The alchemical concept, the "philosophical mercury," on the other hand, had become all things to all men. It might appear in almost any guise in the alchemical literature and with almost any characteristics attached to it. But there was still some commonality between the two ideas and the swing towards experimental-ism in alchemy in some circles in the seventeenth century would have served to bring the concepts closer together in those circles. Thus it might come about that a concrete "philosophical mercury" would be sought by ordinary or extraordinary chemical means.

By far the most common techniques in use for extracting a "mercury" from metals at the time Newton began his alchemical studies were some which involved the use of heat. Two types of approach may be stated briefly in terms of modern chemistry. A metallic ore or compound, usually of lead or antimony, might be distilled from a reducing agent, and, when the molten metal appeared in the receiver, it was called a "true running mercury."[18] Actually of course it was only the metal itself. That should have been apparent as soon as it cooled, but impurities might have masked the fact. Another process sometimes used was the treatment of a metal with the volatile sublimate of mercury, mercuric chloride. When the two substances were heated together, a substitution reaction occurred, a

theory of smoky and moist exhalations in the bowels of the earth; Holmyard suggests the germs of it might have been found in Apollonius of Tyana.

[16] Allen G. Debus, "Fire analysis and the elements in the sixteenth and the seven-teenth centuries," *Annals of Science* **23** (1967), 127–47; Thorndike, *Magic*, VIII, 104–69 (3, n. 5).

[17] Digby, *Secrets*, pp. 72–83 (3, n. 69), contains many recipes for the preparation of "mercuries" of different metals and many of Digby's other recipes involve the use of specially extracted "mercuries." Boyle generalizes his remarks, saying, "That there may be extracted or obtained from metals and minerals a fluid substance, in the form of running mercury, is the common opinion of chymists; in whose bodies [works?] we may meet with many processes, to make these mercuries...." Boyle, *Works*, I, 630 (3, n. 68).

[18] For example, see Digby's "The best way to extract the ☿ of ♄," in which antimony oxide is reduced: Digby, *Secrets*, pp. 72–73 (3, n. 69).

chloride of the original metal was formed, and the mercury released from its initial compound appeared "running" at the bottom of the apparatus. The mercuric product in that process was simply ordinary quicksilver, but was frequently considered to be the "mercury" of the original metal.[19] If the original metal had been in excess of the quantity needed to decompose the sublimate, the actual product would have been an amalgam of quicksilver and the metal. In that case, the product would have had some of the properties of the original metal and the operator would have been reinforced in his belief that the product was truly the "mercury" of the metal used because it resembled both the metal and ordinary mercury.

One of Newton's very early manuscripts includes recipes of this same sort. There are actually two of his compilations of recipes extant in the Keynes Collection, as noted above, both probably datable to 1668. Keynes MS 62, however, which Newton drew from the sixteenth- and early seventeenth-century collection of manuscripts in Keynes MS 67, does not contain any discernible theoretical base, even that of the "mercury" of metals. The recipes Newton selected from Keynes MS 67 would have been worse than useless as practical directions. They merely allow one a startling glimpse of the old "empirics," probably pursuing their alchemical labors in their wives' kitchens and throwing in the odd ingredient that came to hand: child's urine, a lock of woman's hair, the grease of a roasted eel, all of which are required in this set of recipes.

But Keynes MS 31, entitled "Liber Mercuriorum ⟨*sic*⟩ Corporum" – the "Book of the Mercuries of Bodies" – offers a set of recipes for extracting "mercuries," some of which were quite in line with the best type of recipe collected by the Hartlibians. And of course that group may even have been the source for Newton's manuscript compilation.

In Keynes MS 31 one finds, for example, clear directions for making a "mercury" from lead.

℞ old ♄ yt hath layn on church windows or such like (for yt is whitest & best for this purpose) melt it in an Jron dish, into wch, when molten, cast ⊖ calcined to white, & work & stirr it well wth a spatula till it bee calcined, then wash away ye salt wth ↗ hot or ↙ boyling water untill it bee fresh & sweet; put this pouder into a glasse vessell & put theron ⍹ & ⍦ & stop it close & set it in dung for 20 days & most of it shall bee ☿ .[20]

---

[19] Boyle criticized that sort of process in quite modern terms in 1680 in his "Experiments and Notes about the Producibleness of Chymical Principles; Being Parts of an Appendix, designed to be added to the *Sceptical Chymist*" (Boyle, *Works*, I, 631–32) (3, n. 68). But earlier he had himself used mercuric chloride to "open" metals, as will be seen below.

[20] Keynes MS 31, f. 1v. On f. 4v of this manuscript Newton prepared a neat table of definitions for all the unusual symbols in this set of recipes (Plate 2). There ⍹ is "Salarmoniack water" and ⍦ is "A solution of Tartar."

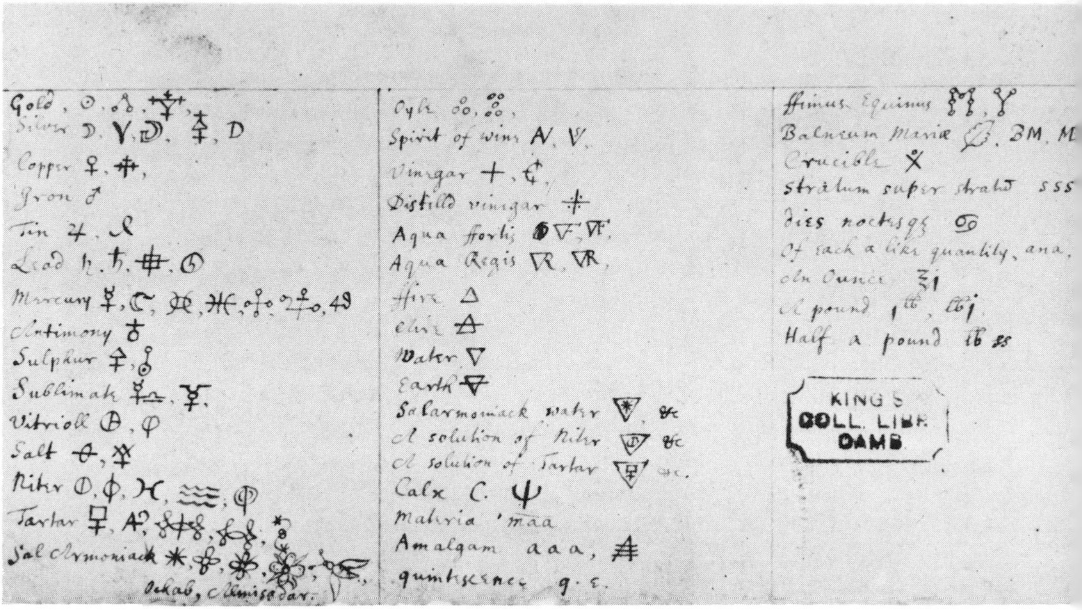

Plate 2. ALCHEMICAL SYMBOLS FROM KEYNES MS 31.
One of Newton's very earliest alchemical manuscripts, Keynes MS 31, f. 4v, bears this table that Newton prepared for himself to explicate the symbols used in the recipes he had copied. (Reproduced with permission from the Provost and Fellows of King's College, Cambridge.)

Or if one prefers the "mercury" of gold, he might obtain it by the following process, although in this case he might have trouble preparing the "water of life" he would need to use.

⚬ of ⋀. ℞ fflorentine ⋀ for yt needs little or rather no purging, calcine it to a most subtile pouder, & pour ⚬⚬ of ⊔ to it yt it may swim two fingers above it, digest it in ⧠ 8 days & nights, pour away ye ⚬⚬, dry it; put upon it ye ▽ of life, or our menstruall, lute it close, give it a lent △, & yt Calx will visibly turne to ☾, & put it in a glass wth ye ▽ of ☿ prepared, yn strein it through a linnen cloth or sheeps skin, & wt will not passe through dissolve again wth more of your blessed ▽ till it will passe through; soe have ye ☾ of ⋀ wch will bee heavy of motion, & volatile.[21]

Whether Newton actually experimented in 1668 with those recipes he had collected probably can never be determined now, for if he did, no record of his results seems to have survived. But in April of 1669 he laid in a supply of glassware and chemicals, materials for two furnaces, and the six-volume *Theatrum chemicum*,[22] and set to work in earnest.

[21] *Ibid.*, f. 4r. On f. 4v, ⚬ and ☾ are defined as mercury, ⋀ as gold, ⚬⚬ as oil, ▽ as water, △ as fire, ☿ as salt, and ⧠ as "ffimus Equinus"—horse dung.
[22] Brewster, *Memoirs*, I, 31–32, n. 1 (1, n. 2).

138

"Mercury" extracted the wet way: the first experiments

The technique with which Newton began his first recorded chemical experiments on the "mercury" of metals was not so common as those requiring the use of heat; rather than heat processes, it utilized solutions of different metals in *aqua fortis* (nitric acid, HNO_3). However, Boyle had published an experiment in the second edition of his *Physiological Essays* of 1669 which was remarkably similar to Newton's initial one. Here is Boyle's very suggestive report, reduced somewhat in volume.

> Dissolve one Ounce of clean common Quick-silver in about two Ounces of pure *Aquafortis*...pour into it by degrees...half an Ounce or one Ounce of Filings of Lead, and...the Lead will be in a trice praecipitated into a white Powder, and the Mercury reduc'd into a Mass (if I may so speak) of running Quick-silver.... And though this be far from being the true Mercury of Lead, as I may elsewhere shew you; yet some Inducements, not here to be named, incline me to look upon it, as somewhat differing from common Mercury, and fitter than it for certain Chymical uses.[23]

Newton's first recorded experiments, entered in MS Add. 3975 directly after his notes on Boyle's *Of Formes*, which experiments are probably datable to 1669, begin in a fashion almost identical with Boyle's experiment in the *Essays* of 1669.

> Jn Aqua fortis 2 ℥ dissolve ☿ 1 ℥ or as much as it will dissolve. Then put an ounce of Lead laminated or filed into it by degrees & ye lead will bee corroded dissolving by degrees into ☿ & besides there will

[23] Robert Boyle, *Certain Physiological Essays And other Tracts, Written at distant Times,* and on several Occasions. The Second Edition. Wherein some of the Tracts are enlarged by Experiments, and the Work is increased by the Addition of a Discourse about the Absolute Rest in Bodies (London: Printed for *Henry Herringman* at the *Blew Anchor* in the Lower Walk of the *New-Exchange*, 1669), pp. 202–03; Boyle, *Works*, I, 399 (3, n. 68). Newton owned this second English edition and it is still extant in the Newton Collection, Trinity College, NQ. 8. 48. This experiment was not in the first edition: Robert Boyle, *Certain Physiological Essays, Written at distant Times, and on several Occasions* (London: Printed for *Henry Herringman* at the *Anchor* in the Lower walk in the New-Exchange, 1661). According to Fulton, the Latin editions of 1661, 1667, and 1668 are all only translations of the first English edition. See Fulton, *Bibliography*, pp. 20–25 (3, n. 99). Therefore if Newton's first recorded experiment on obtaining the "mercuries" of bodies did derive from this experiment of Boyle's–and Newton's beginning is so nearly identical to Boyle's that it seems likely that it did – then Newton's experiments must be dated in 1669 or after. But all of the experiments of Newton's recorded in MS Add. 3975 which are to be considered in this chapter could have been done later than that. They are all followed by or are surrounded by notes from Boyle's works, one of which works only appeared in 1671, as will be discussed below. It is entirely possible that Newton had left a few pages blank in MS Add. 3975, for experiments he thought he might do, and only filled those pages with experimental notes after 1671. Although it is not the case with the first experiments, to be considered in the present section, some of the handwriting in the experimental notes does indeed seem to indicate a later date, even though one still in the 1670's.

> fall downe a white praecipitate like a limus being ye ☿ praecipitated by ye 🜍 of ♄. Out of an ounce of ♄ may bee got ⅓ ℥ of ☿ Jf the remaining liquor bee evaporated there remaines a reddish matter tasting keene like sublimate.[24]

Newton is obviously thinking in terms of the mercury–sulfur theory of metallic composition, since he speaks of the sulfur of lead as precipitating the mercury into a white "limus." The mercury in the "limus" at the end of the reaction seems to him to be the ordinary mercury which he had originally dissolved in the *aqua fortis*. The *other* mercury, which would have appeared free and "running" in the bottom of the vessel, was considered by Newton to be the "mercury of lead," for he said the lead he added to the solution was corroded and dissolved *into* mercury.

Passing then beyond Boyle, he tried tin and reported: "The same liquor will extract ye ☿ of ♃." Then he tried copper and recorded, "Jf ♀ bee put into it, it is presently covered wth ☿."[25] But he began to have doubts and noted:

> J know not whither yt ☿ come out of ye liquor or of ♀ for ye liquor dissolves ♀. Also ♀ will draw ☿ out of ye limus wch falls down in dissolving ♃ or ♄ & also out of ye liquor both during ye solution & afterward.[26]

What Newton had seen in these experiments, and brought his critical judgment to bear upon, may be easier to understand in modern terminology. Mercury dissolved in *aqua fortis* or nitric acid yields a mercuric nitrate solution.

$$Hg + 4HNO_3 \rightarrow Hg(NO_3)_2 + 2H_2O + 2NO_2\uparrow$$

Lead, tin, and copper all being more active than mercury, however, will, if added to the mercuric nitrate solution, reduce the mercury while they themselves are oxidized and pass into solution. Depending on the experimental conditions, some of the mercury will be set free and will reappear in its original form, and at the same time some would probably be reduced only to the lower oxidation state of the mercurous ion rather than to the free metal. The latter condition would result in an additional precipitate of white mercurous oxide, Hg_2O, Newton's "limus." In the reactions with lead and tin, the original solution would not appear to change and would remain colorless, as it was at the beginning. But in the reaction with copper, the solution would turn blue, thus signaling that some copper was going into solution.

It must have been the appearance of the blue color which warned Newton that all was not occurring as he first thought, for he said, "J know not whither yt ☿ come out of ye liquor or of ♀ for ye liquor dissolves ♀."

[24] MS Add. 3975, f. 41v (Newton's p. 80) (3, n. 151).
[25] *Ibid.* [26] *Ibid.*

At that point he apparently decided that the solution method was not a good one for the extracting of "mercuries" and abandoned his first technique for another.

Eventually the data from this first set of experiments appeared in the *Opticks*, first added to the Latin edition of 1706 and kept in all subsequent editions. In the *Opticks* there was no hint that Newton had first explored the reactions when searching for the "mercury" of lead, tin, or copper, and the statement is quite a modern one, setting out the relative activities of the metals.

...[A] Solution of Mercury in *Aqua fortis* being poured upon Iron, Copper, Tin, or Lead, dissolves the Metal and lets go the Mercury; does not this argue that the acid Particles of the *Aqua fortis* are attracted...more strongly by Iron, Copper, Tin, and Lead, than by Mercury?[27]

"Mercury" extracted the dry way: the second experiments

Newton's second set of recorded experiments was closely related to the common use of mercuric chloride to obtain "mercuries." He had read some processes of that sort in Boyle's *Of Formes*,[28] in which Boyle talked about "opening" the bodies of antimony, silver, and tin with mercury sublimate. Newton diligently took notes on Boyle's experiments[29] and then entered the information in a condensed form in his chemical dictionary.

The dictionary seems to have already had this brief entry for mercury sublimate in place:

Mercurius Sublimatus is ye flowers of Mercury ⚴ well mixed ♀ with Vitrioll (calcined twixt white and red) & Salt Peeter (decrepitated) wch in Sand sublime to a white corrosive salt wch is ye Mercury joyned wth the spirits of the salts in a dry forme.[30]

Then when Newton decided to add the note he had drawn from Boyle, he put as much of it as he could after that earlier entry, but since there was not room for it all, he made a continuation indication and finished it on another page. What he had taken from Boyle follows. Newton is referring to mercury sublimate ($HgCl_2$) and summarizing Boyle's experiments very succinctly.

Jts fumes strangly open & volatize minerall bodys as of Antimony in making it Butter, and ↗ grosly beaten Venetian Sublimate opens ↙ Copper cemented wth it so as to make it brittle transparent & easily

[27] Newton, *Opticks*, p. 381 (1, n. 9).

[28] Boyle, *Of Formes*, pp. 283–302 (Experiment II) (4, n. 78); Boyle, *Works*, III, 78–83 (3, n. 68).

[29] MS Add. 3975, ff. 38v–39r (Newton's pp. 74–75) (3, n. 151).

[30] MS Don, b. 15, f. 4r (4, n. 77).

fluxible & inflamable too being lay ⟨*sic*⟩ upon blown coales or put to
ye flame of a Candle. So also of Silver, Tin &c but not of gold. Yet
perhaps there may bee Sublimates made (as by subliming common
Sublimate & Sal Armoniack ↗ well poudered ✗ together) wch
besides notable operations on other metalls, may act upon Gold too.
☿ may be sublimed from common salt.[31]

What had especially interested Newton was Boyle's suggestion of improving the operation of ordinary sublimate by the use of various additives
to it. Boyle had said towards the end of his report on the action of ordinary
sublimate on metals,

> That this Experiment may probably be further improv'd, by imploy-
> ing about it various and new kinds of Sublimate, and that several
> other things may be sublim'd up together either with crude Mercury,
> or with common Sublimate, he that considers the way of making
> vulgar Sublimate, will not, I suppose, deny.[32]

Boyle continued with an example of what had happened to copper
when he treated it with common sublimate to which had been added
"Sal Armoniack" (ammonium chloride, NH_4Cl). The results had been
much more impressive than when copper was treated with ordinary sublimate alone, "so that we suppos'd the Body of the Venus to have been
better wrought upon by this, then by the former Sublimate."[33] Boyle then
ended his discussion with further suggestive hints:

> And yet I judg'd not this way to be the most effectual way of improv-
> ing common Sublimate, being apt to think, upon grounds not now to
> be mention'd, that it may, by convenient Liquors, be so far enrich'd
> and advanc'd, as to be made capable of opening the Compact Body
> of Gold it self, and of producing in it such Changes, (which yet per-
> haps will enrich but mens Understandings,) as Chymists are wont
> very fruitlesly to attempt to make in that almost Indestructible Metal.
> But of This, having now given you a Hint, I dare here say no more.[34]

It was the union of common sublimate and sal ammoniac which Newton
proceeded to use in his second set of experiments, applying the mixture
apparently in an attempt to "open" the bodies of the metals and get out
their "mercuries." He was later to return to Boyle's last suggestion and
make something he called a "philosophical sublimate," but at this time –
probably still in 1669 – he stayed with the relatively simple combination
of $HgCl_2$ and NH_4Cl.

Newton's brief report follows. It obviously includes the results of a
multitude of detailed experiments.

[31] *Ibid.*, ff. 4r and 7r.

[32] Boyle, *Of Formes*, pp. 299–300 (4, n. 78),; Boyle, *Works*, III, 82 (3, n. 68).

[33] Boyle, *Of Formes*, pp. 300–01, quotation from p. 301 (4, n. 78); Boyle, *Works*, III, 82 (3, n. 68).

[34] Boyle, *Of Formes*, pp. 301–02 (4, n. 78); Boyle, *Works*, III, 82–83 (3, n. 68).

> Sublimate 1 ℥, ✱ ℥½, ♂, ♃, ♀, ☽, or ↗ their ↙ Reguluses or ye Reg of ♁ ℥½. ffirst bake ye ☿ & ✱ together & put ye mettall poudered of ⟨*sic*⟩ filed into a crucible & ye salts upon it & in a gentle fire ye salts will act upon ye metalls & you shall have their ☿ running at ye bottom. You must but just let them begin to boyle on ♂ because his ☿ is very volatile, but for other metalls let them stand longer.[35]

A discussion of what Newton meant by "their Reguluses or ye Reg of ♁" will be reserved until the next section. The primary thing to be noticed in Newton's experimental notes here is that he thought he had obtained the "mercuries" of all those different metals by this method of treating them with mercuric chloride and ammonium chloride. His approach was systematic, as in each case he used the same quantities of his special sublimate and metal, varying only the nature of the metals. And he duly noted the variation in procedure which had to be applied in the case of iron. But there was one very important experimental variation which he needed to apply but which he could not because it was a concept totally unsuspected at the time, the concept of equivalent weights. Newton used equal weights of the metals just as if they were all "worth" the same chemically. But in fact that is not correct. Using *equal* weights of the metals rather than *chemically equivalent* weights, he would have obtained products which were "mercuries" contaminated in varying degrees with the original metals. There was nothing incorrect about his experimental approach by modern standards, systematic and quantitative as it was, except that his work plan did not and could not take into account the factor of equivalent weights. If he had known to use the proper proportions of equivalent weights in each case and so been able to obtain pure mercuric products, one wonders if his keen eye would not have seen their chemical identity.

Here a question may be raised about Newton's interpretation of his results. He had actually obtained only ordinary quicksilver in each case – deriving from the quicksilver compound, mercuric chloride, which he had put in at the beginning – or perhaps an amalgam of quicksilver and the original metal used in each instance. Did he in any sense realize that fact?

In the first experiments, the extraction of "mercury" the wet way, he seems to have received a tell-tale sign which he correctly interpreted, when the solution began to turn blue with dissolving copper. But in these second experiments there would have been no such convenient indicator and he probably did not realize his mistake. One would think that he should have suspected it, for, immediately following the summation of his experiments, he set down the recipe for "Venetian sublimate" (another name for the mercuric chloride he had just been using), and in the recipe he quite clearly states the use of mercury in the preparation of the sublimate.

[35] MS Add. 3975, f. 41v (Newton's p. 80) (3, n. 151).

Venetian Sublimate is made of ☿ 2 pts, refined ☽ 2 pts ⊕ calcined to red 1 pt & salt decrepitated 1 pt. The Hollanders sophisticate it wth Arsnic. The sophisticated is in long splinters & turns black wth oyle of ♄ dropt on it, but ye true turns yellow & is in little grains like hempsed.[36]

When the modern reader, with the benefit of an additional three hundred years of chemical studies intervening, looks at that recipe Newton wrote down about 1669, he immediately thinks Newton should have suspected the presence of ordinary mercury in his product since he himself had put it into his reaction at the beginning. And there were some seventeenth-century chemists who would have done so. Even without the modern concept of mercury as a chemical element, always recoverable in its original form from its compounds, it was possible for a person to show that level of sophistication in his chemical reasoning. Boyle in fact did. Writing about 1680 on "The Producibleness of Mercury," a part of an appendix being prepared for the second edition of *The Sceptical Chymist*, Boyle recalled that he had "long since" explained that matter to some of the *virtuosi*. In their presence, he said,

> having mingled the filings of copper with a certain salt, and put them in a conveniently shaped vessel of glass, I warily held it over a competent fire of well kindled charcoal, till the salt was thoroughly melted, and in part sublimed; by which operation the copper seemed to be quite changed, especially in color, and was really pretty inflammable: and there remained in the lower part of the glass a pretty deal of running mercury, so that they would have gone away persuaded, that they did see me make the mercury of Venus, if I had not been careful to undeceive them; which I did by telling them, that this quicksilver was only the common mercury, that lay disguised in the compounded sublimate I had employed, together with the copper, which set the mercury at liberty from the corrosive salts it lay concealed in before, by presenting them a metal more disposed to be wrought on by them, than quicksilver is.[37]

Newton was evidently not one of the *virtuosi* privileged to hear Boyle's explanation, but although the critisicm of Newton for not suspecting that he was obtaining his mercuric products from the mercury sublimate which he was putting into the reaction is perhaps justified, it may miss the point of the presence of his recipe for Venetian Sublimate right there. Newton had already put that same recipe in his chemical dictionary in two places, where it varied only slightly from the one just quoted,[38] besides which there was his original dictionary entry on "Mercurius Sublimatus," quoted above, which is also quite similar. He had known all along that

---

[36] *Ibid.*   [37] Boyle, *Works*, I, 632 (3, n. 68).
[38] MS Don. b. 15, ff. 7r and 8r (4, n. 77).

ordinary mercury was going into the reaction and that probably was not what concerned him when he wrote out the recipe yet again. More likely, he was taking occasion to consider the preparation of more sophisticated sublimates, acting out Boyle's suggestion, also quoted above, that those who knew how to make "vulgar Sublimate" might also devise various new ones. At least, since Newton did make very complicated sublimates later on, that seems a reasonable suggestion.

Returning to the question about Newton's interpretation of his results in these second experiments, there seems to be no reason to doubt that he believed he was obtaining the "mercuries" of the metals he worked with. But they were not good enough for his purposes. He was later to say, in an alchemical manuscript of the "middle" period:

> this ☿...drawn out of bodies hath as many cold superfluities as common ☿ hath, & also a special form & qualities of ye metals from wch it was extracted, wch makes it more remote from ye philosophick ☿ then ye common ☿ is.[39]

Three things may be deduced from Newton's later comment. First, he did not doubt that he had obtained "mercuries" from his metals, but, second he had realized – probably from some more experiments which he never recorded – that they resembled the metals he got them from and so were unsuitable for further use. Third, and perhaps most important, what he was really looking for was the "philosophick" mercury.

The clue to Newton's intent in using this method of obtaining the "mercury" of metals lies in the Boylian concept of "opening" metals. As will be discussed in more detail in Chapter 6, Newton's concept of the "mercury" of metals probably never was *exactly* based on the "chymical" theory that the principles of metals were "mercury" and "sulfur" – or at least his conception was not *limited* to that approach. In later years he published material indicating that he thought that the large corpuscles, which by cohering formed massy bodies, were themselves composed of complex arrangements of smaller particles. There are indications in his later writings that his conception probably included a "mercurial" particle within the largest corpuscle of the metal, one step down the hier-

---

[39] King's College, Cambridge, Keynes MS 55, f. 3r, hereinafter referred to as Keynes MS 55. It is instructive to compare Newton's statement on the "mercury" of various metals with some statements by Eirenaeus Philalethes, for there are many similarities: "... only this *Mercury* hath specificated qualities according to the nature of the *Metal* or *Mineral*, from which it was extracted; and for that reason, (as to our work which is to dissolve perfect Species of *Metals*,) it hath no more virtue than common *Argent Vive*.... So then, if a *Mercury* drawn from the Bodies, have not only the same deficiency of heat and superfluity of *faeces* as Common *Mercury* hath, but also a distinct specificated form, it must (by reason of this its form) be so much the farther remote from our *Mercury* then common *Argent Vive* is." Eirenaeus Philalethes, "Sir George Ripley's Epistle to King *Edward* the Fourth, Unfolded," in Philalethes, *Ripley Reviv'd*, pp. 11–12 (2, n. 9). Newton appears to have made the concept his own, however.

archical ladder from the largest particles, and *inside* them. Thus the "mercury" of which he spoke, although having mercurial chemical characteristics certainly, was made of relatively complex particles of matter and was obtained when the larger particles of the metal were "opened," allowing the smaller component parts to exit. In various aspects of his work, Newton seems to have been able to hold divergent concepts in his mind in a powerful tension fruitful of new ideas. In this case, the tension lay between the new corpuscularianism of particles and the old "chymistry" of substances. From the tension were born his complex hierarchical ideas on the sub-structure of matter, which comprehended both.

Although he was later to return to it, he dropped the particular line of inquiry of the second experiments at that time – probably still in 1669 – and took up another one, which was much more distinctively alchemical, one which was to occupy him for a long, long time with all its variations and ramifications. The new one was related to the "Reguluses" which he had already mentioned, and apparently was designed to draw a more "philosophick" mercury out of metals. Some chemical and alchemical background to the "Reguluses" is in order before the next experiments are considered.

The star regulus of antimony

Antimony is a semi-metal which most commonly occurs naturally as antimony sulfide, Sb_2S_3, a mineral now called stibnite. Stibnite may be reduced to antimony metal simply by heating it with charcoal or other mild reducing agent under the proper conditions, but the standard modern laboratory and commercial procedure for the preparation of the metal involves heating the Sb_2S_3 with iron filings and a flux, such as borax. The iron combines with the sulfur of the stibnite, forming an iron sulfide which floats to the top, and the metallic antimony sinks to the bottom. The borax facilitates the separation of the liquid layers but does not enter into the reaction.

$$Sb_2S_3 + 2Fe \rightarrow 2Sb + Fe_2S_3$$

When the whole mass has cooled, the layers may easily be separated and the upper slag discarded. Then the antimony will appear with a metallic luster. If conditions have been suitable, and if the antimony has been well purified, metallic crystals will have formed. The crystals of antimony are long and slender and sometimes arrange themselves in patterns on a sort of stem and so resemble the fronds of ferns. If certain very special conditions prevail in the purification and cooling of the metal, the crystalline "branches" may be arranged around a central point and so take on the appearance of a star. The star of antimony fascinated the alchemists, and especially Newton.

146

But in the seventeenth century neither the nomenclature nor the chemical understanding of the star of antimony was quite the same. In the first place, the name "antimony" was then applied only to the ore, while the terms "regulus" or "regulus of antimony" indicated the antimony metal. In the second place, it was thought that the iron, or any other metal, used in the reduction of the ore remained in the metallic product, whereas actually it did not if the right proportions had been used. That belief, however, gave rise to a variety of designations for metallic antimony: "regulus per se," if the metal had been formed by the heating of the ore with a non-metallic reducing agent; "Martial Regulus," if iron had been employed in the reduction (Mars being identified with iron); "Venereal Regulus," if copper had been used (Venus being equivalent to copper); and so forth. When the star appeared in the refining, the antimony was given the special designation of *regulus antimonii stellatus*, the "starred regulus of antimony."

Nevertheless the common method of preparation in the seventeenth century was identical with the modern one and employed the stibnite ore and iron. Newton was familiar with it when he wrote his chemical dictionary, as well as with a method of making "regulus per se," but seems to have thought the star only appeared with the use of the iron. Here are his dictionary entries on the subject.

Regulus of Antimony is made of Antimony Salt Peeter & Tartar ana: put by spoonfulls into a red hot crucible & fluxt when all is in. When tis cold in the crucible you will find a black Sulphur on ye top layd hold on by fixed Salts, & a metalline body below like lead but brittle wch is ye Regulus & of this they make ye Antimoniall Cups.

Regulus Martis Is made by casting two pts of Antimony upon one of Jron heated white hot in a Crucible & melting them ⌐ well ⌐ together wth a little Saltpeeter to promote the fusion. When tis cold ye Regulus will bee in the bottom; which being againe 3 or 4 times melted wth Salt Peeter is thereby purifyed & when cold hath an upper surface (under ye Saltpeter wch is then of a cleer Amber colour) wth stellar figures & is called

Regulus martis Stellatus[40]

No alchemical implications at all appear in Newton's dictionary entries: if one modernized the spelling and the nomenclature a bit, Newton's descriptions could go straight into a modern chemical text. But he got involved with the alchemical meaning of the star of antimony early on, and already in the second set of experiments examined in the last section – which probably date within a year or two after the dictionary – he was experimenting with "Reguluses" in an alchemical context.

It will be apparent by now that when Newton mentioned the reguluses

[40] MS Don. b. 15, f. 4v (4, n. 77).

147

of iron, tin, copper, and silver in those experiments that he meant metallic antimony prepared by the use of those different metals. Likewise when he mentioned "ye Reg of ☿" he meant "regulus per se," or antimony metal prepared "by itself," i.e., with a non-metallic reducing agent.

The term "regulus" is another word which has changed its meaning since Newton's time. Although now it refers to any metallic product which forms under the slag when ores are refined, then it was applied *only* to metallic antimony. Thus the "regulus of iron" did not then mean metallic iron, as it would today; rather it meant metallic antimony prepared by the use of iron. To the seventeenth-century user of that designation it would also have meant that iron was present in the regulus, as well as antimony metal, although that would not always have been the true state of chemical affairs. On occasion, it might possibly have meant to the seventeenth-century "chymist" that some particular portion of the original iron, such as its "mercury," was present in the regulus.

The word "regulus" means "little king," as it is the diminutive of the Latin word for "king," *rex*. It has sometimes been suggested that the word came to be applied to metallic antimony because of the special chemical relationships that antimony has with gold, "the king of metals," one of which special relationships will be considered below. It has also been suggested that the term "regulus" was used for the metal because the metallic regulus was something of special value obtained from the ore. But it is here suggested that perhaps metallic antimony got its name "regulus" from its ability to form a star, because there was, and is, a prominent star by that name: Regulus, a star of first magnitude, the brightest star in the constellation Leo, and also known as *cor leonis*, the heart of the lion. At any rate, Newton saw a relationship between "the regulus of antimony" and the "regulus of Leo," as will appear later, so, whether the relationship originally existed or not, it is important for the present discussion. Newton – and it is at least possible that others did also – seems to have interpreted the lion of alchemical symbolism as antimony ore. The starry metallic antimony at its heart then became *cor leonis* or Regulus (Plate 3).

With the star of antimony one steps out onto the quicksands of alchemy and becomes willing to accept assistance from any source, however unlikely. What did the star of antimony mean to Newton? The first step in finding out involved for the present writer help from a very unlikely source indeed, Mary Anne Atwood.

Mrs. Atwood, born in 1817, was the daughter of Thomas South, a scholarly English theosophist who trained his daughter to become his intellectual comrade. Their mutual studies proceeded through the mystical literature of all ages and included the alchemical philosophers. About 1849 father and daughter decided that they had reached mature conclusions from their long researches, and he retired to one room to cast his knowledge

148

Plate 3. THE STAR REGULUS OF ANTIMONY.
The terrestrial "star" at the heart of antimony ore was symbolically designated as
cor leonis, the heart of the lion, after that celestial Regulus, the star at the heart of
the constellation Leo. (Crown copyright. The Science Museum, London.)

into verse while she occupied another and composed in prose *A suggestive
inquiry into " The Hermetic Mystery,"* published in 1850. A change of heart
overtook Mr. South about the time the books began to be publicly distributed: he seems to have decided that the mystery should not be laid
open to the world after all. The remainder of the edition was called in at
no little expense and burned on the front lawn along with the manuscript
poem. Only a few copies escaped and it was from one of these that the
recent edition came.[41]

Mr. South might saved himself all that trouble, for his daughter's book
is not the sort of work which will ever greatly enlighten the world, either
for good or for evil. There is no doubt that Mrs. Atwood was a very
erudite woman, however, and she did convey a hint which has been invaluable in this study of Newton. Before Mrs. Atwood's theosophy was

[41] Mary Anne Atwood, *Hermetic Philosophy and Alchemy. A suggestive inquiry into " The
Hermetic Mystery"* with a dissertation on the more celebrated of the alchemical philosophers, introd.
by Walter Leslie Wilmhurst (revised edn.; New York: The Julian Press, 1960), "Introduction," pp. 1-96.

149

read, it had always seemed to the present writer that the lines of crystals in a star regulus radiated *out* from the center, as the rays of light radiate out from a star. Suddenly in a shift of perspective it appeared that the lines might just as well be considered as radiating *in* to the center, which gives them the character of *attraction* rather than the character of emission, for that, Mrs. Atwood said in an indirect way, was what the alchemical philosophers had seen in the star of antimony.[42]

When the star regulus is seen as having lines radiating in to a central point, it assumes a whole new dimension of meaning, especially considered in relation to Newton. After all, he is the man who is most famous for working out a law for the attraction of gravity, in which the lines of attraction run in to and converge in a central point.

Newton would have been alerted to the alchemical potentials of the star of antimony quite early in his alchemical studies, for probably one of the first things he did was to read and take notes on *The Triumphal Chariot of Antimony* by Basilius Valentinus, which notes survive in Keynes MS 64, as noted above. Basilius Valentinus was primarily concerned in this work to delineate the important chemical medicines which might be obtained from antimony, but he had the following to say about the star.

Many have esteemed the Signed Star of Antimony very highly, and spared neither labour nor expense to bring about its preparation. But very few have ever succeeded in realizing their wishes. Some have thought that this Star is the true substance of the Philosopher's Stone. But this is a mistaken notion, and those who entertain it stray far afield from the straight and royal road, and torment themselves with breaking rocks on which the eagles and the wild goats have fixed their abode. This Star is not so precious as to contain the Great Stone; but yet there is hidden in it a wonderful medicine. . . .[43]

Basilius gave the standard preparation for the star with stibnite and iron and added the method of making a medicine from it of great "surgical utility." Then he offered another alchemical hint.

This Regulus, or Star, may be very often carried through the fire with a stone serpent, till at length it consumes itself, and is completely joined to the serpent. The Alchemist has then a hot and ignitable substance, in which wonderful possibilities are latent.[44]

Newton made notes on those passages – even though his chemical dictionary indicates he already knew how to make the star regulus.

Furthermore, there is a clear indication in Newton's notes that he was

[42] *Ibid.*, pp. 147–59.

[43] Basilius Valentinus, *The Triumphal Chariot of Antimony. With the Commentary of Theodore Kerckringius, A Doctor of Medicine. Being the Latin Version Published at Amsterdam in the Year 1685 Translated into English. With a Biographical Preface by Arthur Edward Waite* (Reprint of the London edn. of 1894; London: Vincent Stuart, 1962), p. 175, hereinafter referred to as Basilius Valentinus, *TCA*.

[44] *Ibid.*, pp. 176–77.

relating Basilius' formulation of the regulus being "carried through the fire with a stone serpent" with his own experiments on obtaining "mercuries" the dry way. For Newton, in one of his few personal additions to the manuscript, identified "the stone serpents" as components of mercury sublimate and said that the regulus or star might be carried through the fire with mercury sublimate. In the following brief translation from Keynes MS 64 the material within the square brackets is Newton's own.

> Still more often this Regulus or star may be led through the fire with stone serpents [i.e. with mercury sublimate, for stone serpents i.e. vitriol and Salt peter are in the sublimate] so that at length it is completely consumed and joins itself with the serpent, by accomplishing which the artist has a matter most excited and wholly fiery in which many works are latent....[45]

In that passage in square brackets – which is definitely Newtonian since that material is not in *The Triumphal Chariot of Antimony* – Newton's understanding of the stone serpents is conditioned by his understanding of the composition of mercury sublimate. As was seen in his dictionary entry on "Mercurius Sublimatus," cited at note 30 above, Newton made mercury sublimate by subliming "flowers of mercury" with vitriol and salt peter and said that mercury sublimate consisted of "Mercury joyned wth the spirits of the salts in a dry forme." Actually the "flowers of mercury" Newton started with must have been mercurous chloride and the vitriol and salt peter only served to oxidize that material to the mercuric chloride, but he thought the acid spirits from the vitriol and salt peter were present in the mercury sublimate also. Those "spirits" he then identified as stone serpents because they came from stony salts and interpreted the passage from Basilius Valentinus to mean that the regulus might be treated with mercury sublimate to obtain an alchemically interesting product. That treatment he then proceeded to undertake in connection with his second set of experiments, combining that idea drawn from Basilius Valentinus with the use of the enriched sublimate drawn from Boyle.

Newton also abstracted another passage from Basilius Valentinus that followed close on the heels of the ones above. In the later section that Newton noted, Basilius explained that the "lead of antimony," the "philosophical lead," should be identified with the regulus of antimony. Newton drew a little hand in the margin of his manuscript notes, pointing to that place, to indicate its special importance.[46]

---

[45] Keynes MS 64, f. 4r: "Ducitur etiam hic Regulus vel stella cum serpentibus lapideijs [i.e. cum ☿o sublimato, serpentes enim lapidei i.e. vitriolum & Sal petrae insunt sublimato] saepius per ignem ita ut tandem omnino consumptus serpenti se associet, quo peracto artifex habit materiam fervidissimam & totam igneam in qua multa artificia latent...."

[46] Basilius Valentinus, *TCA*, pp. 178–80 (5, n. 43); Keynes MS 64, f. 4r. Another marginal "hand" on that same folio points to a discussion of the "mercury of the philosophers," *Mercurius Philosophorum*.

Basilius said one more thing about antimony which Newton did not note at the time but may have remembered later: "Know that in Antimony also there is a spirit which is its strength, which also pervades it invisibly, as the magnetic property pervades the magnet."[47] It was the "magnetic" or attractive property which was attributed to the star which eventually riveted Newton's attention upon it and led him to think that it would draw the "philosophical mercury" out of other metals.

Sometime early in 1669 Newton took up the study of alchemical works by Sendivogius and d'Espagnet and an interesting set of notes on the two of them survive in Keynes MS 19. Newton had divided each page of the manuscript into parallel columns. The column on the left he entitled "Collectiones ex Novo Lumine Chymico quae ad Praxin spectant," and it contains excerpts from *A New Light of Alchemy*, the *Philosophical Riddle*, and the *Dialogue of Mercury, the Alchemist, and Nature* by Sendivogius, and *Arcanum Hermeticae philosophiae opus* by d'Espagnet. The column on the right was entitled "Collectionum Explicationes," and it carries Newton's own explanatory notes, keyed to the snippets of textual material on the left.[48]

The manuscript may be dated with some degree of confidence – more confidence in this case than one can muster with many of the manuscripts – because of the following items. (1) The handwriting and Newton's failure to cross his symbols for lead place it sometime in the 1660's. (2) The only reference in it to a book other than the ones being noted is to *Secrets Reveal'd* by Eirenaeus Philalethes. That work, which will be referred to again below, was not published until 1669. (3) In April of 1669 Newton purchased the six-volume *Theatrum chemicum*, as noted above. Virtually every alchemical manuscript which Newton seems to have written after the "very early" period contains a reference to some item in that enormous collection, and since Keynes MS 19 does not, it seems probable that it was composed before he plunged into the *Theatrum chemicum*, that is, fairly early in 1669. The dating of Keynes MS 19 is important because its internal characteristics seem to place it between the second set of experiments, where Newton is still more or less following Boyle's lead in his search for the "mercury" of metals, and the third set of experiments, where Newton addresses himself to the difficult and hopeless attempt to extract a more "philosophick" mercury by way of the star regulus. Although Newton utilized various "reguluses" in the second set of experiments, there they are mentioned in series with the common metals, and there is no indication that he considered a star regulus to have any special or "magnetic" properties. He seems to have acquired his ideas on the "mag-

---

[47] Basilius Valentinus, *TCA*, p. 30 (5, n. 43).

[48] It has not been possible to identify the editions Newton used for Keynes MS 19 because he included no page numbers in this manuscript and all of the works had had more than one Latin edition before 1669.

netic" effects of the star regulus from Basilius Valentinus, Sendivogius, d'Espagnet, and Eirenaeus Philalethes, and it was only after he had read all those authors that he began to focus and concentrate on the star regulus.

It will be recalled from Chapter 3 that Sendivogius was one of the most popular of the alchemical writers in the seventeenth century and that both he and d'Espagnet had strong affinities with the philosophical position of Neoplatonism. D'Espagnet's *Enchiridion physicae restitutae*, in fact, was a fairly important restatement of Neoplatonism at the time, although his alchemical work, *Arcanum Hermeticae philosophiae opus*, being considered by Newton in Keynes MS 19, seems to have been somewhat less popular. Both Sendivogius and d'Espagnet firmly believed in "magnets." They conceived them as matrices which drew other things – bodies or spirits – to themselves by virtue of an attractive power and then somehow made manifest and substantial a new form for what had been drawn in.

In more than one place in Keynes MS 19 Newton excerpted a passage from Sendivogius in which the "Magnet" or the "Chalybs" was mentioned. In his own notes keyed to those passages Newton then identified the attractive body as antimony, even though Sendivogius said never a word about antimony. In Tractate 9 of the *New Light*, Sendivogius had said:

There is another Chalybs which is made like this, created of itself from nature, which knows how to draw from rays of the sun that which so many men have sought, and it is the beginning of our work.[49]

Newton's response to that was as follows.

That other (and properly named) chalybs is antimony for it is created from nature of itself (without art) and it is the beginning of the work; neither are there more than two principles, Lead and Antimony.[50]

Perhaps it was at that point, when Sendivogius began to emphasize "magnets," that Newton first began to see the significance of Basilius Valentinus' idea about the "magnetic" property of antimony and so identified the Sendivogian "Chalybs" with antimony.

Another example in Keynes MS 19 may be drawn from the section on the *Aenigma* or the *Philosophical Riddle*. There Sendivogius spoke of "our water."

Our water is wondrously drawn, but that is the best which is drawn by the power of our Chalybs which is found in the belly of Aries.[51]

[49] Keynes MS 19, f. 1r; "Est et alius Chalybs qui assimulatur huic, per se a naturâ creatus, qui scit ex radijs solis elicere illud quod tot homines quaesiverunt, & operis nostro principium est."

[50] *Ibid.*: "Jste alius (& proprie dictus) chalybs est antimonium nam per se a natura (sine arte) creatur & est operis principium; nec plura sunt quam duo principia, Plumbum & Antimonium."

[51] *Ibid.*, f. 3r: "Aqua nostra hauritur miris modis, sed ista est optima quae hauritur vi Chalybis nostri qui invenitur in ventre Arietis."

Newton's comment applies to the "power of our Chalybs which is found in the belly of Aries," and in it one may see the effects of the *prisca sapientia* doctrine in operation. Newton said that the best water was drawn

> by the power of our sulphur which lies hid in Antimony. For Antimony was called Aries with the Ancients. Because Aries is the first Zodiac Sign in which the Sun begins to be exalted and Gold is exalted most of all in Antimony.[52]

It is a chemical fact that gold can be refined or "exalted" by heating it with antimony ore. In such a treatment, any metals contaminating the gold combine with the sulfur of the stibnite and all rise to the top of the molten mass as a sort of scum. The gold sinks to the bottom, along with metallic antimony, from whence the gold may be recovered in an extremely pure state. That is one of the special chemical relationships between antimony and gold, the "king of metals," which lend credence to the idea that the use of the word "regulus" for metallic antimony derives from its meaning of "little king." Newton knew that gold could be refined in that manner, evidently from his general chemical reading, for it was common knowledge in the seventeenth century. In his chemical dictionary he described the process, calling it a form of "probation."

> Probation by Antimony is when you ↗ ⟨ ⟩ & ⟨melt⟩ ↙ one pt of impure ☉ with 3 pts of Antimony together in a strong fire & poure them forth into an Jron cone, first a little heated. And wn tis cold you will find a Regulus in the bottom in wch is all ye ☉, ye Antimony having drunk up ye other Mettalls. This Regulus being set in ye fire till yt Antimony ↗ in the Regulus ↙ is all driven away you will have the gold remaining very pure[53]

Newton then attributed that knowledge to the "Ancients," in accord with his belief that all wisdom was anciently held by at least some wise men, and then interpreted the mystical Sendivogian phrase, "in the belly of Aries," in terms of that supposed antique knowledge. So in Keynes MS 19, in his explanatory comment on the Sendivogian passage just quoted, Newton said that antimony was called Aries by the ancients because the sun begins to be "exalted" in Aries (meaning it begins to rise towards its summer zenith, for Aries is the first spring zodiacal sign) as gold, always symbolized by the sun, is "exalted" or refined in antimony.

But the really important part of that passage for the purposes of the present study is the fact that Newton identified antimony with the draw-

---

[52] *Ibid.*: "vi sulphures nostri quod latet in Antimonio. Antimonium enim apud Veteres dicebatur Aries Quioniam Aries est primum Signum Zodiaci in quo Sol incipit exaltari & Aurum maxime exaltatur in Antimonio."

[53] MS Don. b. 15, f. 4v (4, n. 77).

ing power of the "Chalybs," and from Keynes MS 19 it may be seen that Newton was considering antimony as a material operating by attraction in 1669. He does not mention the regulus of antimony, but immediately following the note just quoted he conveys the idea that although crude antimony is a great poison, after coction or heating it is a great medicine. So he may have been thinking in terms of the regulus as being the operative "magnet" rather than the unrefined ore. In any case, he had decided that antimony was the beginning of the work by this time, and in the laboratory he embarked on an exhaustive study of the star regulus.

The star regulus five different ways: the third experiments

"To make Regulus of ☿, ♂, ♄, ♀ &c.," Newton said, you must take certain specific proportions of antimony ore and metal in each case. After he had worked through the directions for those four, he added the proportions for a regulus of tin. The proportions were of great importance to him because they determined whether one would get a good star, the star being the desired product in each case. He said,

> The better your proportions are the brighter and britler will ye Reg bee & ye darker ye scoria & the easier will they part: And also ye more perfect the starr, unlesse the salts on ye top worke & bubble in the cooling to disturbe ye sd superficies. The work succeeds best in least quantitys.[54]

Newton continued with mention of every conceivable variable which might affect the success of the work – the degree of fire, when the fluxing salts should be added, the order of addition of the materials to the hot crucible, when they should be poured off to cool. Much of this practical data appears under two headings: "These rules in generall should bee observed," and "Also these signes may bee observed in generall." The whole of his discussion about the star reguluses really comprises a little experimental essay, and it presents the results of perhaps hundreds of individual trials.[55]

The star regulus was of fundamental importance to Newton. He set out his information about its preparation meticulously, and he continued to use it in some of his experiments for many, many years. Its significance for him, in fact, seems to warrant the inclusion of his experimental essay on it in its entirety, and it may be found in Appendix B.

Because they may all be studied directly in Appendix B, the details of

---

[54] MS Add. 3975, f. 42r (Newton's p. 81) (3, n. 151).

[55] The plural of "regulus" ought properly to be "reguli" since it is a Latin noun, second declension masculine. However, since Newton chose to give it an English formulation as "reguluses," his usage is continued here.

his experimental work in making the reguluses need not be set out here, and the significance of the star may be considered. If one may judge by a number of later manuscripts, which will be considered in their turn, Newton thought from the first that the star regulus had some special relationship with the "philosophical mercury" but as a practical matter he did not know exactly what the relationship was or what other substances might be involved. As a result, he entered into a great deal of alchemical reading in the next few years to try to solve those problems, as well as performing what was probably a large number of experiments which led him nowhere.

The statement that he performed a large number of experiments is partly surmise, because there are not really that many experiments recorded in MS Add. 3975 which seem to date before he wrote his "Clavis" in Keynes MS 18. But three factors make it seem reasonable that all his laboratory work was not recorded.

One is that the first experiments – those already considered – are not by any means running laboratory records. They are summations of what Newton considered significant after he had mentally digested and organized his experiments. Therefore, if he concluded that a chain of investigation had no significance, he might well not have written it down at all.

Another factor which makes it seem that MS Add. 3975 does not contain all his experiments is that the ones which are recorded frequently seem to start in the middle of a complex situation. For example, in one case a "double salt" is suddenly introduced as a reagent. No "double salt" existed as a standard chemical in the seventeenth century; therefore there must have been an unrecorded series of experiments in which Newton prepared it himself. Such preparatory work he did not consider important enough to enter in the laboratory notebook, and it is only through his mentioning the "double salt" in one of the Keynes manuscripts that its composition may be estimated at all.

Yet another factor is the fact that the "Clavis," in which Newton wrote out a whole lengthy process with which he was well pleased, does not directly reflect experimental results recorded in MS Add. 3975. Bits and pieces of the process may be found in the notebook, but not the whole and not even all the pieces.

Nevertheless, virtually all of the experiments which are recorded for this early period involve the use of antimony in some form or other, and it may definitely be said that Newton was seeking to make the "philososophical mercury" with it during the period of the early 1670's. The general rationale with which he approached the problem seems to have been Sendivogian and is best understood by dipping into the Sendivogian treatises themselves.

The Sendivogian theory of the formation
and growth of metals

Sendivogius was more than an alchemist; he was also a natural philosopher. In the *New Light of Alchemy* and also in the *Treatise on Sulphur*, what Sendivogius expounded was an alchemy writ large and applied to all manner of physical and biological changes in the natural world. In his hands God became the Great Distiller and the central fire in the earth worked the same way as the fire under the philosopher's distillatory vessel. It was a very naturalistic natural philosophy which Sendivogius wrote, and he emphasized again and again nature's consistency and lawful behavior. Let the philosopher consider the possibility of nature, Sendivogius said, and follow her example. Then he will be able to do the great work of alchemy.

In the twelve treatises of the *New Light* Sendivogius first set up his theory of generation of all sorts in the world at large. Nature is divided into four places, he said in Treatise I, and these are the locales of the four elements.[56] In Treatise II the constant motion of the elements is established, and their mutual interaction. Each element sends out "his owne thinness, and subtlety" which Sendivogius calls the element's seed or sperm. As a man's seed is cast into a woman's womb, so is the seed of each element cast into the center of the earth. The center is empty and nothing can rest there, but the Archeus, "the servant of nature," mixes the different sperms there and sends them out again. From diversity of places through which the mixture passes, or in which it lodges, different things are made. But the mixture of sperms from the elements is the same for all: it is

> the Elixir of every thing, or Quint-essence, or the more perfect decoction, or digestion of a thing, or the Balsome of Sulphur, which is the same as the Radicall moisture in metals.[57]

From Treatise III it is learned that two matters are required for metals. The more important, equated with the "philosophical mercury," is a "certaine humidity mixed with warm aire." In the making of metals, this first matter is not joined to anything except that it must have a "covering or shadow," which is Sulphur or "the dry heat of the earth."[58] In Treatise IV it is explained how metals are generated in the bowels of the earth. The mixture of sperms sent out by the Archeus sublimes through the pores of the earth, and by a continuously reiterated circula-

[56] Sendivogius, *New Light*, pp. 1–5 (2, n. 8).
[57] *Ibid.*, pp. 5–8, quotation from p. 6. A general treatment of the concept of the "radical moisture," which derives from medical theory, may be found in Thomas S. Hall, "Life, death and the radical moisture, A study of the thematic pattern in medieval medical theory." *Clio Medica* **6** (1971), 3–23.
[58] Sendivogius, *New Light*, pp. 9–10 (2, n. 8).

tion becomes more subtle and perfect.[59] Treatise V similarly describes the generation of stones.[60]

In Treatises VI, VII, and VIII, Sendivogius establishes that nature must act through seeds. Of these there are three kinds: mineral, vegetable, and animal. The vegetable seed is common and vulgar and demonstrates how nature makes seeds from the four elements, but metals must have seeds too, for "Are not Metalls of as much esteem with God as trees?" The natural virtue of a seed is to join itself to another within its own kingdom and then congeal in an appropriate matrix. The seed is not worth anything unless by art or by nature it is put into its proper matrix. Nature has sometimes failed to perfect developing minerals because "crude air" in the pores, bowels, and superficies of the earth prevented it. Art can complete the perfecting process.[61]

Having established the larger scene, Sendivogius moves in Treatises IX, X, XI, and XII to the crucial alchemical question of perfecting metals, a process which he holds to be parallel to the natural generation of metals in the earth. The first step is to draw out the metallic seed, and it is here that the "magnet" enters into prominence, for it is the magnetic "Chalybs" which may draw out the seed from gold, and presumably from other metals also. The passage quoted above which Newton abstracted in Keynes MS 19 is here given in its entirety. In it Sendivogius said:

[T]here is granted to us one Metall, which hath a power to consume the rest, for it is almost as their water, & mother: yet there is one thing, and that alone, the radicall moisture, viz. of the Sunne, and Moon that withstands it, and is bettered by it; but that I may disclose it to you, it is called Chalybs, or Steel. If Gold couples eleven times with it, it sends forth its seed, and is debilitated almost unto death; the Chalybs conceives, and bears a son, more excellent then his father: then when the Seed of that which is now brought forth is put into its own Matrix, it purifies it, and makes it a thousand times more fit, and apt to bring forth the best, and most excellent fruits. There is another Chalybs, which is like to this, created by it selfe of Nature, which knows how to draw forth the vertue of the sun beams (through a wonderfull power, and vertue) that which so many men have sought after, and is the beginning of our work.[62]

Then Sendivogius set up a new factor which was to be important in his conception of the "magnet." Reverting to a natural-philosophical discussion, he considered how two heats are utilized by nature in her productions, the heat of the sun and the central heat of the earth. Nature knows how to produce wonderful fruits out of the water of the earth, he

[59] *Ibid.*, pp. 11–14. [60] *Ibid.*, pp. 14–16.
[61] *Ibid.*, pp. 17–25, quotation from p. 19.
[62] *Ibid.*, pp. 26–38, quotation from pp. 27–28.

says, and how to give them life from the air. The fire or heat is the cause of the motion of the air and the life of all things. The earth is the nurse of all things and their receptacle. Without the water and air to cool and temper the two heats, however – the "Centrall Sun" and the "Celestiall Sun" – the earth would be dried up. This can and does happen: when the pores of the earth are obstructed so that the water can not penetrate, then the earth will be inflamed by the sun and great chops or furrows will appear in the earth. And the reason the earth may be so inflamed without water is that the central and celestial suns have a correspondency or "magnetick vertue betwixt themselves."[63]

The climax of Sendivogius' alchemical use of the "magnet" comes in the Epilogue which follows the twelve treatises of the *New Light of Alchemy* and only there does it become clear why he interjected those comments dealing with the central and celestial "suns." In the Epilogue he says it is the "loadstone" which answers "to each Center of the beams, *viz.* of the Sun and Earth," and he expands on that with a short rhapsody on "aire." The "aire" of Sendivogius is fully comparable to the "universal spirit" of d'Espagnet or of le Fèvre: it is the repository of all the circumambient celestial and terrestrial virtues, and the "Loadstone" is the matrix which is to draw those virtues in.

> [The aire] is the Matter of the ancient Philosophers.... [I]t is the water of our dew, out of which is extracted the Salt Petre of Philosophers, by which all things grow, and are nourished: the matrix of it is the Center of the Sun, and Moon, both celestiall, and terrestriall: and to speak more plainly, it is our Loadstone, which in the foregoing Treatises I called Chalybs, or Steel: The Aire generates this Loadstone, and the Loadstone generates, or makes our Air to appear, and come forth.... And so in this place thou shalt have the true, and right explication of Hermes, when he saith, that the father of it is the Sun, and its mother the Moon, and that which the wind carried in its belly, *viz. Sal Alkali*, which the Philosophers have called *Sal Armoniacum*, and vegetable, hid in the belly of the *Magnesia*.[64]

Newton's comment on that climactic passage, in the "middle" period manuscript, Keynes MS 55, again links the "magnet" unequivocally with antimony:

> The air generates ye Chalybs or magnet & this makes ye air to appear. So ye father of it is ⊙ & ye mother of it ☽ This is yt wch bears ye wind in its belly, yt is, ye vegetable ⊖ Alcali or Armoniac hidden in ye belly of ye Magnesia or ♁.[65]

Sendivogius had not specified that the "magnet" was made from

[63] *Ibid.*, pp. 30–36.
[64] *Ibid.*, pp. 39–46, quotation from p. 41.
[65] Keynes MS 55, ff. 11v–12r (5, n. 39).

antimony, but Newton was quite sure that it was. And what was "hidden in ye belly of ye magnesia or ☿" was of course the star regulus.

Keynes MS 19, datable to early in 1669, was entitled "Extracts from the New Light of Alchemy which look towards practice," it will be recalled. In it Newton identified the Sendivogian "Chalybs" with antimony. Keynes MS 55, entitled "Sendivogius Explained," apparently dates from the "middle" part of Newton's career, probably from the mid-1680's, since the handwriting of the most important English section of it shows the confident boldness of the *Principia* period. In it Newton again identified the Sendivogian "Chalybs" with antimony. Between them, Keynes MS 19 and Keynes MS 55 neatly bracket the experimental work of the 1670's, and the conclusion is inescapable that when Newton experimented with antimony during the 1670's, he had at least a tentative concept of antimony acting as a Sendivogian "magnet."

Was the star regulus then supposed to draw the "philosophical mercury" directly out of the "aire"? Perhaps Newton thought in one sense that it could, but his use of the star regulus in his experiments indicates that his laboratory work was directed to the *inverse* of the process of metallic generation treated by Sendivogius in his natural-philosophical approach.

Evidently Newton thought that antimony was the matrix or receptacle of something uniquely metallic – that idea is expressed both in Keynes MS 18, the "Clavis," and in Keynes MS 55, "Sendivogius Explained," and will be explored below. But, when he worked with antimony in the laboratory, he was not concerned to draw that unique metallicity directly from the "aire." He did not need to draw what was uniquely metallic in antimony from the "aire" because it seemed to him to be already present in the antimony. From thence it was able to draw something – the metallic seed, or the radical moisture, or the philosophical mercury – out of other individual metals.

In the laboratory he was not attempting to operate at a cosmic level and create metals *de novo* by drawing down the "universal spirit" into a metallic matrix of antimony and there forging it into a new material metal. Rather he was thinking in terms of decomposing metals which already existed in the natural world, of analyzing them into their constituent substantial principles of "sulfur" and "mercury." That attempt – and it was one which was fully in line with his first Boylian experiments – involved him in a way of using antimony as a matrix which was rather different from the Sendivogian natural-philosophical one but much like the parallel alchemical one which Sendivogius offered to his readers in the passage cited at note 62. For Newton, the antimony was perhaps able to draw the "philosophick" mercury, or the metallic seed, back out of the metals in which it had already been "specificated," i.e., out of quicksilver or iron or lead or gold, and it was to that end that his next experiments seem to have been directed.

The net: the fourth experiments[66]

In his *Arcanum Hermeticae philosophiae opus* d'Espagnet had spoken of the net to catch the philosophical fishes in, and Newton had duly noted it in Keynes MS 19.

> Also Philosophers have their sea in which are generated fat and scaly silvery flickering little fishes; he who learns to envelop them with a finely woven net and to extricate them deserves to be considered a most skillful fisherman.[67]

Newton seems to have considered the fishes as being of two sorts, the one fat and the other silvery, although that is not a necessary interpretation of d'Espagnet's symbolism, and then to have identified the fat ones with the "sulfur" of metals and the silvery ones with the "mercury" of metals. Whether that interpretation was inherent in d'Espagnet's words or not, it was a perfectly reasonable way of giving "signification" to the passage, for in doing so Newton was merely applying conventional chemical theory. It was quite customary to attribute the properties of fatness, color, and inflammability, for example, to the chemical principle "sulfur," and to consider that the principle "mercury" was the bearer of the property of silvery metallic luster as well as other metallic properties. In addition, it may have been conventional alchemical symbolism as well: for example, Lambsprinck's first figure shows *two* fishes swimming in the philosophical sea.[68] In any event, in his explanatory note on the passage just quoted, which Newton had drawn from d'Espagnet's Canon 54, Newton said:

> Just as purple violets are plucked, so fat (i.e., Sulphureous) & silvery fishes are extricated: certainly Mercury grows white in the last sublimations.[69]

The fat fishes are the sulfureous ones and Newton seems to emphasize their sulfureous nature by comparing their extraction to the plucking of the purple violets, a metaphor which d'Espagnet had employed in his

---

[66] A discussion of the experiments which follow directly after the experimental essay on the star reguluses in the laboratory notebook is postponed until the next section because the notes on them appear by handwriting to have been added later. Also they must have been entered after the "formula" for the net was established, since the net itself was used in the experiments.

[67] Keynes MS 19, f. 4v: "Mare suum etiam habent Philosophi in quo pingues squammisque argenteis micantes generantur Pisciculi, quos qui subtili rete involvere & extrahere noverit, piscator peritissimus habendus erit."

[68] Burland, *Arts*, p. 145 (2, n. 16).

[69] Keynes MS 19, f. 4v: "Sicut violae punicae avelluntur sic pingues (i.e. Sulphuri) & argentei pisci⟨cu⟩li extrahuntur: nempe Mercurius in postremis sublimationibus albescit."

immediately preceding Canon 53.[70] Newton's further comment that the mercury certainly grows white in the later sublimations, taken in this context, seems then to imply an idea of separation of the "sulfur" from the "mercury," for it could only be when the "sulfur," the color-bearer, was taken away that the "mercury," left alone, could become white. If that was indeed his thinking on the matter, then in 1669 Newton was considering the philosophical net as a sort of filter which would help to separate the "sulfur" of metals from their "mercury." Interposed between the antimonial "magnet" and a specific metal then, the philosophical net might hold back the "sulfur" – the fat fishes – and allow the "mercury" – the silvery fishes – to pass through to the natural matrix of the star regulus of antimony. Or, conversely, it might hold back the "mercury" and allow the "sulfur" to pass through.

In any event, Newton set about making the philosophical net with his usual thoroughness. It was to be formed from copper and from the star regulus of antimony (prepared with iron). Various proportions were tried out for the new alloy and Newton's choice of the best evidently arose from the character of the literal, physical network formed when the metals cooled.

R ♂ 9¼, ♀ 4 gave a substance wth a pit hemisphericall & wrought like a net wth hollow work as twere cut in.

R ♂ 8½, ♀ 4 noe pit but a net work forme spread all over ye top yet more impressed in ye middle

R ♂ 2 ♀ 1 gave net worke but not so notable as ye former, & so did R ♂ 5 ♀ 2

The best proportion is about 4, 8½ or 9.[71]

There is no indication anywhere that the present investigator has been able to find as to what drew Newton's attention to the possibility of making a net-like structure out of antimony and copper. Perhaps he systematically investigated numerous alloys until he stumbled onto the particular combination of antimony and copper. It is possible, however, that he was attracted to the use of a copper–antimony alloy because in many proportions those two metals produce alloys which exhibit various shades of purple. Could it be that he interpreted d'Espagnet's purple violets as literally as he did d'Espagnet's net to catch the philosophical fishes in? If so, then the disappearance of the purple color as the new net-like alloy was combined with other metals could have meant to him that the "sulfur" was being "extricated," leaving the silvery fishes – the "mercury" – to stand alone and grow white.

Newton, however, does not seem to have worked out a process utilizing the net during the early period under consideration here. The alloy of copper and antimony disappears from the laboratory notes of the early

[70] Ibid., f. 4r.   [71] MS Add. 3975, f. 43r (Newton's p. 83) (3, n. 151).

1670's except for the one entry to be considered in the next section, and in the "Clavis" of Keynes MS 18, to be discussed later in the present chapter, Newton interposed an entirely different substance between the star regulus and another metal, a substance with an entirely different alchemical rationale behind it. The concept of the net evidently was not rejected by Newton nevertheless, for although his interpretation of it seems later to have been modified somewhat, it shows up again in "Sendivogius Explained," and also in later experimental notes. In "Sendivogius Explained" Newton went over the now familiar preparations of the star and the net once again in his "Jntroduction to ye 9th Treatise" of Sendivogius.

> To understand ye following Treatise know ye Author intends you should melt by spoonfulls ye pouders of ♁ ⚲ & ⦶ in an hot crucible. Then shake it yt ye ℞ may fall to ye bottom: Which freed from its dross shines like ♃ after ♄ expulsion. Then melt it four times with half so much ♂, still separating it from ye gross dross. Then you will have ye ⚹ of ♂ which melt wth so much ♀ till both are catched in a fine net, & you have ye Philosoph. flying ☉ & ⚹ of ♀. . . .[72]

Newton's concept of the philosophical fishes appears to have undergone a change also by the time he wrote "Sendivogius Explained" in the 1680's,[73] and the curious episode of the net and the fishes is in a way a fairly good exemplar of the manner in which Newton apparently operated in his alchemical studies. Taking hold of an obscure passage in which an alchemical author had couched a supposedly real secret in symbolic language, Newton first reduced it to terms with real meaning for himself. In the case of the fishes, he found that meaning in current chemical theory, but presumably he had to create a meaning for the net. Treating the symbol as a literal description, he then took his hint into the laboratory and began to work with it. There is now no way of telling how many trials he made before he succeeded in making a physical metallic net which satisfied him, but when he was satisfied, he entered the best proportions for it in his laboratory notes.

One can only guess what happened next, but it seems reasonable to assume that he then attempted to apply his net in whatever larger scheme he then had in mind, operating with a tentative notion of what it might do. When it failed to perform according to his expectations, he dropped it temporarily but did not totally reject it. Eventually, he picked it up again, presumably applied it in a new working hypothesis, and gave it a new interpretation according to its new performance.

[72] Keynes MS 55, ff. 7v–8r (5, n. 39). [73] Cf. *ibid.*, f. 13v.

Amalgamations: the fifth experiments

It is difficult to tell from the laboratory notebook exactly which experiments followed directly after the preparation of the net, but it may have been the set noted in the small spot between the end of the experimental essay on the star reguluses and the beginning of the experiments on preparing the net. These interposed experimental notes were evidently added at some time after the net experiments but, since the symbols for lead in them are not crossed, they could not have been written much later than the net experiments. Probably both were done in the first part of the decade of the 1670's. Since the interposed notes do contain references to the *rete* or net, they will be considered here as following fairly soon after Newton first successfully prepared the net.

These experiments seem to have consisted of a large number of attempted amalgamations at elevated temperatures, amalgam being a specific, technical term for an alloy of mercury with some other metal or metals. Since mercury is extremely volatile compared to other metals, it is not always possible to form an amalgam by melting another metal or metals and then adding the already liquid quicksilver, as the heat of the melt may be sufficient to volatize the mercury. Newton, by inference, made some amalgamation trials in which that happened, for in these notes he listed several sets of reagents which he said "will amalgam before ye ☿ fly."[74]

One series of experiments which he seems to have done was an attempt to make amalgams of all the different types of reguluses he had detailed in the experimental essay on reguluses. Only the trial with the "regulus of lead" was successful: "If Reg ♄ melted be dropped upon ☿ it will amalgam but noe other Reg."[75] Other successful trials all involved more complex compositions, and their proportions are presented in Table 6.[76]

A few generalizations about the meaning of these experiments may be ventured, but before turning to that it will be well to discuss briefly the peculiar symbol " ♈ ." This symbol appears for the first time here and no explanation of it is given. Neither is it by any means a conventional symbol. It is possible, however, to attach a name to it from a manuscript of the "late" period entitled, "Praxis." That name is "quintessentia,"[77] but the "Praxis" does not provide an unequivocal reading for the actual chemical referent for the quintessence. Some pages of the manuscript seem to identify the quintessence with antimony (presumably the ore),[78]

---

[74] MS. Add. 3975, f. 43r (Newton's p. 83) (3, n. 151).

[75] *Ibid.*   [76] *Ibid.*

[77] Babson College, Babson MS 420 (Sotheby Lot 74), pp. 15 and 14a. This manuscript has a complicated format and the pagination assigned to it by Babson College is followed here.

[78] *Ibid.*, pp. 1–2.

TABLE 6

Antimonial ingredient	Other ingredient	Mercury
Reg ♄	—	+
Reg ♃ 1	♃ 1	+
Reg ♃ 1	♄ ½	+
Reg ♃ 2	♈ 1	+
Reg ♃ 2, Rete 3	♈ 2	+
Reg ♄ 1, Reg ♀ 1	♄ 1⅓	+
Rete 3	♈ 7	+
♁ 2	♃ 5	+

but another renders the symbol " ♈ " as "Bism.," probably meaning an ore of bismuth.[79] Either of these designations would at least make it appear to have been a concrete substance to Newton, yet elsewhere in the same manuscript he defined the quintessence as a "corporeal spirit" and a "spiritual body" and the "condensed spirit of the world."[80] Probably all that can be said about it from the "Praxis" then is that in his later years Newton was doubtful about the nature of the quintessence. In fact, much of the "Praxis" manuscript consists of pages on which Newton wrote out his alchemical ideas and then scratched them out, unsatisfied. But in the first part of the 1670's, when the amalgamation experiments under consideration here were probably being done, Newton evidently had a real chemical substance in mind for the quintessence because he designated real and definite proportions in which it entered into his successful amalgams.

The most important generalization to be drawn from the table of Newton's amalgamation experiments is this: that in each case but one a star regulus of antimony was one of the ingredients, and in the remaining case the raw ore of antimony was used. The presence of antimony in each trial of course places these experiments squarely in line with the others involving antimony, and if one accepts the notion that Newton was employing antimony as a Sendivogian "magnet" at this time, then the whole set of amalgamation experiments must have been aimed at making the philosophical mercury.

Some additional generalizations may be made about the other ingredients. In the first place, all the trials he made involved the addition of common mercury, as by definition mercury is a part of all amalgams.

[79] *Ibid.*, p. 14a.
[80] *Ibid.*, p. 2: "Est spiritus corporalis et corpus spirituale. Est condensatus spiritus mundi...."

Evidently then he considered common mercury as a likely source of the philosophical mercury. The passage from Keynes MS 55 quoted above at note 39 offers a reason for his use of common mercury. There, it will be recalled, Newton stated that the "mercuries" drawn out of other metals had special forms and qualities related to the metals from which they came, and that those idiosyncracies made such "mercuries" more remote from the philosophical mercury than common mercury is. That seems to have been the conclusion Newton drew from his second set of Boylian experiments, discussed above. The *converse* of that statement of Newton's of course would be that the common mercury is closer to the philosophical mercury than the "mercuries" of the other metals are: hence it seems that, after the failure of the Boylian experiments, Newton began to concentrate on producing the philosophical mercury from common mercury.

One further, but very tentative, generalization may be offered concerning the nature of the other ingredients. Both tin and lead, the substances which appear most frequently in Newton's successful trials, are relatively soft metals and are easily melted. It is likewise with the semi-metal bismuth, and, if Newton was identifying the "quintessence" with bismuth during this period, then one might hazard a guess that he was concentrating his work on the softer metals on the theory that they were more similar to the liquid common mercury. Probably, however, that is a false generalization. The recorded results only reflect successful trials, it must be remembered, and it is precisely the softer and more easily fusible metals which would be likely to amalgamate "before ye ☿ fly." So the fact that most successful trials involved the soft metals really does not speak to the question of what type of metal Newton experimented on in general.

Leaving aside the specific chemicals involved in these amalgamations, it must be noted that a very important alchemical concept is built into the structure of these experiments, the concept of mediation. A fuller discussion of alchemical mediation will be given below, when Newton's "Clavis" is discussed, but it may be introduced here in passing. Although Newton did try combining all the reguluses directly with common mercury, in most of the recorded trials he interposed some other substance between antimony and mercury. In the case of the net, as noted above, he perhaps thought of this as a sort of filter which would hold back the "sulfur" of the common mercury and allow the philosophical mercury to pass through into the "magnet," or conversely that it would "extricate" the "sulfur" and leave the philosophical mercury standing alone, and he seems to have tried the net both alone and in conjunction with the simple star reguluses. In the cases of tin, lead, and the "quintessence" of the other trials, he may possibly have been looking for the

"Doves of Diana," whose tender wings were supposed to enfold and meliorate, but that concept will be considered later.

The metamorphosis of the planets: a glimpse of methodology

The relationships of pagination and handwriting in MS Add. 3975 suggest that Newton was continuing his systematic note-taking from Boyle's books both before and during the time of the preparation of the net and the amalgamation experiments. MS Add. 3975 contains references to both volumes of Boyle's *Usefulness of Experimental Natural Philosophy* (vol. I, 1664; vol. II, 1671) and to the volume of *Essays* (1669).[81]

At the same time Newton also continued his alchemical studies, and one product was Keynes MS 58. Keynes MS 58 is in the handwriting of the 1670's and may perhaps be dated more precisely about the middle of the decade since some of the symbols for lead are crossed and some are uncrossed.

There are three distinct sections to this manuscript, all apparently unrelated to each other except in time. F. 1r,v is a single sheet which shows marks of having been folded, as if to fit in a pocket. It contains a single recipe, a cementation process.[82] It would appear that this recipe was someone else's and that Newton merely copied it. Although it is quite definitely alchemical, providing for the "multiplication" of silver, it is inconsequential. If Newton ever tried it out, he never bothered to record his results on it, and the non-theoretical nature of the recipe makes it seem unlikely that it ever captured much of his attention. Its presence in Keynes MS 58 serves, however, to remind one that Newton must have been in contact with groups in which such "secrets" circulated. The last section of the manuscript, ff. 6r–8v, is likewise of little importance. It appears to be notes on an unidentified manuscript. Many are in symbolic form; some are recipes. F. 6v contains the single item: "NB ☿ est sal commune," in which Newton translates one of the symbols used on f. 6r. F. 8v ends with a recipe for a "Febrifugum."

The recipes in Keynes MS 58, the extensive notes from Boyle's works, the numbers of manuscripts Newton seems to have copied verbatim during this same period, and the variable laboratory techniques to which he applied himself, all show how wide-ranging Newton was in his chemical-alchemical activites. Although he perhaps did concentrate on the search for the philosophical mercury, he was not sure where clues to its preparation would be found, and he looked into literature of every sort. Keynes

[81] MS Add. 3975, ff. 43r–51v (Newton's pp. 83–100) (3, n. 151); Fulton, *Bibliography*, pp. 26 and 37–41 (3, n. 99).

[82] On cementation, see Dobbs, "Digby. Part III," pp. 12–14 (2, n. 59).

MS 58 demonstrates his catholic interests, and the center part of it is really of extraordinary importance for a study of his alchemical methodology for it contains notes in which Newton was apparently trying to unravel an alchemical process and translate it into chemical terms so that it could be attempted in the laboratory.

Three preparations especially concerned him: dry water, the eagle of tin (or of Jove), and Jove's or tin's scepter. He went through variants of the preparations in English, adding annotations and queries in parallel columns or at the foot of the page, with many deletions and re-workings.[83] Then he started over in Latin and worked through the same preparations again.[84] The Latin section is followed by a folio which is blank except for a brief note upside down on the back which Newton seems to have made to himself.[85] That note, probably the last thing written in this important middle section of Keynes MS 58, shows Newton moving from the study to the laboratory to resolve a question or two, and it will be quoted below after the nature of the other contents has been indicated.

Many of the annotations in Keynes MS 58 and at least some of the processes derive from John de Monte Snyders' *The Metamorphosis of the Planets*. Snyders wrote other works, and apparently all of them were published in Latin or in German,[86] but *The Metamorphosis of the Planets* had only German editions and seems to have existed in English translation only in manuscript. Newton somewhere acquired a copy of it and made a complete, carefully written transcript of it which included an elaborate title-page and a detailed symbolic frontispiece. Newton also numbered the pages and even the lines, for easy reference.[87] By handwriting, Newton's transcript probably dates from early in the 1670's.

[83] Keynes MS 58, ff. 2r,v and 5r. F. 2r,v is marked off in parallel columns for text and notes, the right hand column being for notes. F. 5r should follow f. 2v as the text from f. 2v continues of f. 5r. F. 5r, however, is not marked off into columns and the notes to the text there are placed below a line drawn all the way across the page.

[84] *Ibid.*, ff. 5r and 3r,v. F. 3r,v, which should follow f. 5r, shows less re-working and seems not to have been completed, as the note columns are blank.

[85] *Ibid.*, f. 4r, v.

[86] *De pharmaco catholico*, often cited by Newton in later manuscripts and attributed by Newton to Snyders, appeared anonymously as the second part of *Reconditorium ac Reclusorium Opulentiae sapientiaeque Numinis Mundi Magni...* (Amstelodami: Apud Joannem Janssonium à Waesberge, & Elizeum Weyerstraet, 1666). Newton's copy is extant still: Trinity College, Cambridge, NQ. 16. 80. There is no sign that Newton used the first part of this Book, but *De pharmaco catholico* has manuscript notes in Newton's hand and *many* pages turned down. *De pharmaco catholico* contains two smaller treatises referred to by Snyders in *The Metamorphosis of the Planets*: *De elementis magicis* and *De chymico alphabeto*. Thorndike, *Magic*, VIII, 355–56 (3, n. 5), mentions a treatise entitled *De medicina universali* published in 1678; that appears to have been a different translation of *De pharmaco catholico*.

[87] Yale University Medical Library, Sir Isaac Newton, "The Metamorphosis of the Planets by John de Monte Snyders," Sotheby Lot 102, hereinafter referred to as Snyders, "Metamorphosis." This work by Snyders was published in German in four editions: Amsterdam, 1663; Frankfort, 1700; Frankfort and Leipzig, 1773; Vienna, 1774. No

Newton's autograph transcript of Snyder's work was one of the items that so horrified Sir David Brewster when he went through Newton's papers in the middle of the nineteenth century, it will be recalled. And truly it is a distressing document to read, being a complicated allegory that rambles on through thirty-one chapters. The whole comprises sixty-four pages, and in Newton's small early handwriting that is a substantial amount of material. Very little of it is couched in rationalistic language.

Nevertheless, Brewster would perhaps not have been so horrified had he looked a little further and seen what Newton did with the material. For the essence of Newton's approach to Snyders was exactly the same as that which he used in the interpretation of prophecy: a rational, matter-of-fact analysis aimed at finding the true "significations" of Snyders' allegorical figures and their actions. The only variation in method in the case of this alchemical study was that Newton, instead of checking his "significa-tions" against actual historical events as in the case of prophecy, in alchemy checked them against experimental results.

So that it may be seen just how great a distance Newton had to travel to get from Snyders to the laboratory, one passage in which Snyders treats of the eagle and scepter of Jupiter (of Jove) will be given here.

The grey-bearded Jupiter understood from a Comet & Signat-Star, that the double nature as a *Monarch* of this world governed his kingdom in peace by the assistance of Mercury, & that from all parts of ye world envoys were arrived to congratulate & complement ye most mighty & invincible. Wherefore the good Jupiter mounted upon the wings of his nimble Eagle, & hastend to ye Palace & having obtained audience steps in, makes his due reverence with his scepter, bows his knees, kisses ye foot of ye Monarch, presents his Eagle to his service, intreats like as all ye rest also for an inheritance, & an eternall kingdome to be communicated unto him, in consideration that the old Dragon his royal Majesties Father, together with his own Father ye powerful Eagle, namely with Jupiter elect, by ye counsel & good advise of the assembled estates & Burgesses of ye Philosophick kingdom, had established an inviolable friendship.[88]

In considering Newton's treatment of Snyders' work, it should be noted that in the seventeenth century, as for many centuries before, the planetary deities were equated with real metals and that symbols and names were used interchangeably for both. With that equation in hand, Newton had little difficulty in translating Snyders' allegorical figures into chemical terms – as a first step – and Jupiter or Jove immediately became tin to him.

English or Latin edition appears in any standard bibliographic reference. Since Newton does not appear to have been able to read German, his source must have been a circulating manuscript of an English translation.

[88] Snyders, "Metamorphosis," p. 11 (Newton's pagination) (5, n. 87).

But the attempt actually to *make* Jove's eagle, presumably a substance derived from tin, which attempt is reflected in Keynes MS 58, must have involved Newton in a fantastic effort to synthesize the relevant, but meaningless, passages in Snyders, and then draw meaning from them. In the notes in Keynes MS 58 one may see him attempting to work out the relationships among Snyders' allegorical figures and translate them into chemical terms. Saturn (lead), Mars (iron), and Venus (copper) led the cast of characters thus translated in the first passage on making Jove's eagle in Keynes MS 58.

> But for ♃'s eagle [e] mix at ye second time some of ye eagle's minera wth ♄ ↗ or els put it in a little after ye second time ↙ & when they are black, ferment anew ↗ one ↙ wth extracted [f] calces [h] of ♂ & ♀ & ye water & then put ↗ in ↙ ye extracted [g] calx of ye eagle & ye eagle will be mercurialized & distill over.[89]

Newton crossed out that passage but not before he had annotated it. His notes keyed to the above quotation are all from Snyders' *The Metamorphosis of the Planets*, except for the last which expressed Newton's doubt about the accuracy of his interpretation. The notes follow.

> [e] the clashing of ye two worlds in ye dark.
> [Sniders
> [f] The son of ye solary world slid down by a chain.
> [g] ye daughter of ye left world.
> [h] perhaps of ♂ alone[90]

Another attempt at working out a process for making Jove's eagle immediately followed, and then another and another. In one instance Newton even attempted a general statement on making eagles and it will be quoted in full.

> 3. ffor mercurializing mercuriall bodies (viz: ye two eagles Venus & ♃) ferment ↗ the aqua sicca ↙ ♄ anew, ↗ wth a grain of ye old putrefied mother ↙ put to him ye calces ⟨of ye mine, *crossed out*⟩ to be mercurialized, & after a very little working distill them over. But note that for ♃'s eagle, ♄ must ↗ perhaps ↙ be otherwise † prepared then for ♀, viz, by adding to ♄ in his first ↗ or second ↙ working a little of ⟨ye two mineras of ♃, ↗ or rather a little of ↙, *crossed out*⟩ ↗ ye eagles salt, or [a] minera, ↙ that all may putrefy together. Thus you shall have ♀ ye daughter of Saturn & ♃'s eagle.[91]

In the first note keyed to that passage Newton said, "†This ♃ vial ♄ ↗ sublimed ↙ is one eagle as ♄ alone is another,"[92] thus bringing the eagles all the way down the scale from the most mystical symbolism to concrete chemical referents. In the other note to that passage, three points of great interest may be seen.

[89] Keynes MS 58, f. 2r.
[90] *Ibid.*
[91] *Ibid.*, f. 2v.
[92] *Ibid.*

ᵃ And rather of ye minera because ♄ (not ye Lyons blood) ate a stone instead of ♃ & spewed it out again. Perhaps ♄ fermented will mercurialize ye stone wthout more ado because he spewed it up again after he had devoured it. Quaere 1. whether ♄ must eat ye stone for ♃ so soon as ye spirit has dissolved ♄ & is satiated but not yet grown to a dry black calx (as is most probable ↗ because otherwise ♄ will distill over wth him ↙) or after ⟨ward, *crossed out*⟩ ye sublimation of ♄? Quaere 3 whether this stone be crude or its calx. Both may be tryed to see wch ☿ is best. Quaere 2 Whether ye aqua sicca will not be ↗ come ↙ more fixt by joyning wth some bodies as ♂ or ♀ & so let ye ☿ come over alone.[93]

The first point to be noticed is that Newton is actually using mythological material associated with Saturn – the tale in which Saturn was given a stone to eat instead of his own offspring Jupiter and vomited it back up as soon as he had swallowed it – to give himself a hint for the alchemical process he is trying to work out. Perhaps, he says, fermented lead will perform this mercurialization "wthout more ado" since Saturn is said to have regurgitated what he devoured.

Whether Newton followed up that lead in the laboratory cannot now be determined, but in his own response to his third query it is apparent that he did rely on experiment to settle his speculative questions. "Both may be tryed," he says, and that is perhaps the most important point to be gleaned from Keynes MS 58. It is borne out and emphasized by the final brief inverted note at the end of the middle section of the manuscript, which note will be given below.

But before turning to that last note, one further point may be drawn from the queries of Newton's annotation "a" above, which is that one may see him in these notes applying his own considerable technical chemical knowledge to the working out of the alchemical process. In the first query he decides that one tentative process is more likely to work than another because in the second the lead would distill over in an undesirable fashion. In the second query he has a notion to make his "dry water" less volatile or "more fixt" by joining it to iron or copper which are not very volatile.

Leaving aside the rest of the manuscript, which consists largely of variants on the above quotations, one arrives finally at the all-important note at the end.

　　To be tried.

1. Extract ♀ from green Lyon wth Æ diluted & make ye menstrue of this

2. Try if yt menstruum will dissolve Lead ore.

3. Get ye ⊕ s & try ye ferment.[94]

[93] *Ibid.* The words "wth him" in this note spill over onto f. 5r.　　[94] *Ibid.*, f. 4v.

With that note to himself Newton ends his attempt to give "signification" to Snyders by strictly rational means, at least for the time being, and moves into the laboratory to "try" and test by experiment the questions his mind has formulated.

A number of things in the laboratory notes seem to be related to the Snyderian processes of Keynes MS 58, but most of them fall after the time period of the present study. Only two brief paragraphs in Latin in the midst of Newton's notes in English on Boyle in MS Add. 3975, which paragraphs seem to be related chemically to what Newton had to say about Snyders' alchemy in Keynes MS 58, appear to have been written before Keynes MS 18, which is to mark the end of the present study of Newton's earliest alchemy.

The first experiment described in these particular two paragraphs is a "fermentation" process, and, since it utilizes lead, it may represent an attempt to make "the aqua sicca ♄" referred to in the quotation cited at note 91 above. There Newton intended to ferment the material "anew," so he evidently made the dry water by fermentation originally. The experiment is as follows.

The salt of antimony evaporated $3\frac{1}{4}$ ounces; lead 4 or 5 ounces; sal ammoniac about 2 ounces, and through fermentation in heat there will be $2\frac{1}{2}$ ounces calcined from lead and the whole weight of the calx will be $3\frac{1}{4}$ ounces.[95]

The "fermenting" agent in this experiment of Newton's is sal ammoniac, or ammonium chloride, NH_4Cl. Ammonium chloride is thermally unstable and decomposes with heat to give two gases, ammonia (NH_3) and hydrogen chloride (HCl). In Newton's mixture those gases would have bubbled out and would have given the whole mass the appearance of common fermentation, as in the bubbling mass in which beer is being made. The two processes would have been totally unrelated chemically, as Newton's was strictly inorganic and produced ammonia and hydrogen chloride, whereas what is called fermentation now is organic and produces carbon dioxide. But to Newton it was a *type* reaction and its important feature was agitation and not the individual gases being produced. The agitation to his mind probably produced a mechanical rearrangement of the small parts of the substances and so made new substances, in this case perhaps the *aqua sicca* he was looking for.

In Chapter 6 Newton will be seen referring frequently to fermentation, always in a context which presupposes the mechanical philosophy and transmutation by rearrangements of the small parts of bodies, and it is extremely likely that he was applying that same rationale in his laboratory

[95] MS Add. 3975, f. 43v (Newton's p. 84) (3, n. 151): "Sal ♁ij evap ℥iij¼; ♄ 4 vel 5℥; ⚹ ℥ij circiter, & per fermentationem in calido calcinabitur ex ♄ ℥ij½ et calcis pondus totum erit ℥iij¼.

work of the early 1670's. In his early chemical dictionary Newton had defined "fermentation" as follows:

Fermentation or working of liquors, wherby they are further digested & seperated ⟨*sic*⟩ from their faeces &c. Tis impeded by cold. And Must immersed for 6 or 8 weekes in a cold well is soe satled in its constitution that it will not ferment of ⟨*sic*⟩ a long while after.[96]

And a definition of "fermentation" from the contemporary "chymist" George Wilson seems to give a meaning similar to the one suggested here for Newton's usage. Wilson's definition follows.

Fermentation is an ebullition raised by spirits which endeavour to separate themselves from the body, but meeting with earthy parts which oppose their passage, they swell and rarify the liquor, till they find their way out. In this separation of parts, the spirits divide in such a manner, as to make the matter of another nature than it was before.[97]

Also it may be seen that Newton's ideas on the matter were similar to Wilson's around the end of the decade of the 1670's. In his unfinished tract *De aere et aethere*, written about 1679 or 1680, Newton spoke of the generation of air.

... [F]rom these principles the generation of air is easily learned. For this nothing else is required save a certain action or motion which tears apart the small parts of bodies; since when separated they mutually flee from one another, like the other particles of air. And thus it is that every vehement agitation (like friction, fermentation, ignition and great heat) generates the aerial substance which reveals itself in liquids by ebullition....[98]

Thus Newton appears probably to have thought all along that some new sort of substance would be produced by a "fermentation" process in which an "air" was given off, although it cannot be said with absolute certainty how mechanical his thoughts on it were in the early 1670's.

The other experiments in the small set of Latin notes under consideration here look to be directly related to making the eagles of Venus and Jupiter, or "mercurializing" their "mercurial bodies," as quoted from Keynes MS 58 cited at note 91. Both experiments reported here were concerned to volatize the metals and were perhaps conceived as first steps in making the eagles, which no doubt were thought of as being volatile. The experiment which Newton reported with copper was a failure, but his procedure with tin seemed to him to work.

Mercury sublimate 1 ounce, sal ammoniac $\frac{1}{2}$ ounce, copper $\frac{1}{2}$ ounce melted together and evaporated leave in the bottom nearly

---

[96] MS Don. b. 15, f. 7r (4, n. 77).   [97] Lewis, *Chemistry*, p. 10 (3, n. 134).
[98] Isaac Newton, *De aere et aethere*, in Newton, *Unpublished Papers*, pp. 219 and 226 (3, n. 157).

$\frac{1}{2}$ ounce, but they leave nothing of the solution, that dissipates into the sand through the breaking of the glass. But mercury sublimate, sal ammoniac, and selected feminine, arsenical tin wholly fly away, except for about 20 grains. And the glass, though weakening, endures. So mercury sublimate volatizes arsenical tin.[99]

One reason for thinking that these experimental notes do relate to a Snyderian process is not immediately apparent in the translation of the notes because it lies in the Latin word "electus." In the passage from Snyders quoted above, it may be seen that he referred to "Jupiter elect," which is much closer to Newton's "♃ electus" of the Latin laboratory notes than is the rendering in the translation of "selected tin." The double movement from "Jupiter" to "♃" to "tin" and from "elect" to "electus" to "selected" obscures the fact, but the literal meaning seems to be the same in all three versions. The word "elect" was commonly used in pharmaceutical recipes to indicate that one should take the best selected variety of material available. Snyders probably used the word "elect" in just that sense, and then Newton carried it into his notes in his usual custom of adhering closely to his authors' language.

But whether these notes relate to Snyders' eagles or not, the experiments do seem to be a continuation of the Boylian approach to "opening" bodies in the second set of experiments, where Newton attempted to extract "mercuries" the dry way with mercury sublimate and sal ammoniac. Here arsenical tin has been attacked with those same two reagents and made volatile by them.

Snyders continued to be one of Newton's favorite alchemical writers for many years. In fact, Newton once wrote out a list of the more useful authors and included Snyders' name on it.[100] That list certainly dates after 1684 and perhaps several years later than that. Also many experiments recorded in the laboratory notes probably stem from Snyderian ideas.

But to pursue Newton's use of Snyders further at this point would lead too far afield. Neither the eagles nor the dry water seem to have contributed to the production of the important philosophical mercury Newton supposed himself to have made during the early period, and in the structure of the present study the early Snyderian material considered in this

[99] MS Add. 3975, f. 43v (Newton's p. 84) (3, n. 151): "♄ ℥ 1, ⚹ ℥ ss, ♀ ℥ ss simul colliquefacta et evaporata linquebant in fundo ℥ ss ferè, praeter nonnihil solutionis quod per vitri fissuram dilabatur in arenam. sed ☿, ⚹, & ♃ electus, femineus, arsenicalis, penitus avolabant, demptis granis quasi 20. Et vitrum per durabat infractum, Adeóque ☿ volatizat ♃em arsenicalem."

[100] King's College, Cambridge, Keynes MS 13, f. 2r. Although parts of this manuscript are datable as late as the 1690's, and some items may even date from the London period after 1696, the section in which Newton's list of "Authoris magis utilis" appears seems to have been written a little earlier, judging it by the size of the handwriting, and the last internal date that that section bears is 1684.

section is best viewed as a limited case study of Newton's alchemical methodology. Before Newton's report of success in making a philosophical mercury is examined, the structure of his alchemical methodology may be recapitulated briefly.

The surviving manuscripts show three clear stages in Newton's alchemical method. First came the choice of material to be studied. Here one may see Newton turning to the most esoteric and mysterious productions of the alchemists. Although he did sometimes copy other men's practical recipes, neither the bulk of the manuscripts nor any of the laboratory notes reflect deep, continued interest in them. Rather he chose for close study the obscure, the complex, the symbolic, the "theoretical" if you will, the sort of material in which the great secrets of the ancients might lie hidden.

The second step in his method was that of rational analysis to find "significations." During that stage, Newton might draw on cross-references to other alchemical authors, on his own chemical knowledge, on mythology, on anything whatsoever that seemed relevant. For in his eyes all true knowledge was one and all variant statements of it were ultimately reconcilable. The productions of the second stage were in a sense working hypotheses, statements of possible or probable relationships that might be tested.

The third step of course carried out the testing. Usually the failures or the results of no importance must have sent him back to stages one or two again very promptly. But Newton was no more immune than the rest of the world to apparent success, and part of the fascination of Keynes MS 18, to which the next section is devoted, lies in the fact that it reflects a Newton who thought a number of his working alchemical hypotheses had been justified.

The key: Keynes MS 18

Assigning Keynes MS 18, entitled "Clavis," the status of a Newtonian composition has been a procedure beset with many pitfalls, and it is still not entirely clear that the present writer has not fallen into one, for there is considerable evidence that the manuscript is not in fact Newton's own work. In the first place, as will be seen in the discussion of its contents below, some of the concepts used in Keynes MS 18 are redolent of Eirenaeus Philalethes. In the second place, Eirenaeus Philalethes wrote a number of tracts on alchemy which were known to exist but which were lost to view in the seventeenth century. In his 1678 edition of *Ripley Reviv'd*, the bookseller and publisher William Cooper listed thirteen of them, along with a plea to any gentleman who might have one by him to allow Cooper to publish it as he had the others he had obtained. Number 12 in Cooper's list of Philalethes' lost treatises was "A *Clavis* to his Works."[101] To make matters worse, there are two places in Keynes MS 18 which make it appear that Newton was copying some one else's manuscript. On f. 1v there is a short blank space, as if perhaps the original had had an illegible spot in it and Newton had reserved a space for the material in case he should be able to read it later. Then on f. 2r a phrase is repeated, a common enough mistake in copywork. So it appears that the "Clavis" of Keynes MS 18 might very well be Newton's copy of Philalethes' lost "Clavis," the original of which had circulated through the Hartlibians in London and on to those interested in alchemy in Cambridge.

On the other side of the coin is the general character of the manuscript. When Newton solved a mathematical challenge problem anonymously in 1696/97, the continental mathematician who had posed the problem immediately recognized that the solution had come from Newton. One knows the lion by his claw – *tanquam ex ungue leonem*, he said.[102] The same may be said of Keynes MS 18, for Newton's mark is upon it. The proportions for the preparation of the star regulus in Keynes MS 18 are those Newton set out in his experimental essay on that subject.[103] The proportions for mediation with the "doves" of Diana in Keynes MS 18 also appear in the laboratory notes.[104] Then there is the obvious general continuity in the experimental program. By the time Keynes MS 18 was written – by the handwriting, probably around the middle of the 1670's – Newton had definitely engaged himself in a detailed study of star reguluses and had used them in an elaborate series of amalgamation attempts, the purpose of which seems to have been drawing the philosophical mercury out of the common mercury. The latter process is the one at the heart of the procedure in Keynes MS 18. In addition there is in the "Clavis" the use of a mediating agent which is similar to the use of the mediating agents in Newton's earlier amalgamation processes, noted above.

Furthermore, there is that same meticulous, fine-grained quality about the instructions for achieving the best results that one associates with Newton's experimental technique, in optical experiments and in experiments with pendulums as well as in chemistry. In Keynes MS 18 the author warns the would-be practitioner against the practice of throwing nitre on top of the regulus to be purified. The regulus must be beaten and ground and mixed with the nitre before fusion is undertaken, this writer says, and the inconveniences which may otherwise result are spelled out in great detail. Anyone who has read through some of the Philalethes tracts will have difficulty in believing that Philalethes could ever muster that

[101] Philalethes, *Ripley Reviv'd*, unpaginated "Advertisement" following Philalethes' "Exposition Upon Sir *George Ripley's* Epistle to King *Edward* IV" (2, n. 9).
[102] More, *Newton*, pp. 474–75 (1, n. 2).
[103] Compare Appendix B with Appendix C.
[104] Compare Appendix C with MS Add. 3975, f. 53r (Newton's p. 103) (3, n. 151).

degree of precision in description, but would have no difficulty in believing it of Newton.

In addition to the fact that the "Clavis" exhibits a high degree of continuity with Newton's earlier experimental program, there is the fact that it also exhibits a high degree of continuity with his later writings, both in the private alchemical papers and in the published and unpublished scientific papers. The concept of mediation, for example, appears again and again in Newton's later work, where he translated it into mechanical terminology, as will be seen in Chapter 6. Even if Newton did not compose Keynes MS 18, the evidence suggests that he fully accepted the truth of it.

There is one other possibility for the authorship of Keynes MS 18 which perhaps deserves to be explored, which is, that it might originally have been Barrow's. A letter on mathematical affairs written in 1675 points to Newton's great interest in chemical matters at that time. On October 19, John Collins spoke of

> ...Mr Newton (whome I have nott writt to or seene this 11 or 12 Months, not troubling him as being intent upon Chimicall Studies and practises, and both he and Dr Barrow &c beginning to thinke mathcall Speculations to grow at least nice and dry, if not somewhat barren)....[105]

One can imagine that the success of making a philosophical mercury might indeed make mathematics seem a little dry and barren. And Collins' remark, by implication, associates Barrow with Newton in his surge of interest in chemistry. If the "Clavis" had been Barrow's and Newton had taken a copy of it, that might explain the somewhat anomalous evidence mentioned above that seems to suggest that Newton copied the work.

But there is another solution to the problem of the repeated phrase in the "Clavis," a solution which actually strengthens the argument that the work is Newton's own. The passage in question reads as follows in the original: "℞ hujus * ☿ pt j. ☽ae pt ij, funde sumul ⟨sic⟩ donec *[☿ ℥β, ☽ ℥ j.] funde simul donec...." It may be seen that the repeated material comes after the explanatory bracket "[☿ ℥β, ☽ ℥ j.]" in which the author makes more precise the quantitative relationship necessary for mediation with the Doves of Diana (the combination of the star regulus with silver). It is in fact just the sort of error which would be more likely to occur if one interrupted oneself to put in an explanation, rather than the sort of error that might come in copying someone else's work absent-mindedly.

One final point may be made. As noted before, it was Newton's custom to annotate his alchemical manuscripts with references to other alchemical works which seemed to him to convey similar ideas. Many, many of the later ones carry references to the Philalethes tracts which Newton first

[105] John Collins to James Gregory, Oct. 19, 1675, in Newton, *Correspondence*, I, 355–56, quotation from p. 356 (1, n. 52).

obtained in manuscript form and copied into Keynes MSS 51 and 52. Typical of his later references to them are, for example, "Phil. on Ripl. Epist." or "Phil. on Rip. Gates." Similar abbreviated indications of works by other authors appeared in later manuscripts, and, when the work to be cited was anonymous, Newton would often simply use an abbreviated title. So far as the present writer has been able to determine, the "Clavis" is never mentioned later in that manner although it expresses ideas with which Newton dealt later and offers an unusually useful process. For one should make no mistake about the explicit nature of the process in the "Clavis" of Keynes MS 18: alchemical though it is, it is set out so empirically that it should be repeatable even today if anyone cared to pursue it. The fact that Newton does not refer to it by name in later manuscripts seems to be explicable only if it was in fact his own.

There seems to be no way of settling the question definitely with the evidence in hand. If another copy of the manuscript is located elsewhere, a copy not in Newton's hand, then it might be possible to prove that the composition is not Newton's own. But the weight of the evidence, especially the continuity of procedures and ideas, seems to the present writer to fall heavily on the side of its being Newton's, and it will be so treated here, although one should perhaps keep a small doubt in mind. The entire manuscript is set out in Appendix C, both in the Latin original and in English translation, and the interested reader may judge it for himself.

The process Newton described was lengthy and laborious but relatively simple in outline. It required a star regulus of antimony (prepared by iron), some silver, some common mercury, and a little gold. The thing to do, Newton said, was to fuse the star regulus with the silver, then amalgamate that with common mercury. By repeated digestions, grindings, washings, dryings, distillations, and fresh amalgamations, all the "blackness" is removed, and "on the seventh time you will have a mercury dissolving all metals, particularly gold."

I know whereof I write, for I have in the fire manifold glasses with gold and this mercury. They grow in these glasses in the form of a tree, and by a continued circulation the trees are dissolved again with the work into new mercury. I have such a vessel in the fire with gold thus dissolved, where the gold was visibly not dissolved through a corrosive into atoms, but extrinsically and intrinsically into a mercury as living and mobile as any mercury found in the world. For it makes gold begin to swell, to be swollen, and to putrefy, and also to spring forth into sprouts and branches, changing colors daily, the appearances of which fascinate me everyday.[106]

One senses Newton's rapt attention as he watches the play of colors in his

106 Keynes MS 18, f. 1r.

glass, for he is seeing the tail of the peacock, the *cauda pavonis* of ancient alchemy. "I reckon this a great secret in Alchemy," he continues,

and ↗ I judge ╱ it is not rightly to be sought from artists who have too much wisdom to decide that common mercury ought to be attacked through reiterated cohobation by the regulus of leo [that is, of iron or antimony].[107]

The "regulus of leo" is that heavenly star, Regulus, in the constellation Leo, and Newton sees it as a symbol for the earthly star regulus of antimony, prepared either by the use of iron or *per se*, according to his bracketed explanation. The star regulus attacks the common mercury "through reiterated cohobation," that is, by putting the two substances together again and again after repeated distillatory separations. Some artists have had "too much wisdom" to try that method, but it works, Newton says, because "that unique body, that regulus...is familial with mercury...." And with that statement, Newton embraces the alchemical concept of mediation.

That unique body, that regulus, however, is familial with mercury seeing that it is closest to that mercury you have known and recognized in the whole mineral kingdom, and hence most closely related to gold. And this is the philosophical method of meliorating nature in nature, consanguinity in consanguinity.[108]

Alchemical mediation or "melioration" was not an uncommon concept, but it was one which was especially dear to the heart of Eirenaeus Philalethes, that anonymous mid-seventeenth-century alchemist whose works Newton collected and studied with such care. By the time of the writing of Keynes MS 18, Newton had read at least eight of the Philalethes tracts: the *Epistle to King Edward Unfolded*,[109] *On Ripley's Vision*,[110] *On Ripley's Preface to His Gates*[111], *On Ripley's Gates*,[112] the three items in the *Tres tractatus* of 1668,[113] and the *Secrets Reveal'd*.[114]

Eirenaeus Philalethes, the last great philosophical alchemist, wrote mostly in the 1650's. He thus came a generation or more after Sendivogius and d'Espagnet and he relied heavily on concepts from both. Especially was that true of the alchemical "magnet" and the rich, life-giving Neoplatonic "aire," that great repository of all celestial and terrestrial virtues.

[107] *Ibid.* [108] *Ibid.*, f. 1r,v.

[109] Keynes MS 52, dated above in the late 1660's, by handwriting.

[110] Keynes MS 51, dated above in the late 1660's, by handwriting.

[111] *Ibid.* [112] *Ibid.*

[113] Keynes MS 36: *De metallorum metamorphosi, Brevis manuductio ad rubinum coelestem*, and *Fons chemicae philosophiae*, dated above in the late 1660's, by handwriting, probably 1668 by date of publication of the Birrius edition (5, n. 8) and lack of internal cross-references to other alchemical works.

[114] Philalethes, *Secrets Reveal'd* (3, n. 76). Newton seems to have read the work when it was first published in 1669, as witnessed by Keynes MS 19, dated above in 1669, by handwriting and internal evidence.

In the *Secrets Reveal'd*, Philalethes also came very close to stating the concept of alchemical mediation in the exact form in which Newton used it in Keynes MS 18. In his Chapter 11, entitled "*Of the Invention of the perfect* Magistery," Philalethes explained his views on the discoveries the adepts had had to make concerning the mercuries and sulfurs of metals in order finally to achieve success in the Great Work of alchemy. At first, he said, the workers of antiquity sought only for "a simple exaltation of the imperfect Metals to a regal condition:"

> and when they perceived that all Metallick Bodies, were of a *Mercurial* Original, and that ☿ was both as to its weight and homogeneity most like unto Gold, which is the perfectest of Metals, they therefore endevoured [*sic*] to digest it to the maturity of Gold...[115]

But they could not effect that by any fire, Philalethes continued, and so they attempted to achieve their end by applying an internal heat as well as an external one, using "hot" corrosive waters on the mercury. But those waters being merely external agents, as was the fire, the waters could not change or alter mercury's internal proportions either. Neither were salts effective agents, they discovered.

> Wherefore the Wisemen did at length know and consider that in ☿ the watery crudities, and the earthly *faeces*, did hinder it from being digested; which being fixed in the roots thereof, cannot be rooted out, but by turning the whole compound in and out.[116]

But they recognized, Philalethes said, that mercury had in itself "a fermental *Sulphur*" which certainly would coagulate mercurial bodies if only the crudities could be removed. And when they also realized that the reason it did not do so ordinarily was because the sulfur had died and become passive through being detained in the earth, then they had a useful clue.

> [T]here is a passive 🜍 in ☿ which ought to be active; so that it is needful to introduce into it another life of the same nature in the introducing of which it stirs up the hidden life of ☿, So life receives life, Then at length it is fundamentally transformed or changed, and the defilements are voluntarily cast away from the Centre....[117]

The old philosophers first sought the vivifying sulfur in copper but without success. "Then they took the offspring of *Saturn* in hand," Philalethes noted – and from the context Philalethes evidently meant metallic antimony by the "offspring of *Saturn*." That substance, he continued, has absolutely no sulfur in it: "no actual 🜍, but only potential."

Therefore they sought further for an active 🜍, and that most throughly, and at length the said *Magi* sought it and found it hidden in the house of *Aries*. This 🜍 is most greedily received by the son of

[115] Philalethes, *Secrets Reveal'd*, p. 24 (3, n. 76).
[116] *Ibid.*, pp. 25–26.   [117] *Ibid.*, pp. 26–27.

♄; which Metallick matter is most pure, most tender, and most near to the first Metallick *Ens*, void of all actual *Sulphur*, but yet in power or capacity to receive a ♁. It doth therefore draw this to it self like a *Magnet*, and swallows it up in its own belly, and hides it; and the Omnipotent, that he might most highly adorn this Work, hath imprinted his Royal Seal thereon. Then forthwith these *Magi* rejoyced when they beheld the ♁, not only found, but also prepared.[118]

The "magnet," then, according to Philalethes, was not to draw the philosophical mercury out of the "aire" or out of a "specificated" metal, but was to draw that other necessary principle, a "fermental *Sulphur*." That sulfur would then activate and vivify common ordinary mercury and make it into a philosophical mercury. But when the Magi attempted to use their antimonial magnet in that fashion, the event did not answer their desires, Philalethes continued, for there was "an Arsenical Malignity commixt with this ♁" which the "Child of ♄" had swallowed up.

Therefore they assaied to contemperate this malignity of the Air by the *Doves* of *Diana*, and then the event was answerable to their desires; then commixed they Life with Life, and moistened the dry by the moist, and actuated the passive by the active, and vivified the Dead by the Living....[119]

Now returning to Newton's "Key" in Keynes MS 18, it is possible to see his application of Philalethes' ideas in a number of places. Newton opens his remarks with a comment on antimony:

First of all know antimony to be a crude and immature mineral having in itself materially what is uniquely metallic, even though otherwise it is a crude and indigested mineral.[120]

There Newton speaks of antimony ore – "a crude and immature mineral." But in the ore there is materially present something "uniquely metallic," i.e., the regulus, which Philalethes calls the matter "most pure, most tender, and most near to the first Metallick *Ens*." On this point a very useful gloss comes also from Newton's "middle" period manuscript, "Sendivogius Explained." There Newton adds in his "Introduction of ye 3d Treatise" of Sendivogius:

To understand ye following Treatise it is requisite to know that all metals & many minerals have Antimony for their first matter or nearest principle. To which always cleaves an outward ♁ that hinders it from being metal; wch being artificially separated the inward kernell is most pure coagulated ☿ commonly called ye ℞ of ♁.[121]

There Newton has his starting point, a most pure coagulated mercury which is as the first matter of all metals and is the metallic regulus of antimony. He thinks it may be "meliorated" into the true philosophical

---

[118] *Ibid.*, pp. 27–28.
[120] Keynes MS 18, f. 1r.
[119] *Ibid.*, pp. 28–29.
[121] Keynes MS 55, f. 3r (5, n. 39).

mercury by acting upon common mercury, because it is "familial" with mercury as also with the philosophical mercury and with gold. This is the philosophical method of meliorating nature in nature, he says, and of meliorating consanguinity in consanguinity. But it cannot be done directly, and so another term is introduced into the relationship, the mediation of the virgin Diana or of her "doves": Newton seems to use the two terms interchangeably in Keynes MS 18. In his first reference to Diana, he says the following.

Another secret is that you need the mediation of the virgin Diana [quintessence, most pure silver]: otherwise the mercury and the regulus are not united.[122]

Elaborating on the process with specific laboratory directions, Newton says to combine one ounce of silver with half as much regulus and melt them together. Then he notes the following:.

If the regulus is joined with the silver, they flow more easily than either one separately [and they remain fused as long as lead even though there are thus two parts of silver, which is then changed into the nature of antimony, friable and leaden].[123]

Evidently, Newton thought that the silver really had changed its nature into that of antimony, or perhaps into the nature of lead, for he continues, speaking of the mass of silver and metallic antimony which have been melted together, with the instruction to beat "this friable mass, this lead."

Beat this friable mass, this lead, and cast it together with the mercury of the vulgar into a marble mortar.... Grind the mercury for $\frac{1}{4}$ of an hour with an iron pestle and thus join the mercury, the doves of Diana mediating, with its brother, philosophical gold, from which it will receive spiritual semen. The spiritual semen is a fire which will purge all the superfluities of the mercury, the fermental virtue intervening.[124]

Now Newton's theoretical insight into this remarkable process may be further explored. The beginning point was antimony ore. He thought that in that ore there was an "outward" sulfur, which is true enough even according to modern understanding, since the ore is a compound of metallic antimony and ordinary sulfur, it will be recalled. But that "outward" sulfur might be removed, or it might be "digested" by iron – and perhaps Newton thought it was the latter.

Philalethes' discussion implied that the "Metallic matter," which would have been metallic antimony, had been freed of its "outward" sulfur, for he said it contained "no actual ⟨sulfur symbol⟩, but only potential." But Newton him-

---

[122] Keynes MS 18, f. 1v. Newton's "q̄. ē." in the original is here translated as "quintessence" on the basis of his use of that abbreviation for the word in Keynes MS 31, f. 4v.
[123] Keynes MS 18, f. 2r.   [124] *Ibid.*

self speaks of the whole mineral, the antimony ore, as being "digested" by the sulfur of iron to yield the metallic regulus. Then Newton goes on to say that when the star appears, that is a sign that the "soul" of the iron has been made "totally volatile by the virtue of the antimony."[125] Since "soul" was equivalent to "sulphur" in alchemical terminology, Newton seems to be saying that the antimony ore and the iron have acted on each other to effect a mutual elimination of principles. What is left is the star regulus. The star regulus then would appear to have been considered by Newton as compounded from the "mercurial" fraction of the antimony ore and the "sulfur" of the iron.[126]

It was a very special substance, the star regulus. "Signed" as it was with the star, it seemed to Newton to have the power to attract celestial virtues. Exactly what it was capable of attracting perhaps he was not sure. But it looks entirely possible on the basis of Keynes MS 18 that it seemed to him that it attracted a "fermental *Sulphur*" or "fermental virtue" to itself from the circumambient virtues of the surrounding Neoplatonic "aire." In any case, it had in it something uniquely metallic and that something might, by proportionate steps, be "contemporated" or "meliorated" until it became "a mercury as living and mobile as any mercury found in the world."

In Keynes MS 18 it may be seen that Newton described the process of change as taking place in several distinct steps. The goal of the experimental procedure was to prepare a philosophical mercury, and that was to be done by the basic process of "cohobating" common mercury with the star regulus. But it could not be done directly: the "doves of Diana" had to mediate between the regulus and the common mercury. Newton's conception seems to have been that the three substances stood in a proportionate relationship with each other, with the "doves," the silver, serving as a middle term, which might perhaps be expressed as follows.

regulus of antimony : silver :: silver : common mercury

But the three substances were not simply thrown into the pot together: in the first step the common mercury did not enter in at all. Rather there came first the joining of the regulus with silver to yield a leaden mass, which one might call "philosophical lead" in the spirit of the alchemists. Since the silver is said to have changed its nature to become more antimonial, evidently Newton did not think of the process as an additive one to make a new compound. It is rather as if the "natures" of the regulus

[125] *Ibid.*, f. 1r.
[126] The present writer is indebted to Professor R. S. Westfall for calling attention to the fact that the "regulus of iron" might have been considered as containing some special portion of the iron rather than all of it, and also for some clarifying suggestions about the role of the philosophical sulfur in Newton's process.

and of silver move closer to one another and fuse together into a new "nature," as in the following alchemical "equation."

regulus of antimony → "philosophical lead" ← silver

Then taking his creation, the "philosophical lead," as a new starting point, Newton moved on in a second step to another "melioration," fusing the "nature" of common mercury with that of the "philosophical lead." In that process, in which Newton designates the antimony metal as "philosophical gold," and the process itself yields an "actuated" mercury.

"philosophical lead" → "actuated" mercury ← common mercury

In that second step, it seems to Newton that the common mercury is receiving "spiritual semen" from the "philosophical gold" or star regulus of antimony. Presumably, the "spiritual semen" has been drawn into the regulus from the "universal spirit" in the surrounding Neoplatonic "aire." In its union with common mercury, it purges out all the "superfluities." If Newton followed the thinking of Philalethes on that point, the "spiritual semen" acted by vivifying the dead, passive "Sulphur" already present in the common mercury. For Philalethes had said that the passive might be actuated by the active and that then the "defilements" would be "voluntarily cast away from the Centre."

It would appear that it was precisely this process which was indicated by the anonymous designer of one of the most famous of all alchemical illustrations, that of the "greene lyon" devouring the sun (Plate 4). That symbolic picture had first been attached in sixteenth-century manuscripts to a tract called *The Rosary of the Philosophers* and it followed it into print.[127] The original was colored, however, and there the "lyon" itself represented antimony ore, green because of its rawness or immaturity. The life-giving powers of the Neoplatonic "universal spirit" – Newton's "spiritual semen" – were represented by the golden sun, and the red blood that issued from the mouth of the lion represented vivified mercury.

In the final step, Newton took his "actuated" mercury and found that it would dissolve all metals, even gold. Gold was notoriously difficult to dissolve, but the alchemists were forever trying to do it. Their attempts to dissolve gold seem to have been made with the same notion a modern

[127] See H. M. E. de Jong, *Michael Maier's Atalanta fugiens. Sources of an Alchemical Book of Emblems. With 82 Illustrations. Janus: Revue internationale de l'histoire des sciences, de la medécine, de la pharmacie et de la technique. Suppléments, Volume VIII*, rédaction, E. M. Bruins, R. J. Forbes, G. A. Lindeboom, D. A. Wittop Koning (Leiden: E. J. Brill, 1969), pp. 340–41, for information on the first appearance of the "greene lyon" in print in 1572; p. 360, for information on *Rosarium philosophorum*; and p. 448, for a reproduction of the woodcut itself with its German alchemical verse attached. Newton owned a copy that is extant still: Trinity College NQ. 16. 121, *Artis avriferae...* (3 vols. in 1; Basileae: Typis Conradi Waldkirchii, 1610), II, 133–252, "greene lyon" on p. 240. There, in addition to the verse in German, the picture bears this Latin caption: "De nostro Mercurio, qui est Leo viridis Solem deuorans."

Plate 4. THE "GREENE LYON" DEVOURING THE SUN.
Here raw antimony ore, the "greene lyon," draws in vivifying celestial influences,
symbolized by the sun, and emits a vivified mercury, the living or "actuated"
character of which is symbolized by blood. (With permission from The Stadtbibliothek Vadiana, St. Gallen.)

chemist employs when he analyzes a compound before he attempts to
synthesize it: if one knows what a substance is made of, then it is easy
enough to make it.

It was of course known in the seventeenth century that *aqua regia* would
dissolve gold; that had been known for a long time. But it was not commonly thought that that particular dissolution really broke the gold up
into its constituents; rather it was thought that *aqua regia* only dissolved
gold into small particles which were still gold in their essential nature, much
as grinding or filing might do. The dissolution, or analysis, of gold into its
constituent substances was in fact much more difficult, they thought, and
the frustration of the alchemists finally came to be encapsulated in an
alchemical maxim which Boyle was fond of quoting: *facilius est aurum con-
struere, quàm destruere* – it is easier to make gold than to destroy it.[128]

[128] E.g., Boyle, *Works*, I, 513; III, 96 (3, n. 68); cf. Dobbs, "Digby. Part II," pp. 157–
58 (2, n. 58).

Yet in Keynes MS 18 Newton says he has dissolved gold and that the dissolution was clearly not done "through a corrosive into atoms," referring there to the common notion of what happened when gold was dissolved by *aqua regia*. It was the final magnificent success of his series of proportionations, and he had done it with the "actuated" mercury and common gold.

"actuated" mercury → "philosophical mercury" ← common gold

It was the true philosophical mercury that he had finally made and in another useful passage from the "middle" manuscript, Keynes MS 55, Newton indicated something of what this solvent of gold meant to him, in a comment on a passage from Sendivogius.

Jt is needful to search out ye occult matter from wch in a wonderfull manner such an humidity is made as dissolves ⊙ without violence and noyse so sweetly & naturally as ice melts in hot water. Then have you ye same matter of wch ⊙ is produced by nature, to wch ⊙ is friendly & as it were † its mother. ffor no impurity adheres to ⊙.

†yt is ⊙s mother[129]

Now that he has it, it is "a mercury as living and mobile as any mercury found in the world" and its living qualities are evidenced by the fact that it vivifies common (dead) gold and makes it begin to grow again. It is in short the mercury of the philosophers, and something of life and activity has been drawn from the great "universal spirit" and has entered into its composition.

The larger circle

Before the reader too hastily concludes that Keynes MS 18 was the product of a disordered mind, whether Newton's or some other's, it will be well to cast a glance over the larger circle of alchemists with whom Newton had some contact and also to take a hard look at the notion that perhaps Newton's work might be defined as normal science.

The glasses in the fire that contained gold and the philosophical mercury which Newton described as having tree-like formations appearing and disappearing in them, accompanied by a constant changing of colors – was that a real scene he described? No one today puts gold with mercury, philosophical or not, seals it up, and sets it on the fire for many days, and a modern chemist is inclined to say that such reactions could not occur. Gold does not come to life and grow sprouts and branches no matter what one does to it, he might say. Neither does any mercury become "living and mobile."

[129] Keynes MS 55, f. 13r (5, n. 39).

But Boyle had seen it happen too. In tests which he made on "animated mercuries" amalgamed with gold to see whether they would react a certain way more quickly than ordinary quicksilver amalgamed with gold, he sealed up his vessels and set them on the fire for long periods. And about his results he had this to say:

> Nor do six months make the longest term, that the obstinacy of my curiosity has made me keep gold in decoction with animated mercuries without obtaining [the desired result], though in the mean time there were produced very pretty vegetations, and sometimes, which is far more considerable, odd changes of colours, about which it is not necessary to entertain you....[130]

What did Newton and Boyle see? Presumably some sort of unstable intermetallic compounds that underwent rapid shifts in color and form as more or less energy was momentarily supplied by the fire. There is really no reason to doubt the accuracy of their descriptions. But the theories of alchemy that underlay their experiments have been extirpated from chemical thought and practice so thoroughly that now no one considers doing such an experiment, and one is inclined to reject the fact because he rejects the theory.

In T. S. Kuhn's terminology, the experiment itself is no longer normal science. Kuhn sees normal science as a puzzle-solving activity in which quite definite limits are set by a generally accepted paradigm. If the paradigm does not predict or account for or consider as important a class of phenomena, those phenomena will not only not be studied, they will frequently not even be admitted to exist. The paradigm accepted by a scientific community provides the criteria for choosing problems for normal scientific work; it provides a framework in which different sets of phenomena assume greater or less theoretical importance.[131] Within modern chemistry, with a non-alchemical paradigm operating, the reactions between mercury and gold have no special theoretical significance. But in the seventeenth century, within the framework of an alchemical paradigm, they did, and so they were studied.

Can one then say that Newton and Boyle were practising normal science when they explored the reactions between gold and "animated" or "actuated" or "philosophical" mercuries? It would seem that the answer to that should be affirmative, and that in addition they were operating within a scientific community which was fairly well defined in the 1650's, 1660's, and 1670's, although limited in size.

The paradigm which selected and guided research for them was broadly alchemical in the sense that they thought metals might be transmuted. Within that broad framework, a more specific theoretical structure had

---

[130] Boyle, *Works*, I, 649 (3, n. 68).
[131] Kuhn, *Revolutions*, pp. 23–42 (2, n. 45).

grown up and been elaborated in a continuous tradition. Beginning with Paracelsus,[132] the "universal spirit" of Neoplatonism had entered into chemical thought and practice. Elaborated by various Paracelsians in the sixteenth century, by the beginning of the seventeenth century the general idea was probably fairly widely accepted in circles that emphasized chemical medicines. In addition, there had been developed the concept of the matrix in which the "universal spirit" could be "specificated." Early in the seventeenth century, both Sendivogius and d'Espagnet wrote their influential works, and in the 1650's one finds both Eirenaeus Philalethes and le Fèvre building upon the earlier writers.

The research problem which seemed to be of the greatest theoretical importance within that framework, and the one which came to be emphasized and which it seems one might indeed characterize as the puzzlesolving of normal science, was the discovery and use of particular chemical "magnets" in which the "universal spirit" might be captured. As noted in Chapter 2, one of these was considered to be the moisture of the air, and much research was done on various insipid natural waters, such as dew and rainwater. Some of George Wilson's experiments on waters were noted in Chapter 3. Sir Kenelm Digby was involved in many similar experiments in the 1650's and even lectured to the Royal Society on some of his ideas on the matter in January, 1660/61.[133]

One of Digby's French acquaintances gave him several elaborate processes based on a similar approach about 1664 or 1665.[134] In general, Digby's French friend's method was based on the use of deliquescent chemicals that draw water out of the air spontaneously. That Frenchman thought they also drew the "universal spirit" with the water. In one of his processes he made that plain:

> ...what is attracted is the Universal Spirit, the Informing form of the Elements, that of the World, Influence of the Stars, Soul of the World, the vital Nutriment, latent in the Air.... It is attracted by several things, or (to speak plain) there are several things which attract it from the Stars; first, by *Sendivogius* his Magnet, or *Chalybs*.... This most Noble way is clearly and neatly shewn by the Author: But there

---

[132] It is beyond the scope of this study to enter into an exploration of origins, but the reader is referred to the following work as a good beginning place for such an exploration: Walter Pagel, *Das Medizinische Weltbild des Paracelsus seine Zusammenhänge mit Neuplatonismus und Gnosis* (Kosmosophie Forschungen und Texte zur Geschichte des Weltbildes, der Naturphilosophie, der Mystick und des Spiritualismus vom Spätmittelalter bis zur Romantik, Im Auftrage der Paracelsus-Kommission und in Verbindung mit der Paracelsus-Ausgabe, Herausgegeben von Kurt Goldammer; Wiesbaden: Franz Steiner Verlag GMBH, 1962).

[133] Dobbs, "Digby. Part II," pp. 151, 156–59 (2, n. 58).

[134] One of these has been analyzed in detail by the present writer in Dobbs, "Digby. Part III," pp. 17–24 (2, n, 59).

are other ways, which are shorter, by which this Spirit of the World is attracted by several Magnets. . . .[135]

Le Fèvre definitely considered antimony metal to be such a "magnet," and he gave a very interesting "empirical" reason for his belief. Antimony metal was frequently made into cups; it will be recalled that Newton mentioned that fact in his chemical dictionary. The purpose of the cups was medicinal: a little wine was allowed to stand in them overnight and then drunk. The dose served as an emetic, and since the cups could be used again and again, they were considered to have a perpetual inherent power. A small amount of the antimony went into solution in the wine with each use, of course, to provide the emetic action, but it was such a small amount that the cup's loss of weight went unnoticed. Le Fèvre thought that the perpetual power of the cups came from their ability to replenish their virtue by drawing in more of the "irradiation and influence from above" continuously.[136]

Even John de Monte Snyders seems to have been privy to the great "secret" about the "magnetic" powers of antimony. In one passage in *The Metamorphosis of the Planets* he almost lapsed into rationality in explaining it.

The time is not yet come that through the most high in a ⟨smal⟩ moment the weak & imperfect can pass into its highest perfection, but the spirit onely endeavoureth, in the virtue of its magnetique nature to mix it self with the corporeal Planets & by the power of their own soules to carry them gradully ⟨sic⟩ till at length they arrive to the highest degres & equality. Jn wch equality all the Planets finish their cours namely in the point of the heart of ye Lyon, in wch very place nature hath erected a golden Column, whereon stand these words written in the Arabick, Hebrew & many other languages: Go no further, here nature rests in the mineral kingdom, here let the Traveller stop, & the Artist make hast, &c.[137]

There was in addition a new approach to important questions of respiration and combustion which seems to have derived ultimately from that same school of thought based on the concept of "aire" used in a Neoplatonic sense. Involved in research and speculation on these matters were John Mayow, Robert Hooke, Robert Boyle, Malachia Thruston, and George Ent. As Guerlac has demonstrated, the stimulus for their work probably came from that paper of Digby's *On the Vegetation of Plants* delivered to the Royal Society in January 1660/61 and published shortly afterwards.[138] In that work Digby had utilized the idea that "there is in the Air a hidden food of life," which he had drawn from Sendivogius'

[135] Digby, *Secrets*, pp. 179–80 (3, n. 69).   [136] le Fèvre, *Chymistry*, II, 213–15 (2, n. 34).
[137] Snyders, "Metamorphosis," p. 43 (Newton's pagination) (5, n. 87).
[138] Henry Guerlac, "John Mayow and the aerial nitre. Studies on the chemistry of

New Light of Alchemy. The resulting search for a special active principle latent in the air and for "magnets" which might attract it did not lead the group close to an early discovery of material oxygen, as has sometimes been suggested and which myth Guerlac has adequately demolished, but did influence Stephen Hales and therefore did indirectly assist in the rise of pneumatic chemistry in the eighteenth century.[139]

In 1674, just about the same time Newton was having his success with the antimonial "magnet" and the same year in which John Mayow published the *Tractatus Quinque Medico-Physici* which contained his theory on the nitro-aerial spirit that rendered the air fit for respiration, Boyle found occasion to publish a tract entitled *Suspicions about Some Hidden Qualities of the Air.*[140] Attached to that essay was an appendix "Of Celestial and Aerial Magnets." In it Boyle took note of some current research on "magnets," and, although he recognized that some readers might think him "extravagant" in recommending such "unpromising" or even "phantastical" experiments, he suggested a number of experimental variations which might help the operator determine the nature of whatever was in the air.[141] As a conclusion to this brief survey of the larger circle of experimenters who were involved with celestial, aerial, and terrestrial "magnets" in the 1650's, 1660's, and 1670's, Boyle's advice on the matter may be quoted. It seems appropriate for all those who venture into "unknown seas," be they scientific or historical, and in view of the lasting results which seem to have been produced by this area of human endeavor, it may also be recommended for any who are overly committed to remaining inland.

> Those adventurous navigators, that have made voyages for discovery in unknown seas, when they first discerned something obscure near the horizon, at a great distance off, have often doubted, whether what they had so imperfect a sight of, were a cloud, or an island, or a mountain: but though sometimes it were more likely to be the former, as that, which more frequently occurred, than the latter; yet they judged it advisable to steer towards it, till they had a clearer prospect of it: for if it were a deluding meteor, they would not however sustain so great a loss in that of a little labour, as, in case it were a country, they would in the loss of what might prove a rich discovery: and if they desisted too soon from their curiosity, they could not rationally satisfy themselves, whether they slighted a cloud, or neglected a country.[142]

John Mayow – I," in *Actes du VIIe Congrès International d'Histoire des Sciences. Jérusalem* (4–12 *Août* 1953) (Collection des Travaux de l'Académie internationale d'Histoire des Sciences, no. 8; Publiée avec le Concours financier de L'UNESCO; Jérusalem: F. S. Bodenheimer, n. d.), pp. 332–49.

[139] The present writer hopes to pursue some of these questions in more detail in a fourth paper in the series of Digby "Studies," where some of Digby's ideas on life processes will be considered.   [140] Fulton, *Bibliography*, 83–86 (3, n. 99).

[141] Boyle, *Works*, IV, 97 (3, n. 68).   [142] *Ibid.*, IV, 100.

Conclusion

A tentative reconstruction of Newton's early development in chemistry and alchemy may be offered on the basis of the details explored in these last two chapters. It was decidedly a movement from exoteric to esoteric chemistry.

Newton apparently had learned most of the standard operating procedures of ordinary mid-seventeenth-century chemistry, and knew how to prepare the usual chemicals in its repertory by about 1667, at which time he wrote out his chemical dictionary (MS Don. b. 15). Very shortly after that, upon reading Boyle's *Of Formes*, he became interested in transmutation (MS Add. 3975, ff. 32r–41v). It seems most likely to the present writer that that time was 1667 or 1668 and that Newton began then to acquire and copy various circulating alchemical manuscripts.

Those manuscripts fall into two major categories at first. One class consisted of alchemical recipes which were usually completely operational but rested on no theoretical base or on a very slight one (Keynes MSS 31, 62, 67), similar to the recipes collected by many other people in the seventeenth century. The other group was comprised of a more philosophical – but non-operational – variety of alchemy (Keynes MSS 51 and 52), known to have been circulating among the mid-century Hartlibians. The writings of Eirenaeus Philalethes in those latter manuscripts were roughly contemporary and were based at least in part on the early seventeenth-century statements of Neoplatonic alchemy of Michael Sendivogius and Jean d'Espagnet. Newton's interest in the Neoplatonic philosophical position may already have been stimulated by that time by Barrow and More – at least that possibility has been suggested here – but it would certainly have received impetus from his reading of Eirenaeus Philalethes.

At about the same time Newton began to explore printed alchemical literature. Perhaps one of the first of the additional works he encountered was *The Triumphal Chariot of Antimony* by Basilius Valentinus (Keynes MS 64). In a way, that particular work is on the borderline between chemistry and alchemy: it is straightforward enough to serve as an operating manual for the preparation of antimony compounds – and it was so used in the seventeenth century – but at the same time it does carry some alchemical suggestions. Newton's notes on the book reflect the same ambiguity and also show him integrating one of Basilius' alchemical hints with Boyle's procedure for the "opening" of metals, perhaps the earliest surviving evidence of the operation of Newton's great synthetic intellect on the alchemical materials. Perhaps during 1668 Newton may also have tested out some of the recipes he had collected, but if so, no evidence of it has survived.

Late in 1668 and in 1669 Newton moved on to read other works of Eirenaeus Philalethes, then newly published (the *Tres tractates* of 1668 and

191

the *Secrets Reveal'd* of 1669, reflected in Keynes MSS 36 and 19, respectively), and also to explore the foundations of Philalethes' alchemy in Sir George Ripley (Keynes MS 17) and in Sendivogius and d'Espagnet (Keynes MS 19). At approximately the same time Newton seems to have begun to read Michael Maier (Keynes MS 29), for one of the few fixed dates in Newton's alchemical career appears in his letter to Francis Aston of May, 1669, which letter reflects familiarity with the same work of Maier's that is abstracted in Keynes MS 29. His reading of Maier is probably what turned Newton so decisively towards ancient alchemy – for Maier was the chief exponent of the *prisca sapientia* doctrine in alchemical studies – and in April of 1669 Newton purchased the huge collection of antique, medieval, and Renaissance alchemy in the *Theatrum chemicum*.

Glassware, chemicals, and furnaces were purchased at the same time and the experiments began in earnest, whether they had before or not. The first recorded set was probably stimulated by Boyle's *Essays* of 1669, the second recorded set by ideas Newton had synthesized from Boyle's *Of Formes* read in 1667 or 1668 and from Basilius Valentinus' *Triumphal Chariot of Antimony* read in 1668 (MS Add. 3975, f. 41v).

From that time forward the alchemical concepts with which Newton dealt became more and more complicated, and he attempted various synthetic statements of them. There appears to have been at least one of these based on the older alchemy which was written during the early period of Newton's alchemical studies (Keynes MS 12, part A), but in the experimental program of the early period Newton's laboratory notes reflect a certain concentration on the star regulus of antimony (MS Add. 3975, ff. 42r–43r) based on concepts derived from the seventeenth-century Neoplatonic alchemy of Basilius Valentinus, Eirenaeus Philalethes, Michael Sendivogius, and Jean d'Espagnet.

In the meantime Newton's manuscript collection continued to grow as he copied or obtained copies of various unpublished works (Keynes MSS 22, 33, and 58, and Snyders' "Metamorphosis"), made transcripts of published works (Keynes MS 14), and compiled short anthologies of extracts from various writers designed to elucidate some obscure point of procedure (Keynes MS 25). He had in addition some interest in experimenting with Snyders' alchemy (Keynes MS 58; MS Add. 3975, f. 43v) coupled with a continuing concern with Boyle's approach (MS Add. 3975, ff. 43r–51v and esp. f. 43v), but it was the attack on common mercury "through reiterated cohobation by the regulus of leo" – the star regulus of antimony – that finally yielded Newton the philosophical mercury he sought (Keynes MS 18).

Newton's great success seems to have been considered by him to have been a part of "the work in common gold," which many of the later manuscripts refer to. In Keynes MS 20, for example, quoted in Chapter 1, written a few years after Keynes MS 18, one may see him examining "The Hunting of the Greene Lyon" in an attempt to elucidate "ye Regimen of ye work in common gold after ye Pher ☿ is made." Presumably he thought he had the philosophical mercury and was trying to understand what came next in the Great Work.

Newton had found his success in the toils and by-ways of Neoplatonic alchemy, especially in that of Sendivogius, d'Espagnet, and Eirenaeus Philalethes, and their concepts were determinative of his own to a certain extent. But it is entirely possible that Newton had been pre-disposed to an appreciation of Neoplatonism by the ideas of Barrow and More, who found matter–spirit relationships suggestive of needed modifications in Descartes' mechanism, and it may be said that for Newton that larger goal of the re-vision of Cartesian philosophy was probably always part and parcel of his alchemical studies. Did his success in "the work in common gold" give him the vantage point he needed then to attack Cartesianism?

The answer to that question *c.* 1675 has to be a blunt negative, for Newton had not yet come to grips with the great problem inherent in Neoplatonism. The problem was already implicit in the differences be-tween Barrow's and More's positions, and there is no evidence that Newton had resolved it at the time he wrote Keynes MS 18. Stated in terms of "the work in common gold," the problem may be put in a spe-cific way. Was the vivifying principle, which he thought his antimonial "magnet" had drawn out of the "universal spirit," present in his philo-sophical mercury in a spiritual or in a corporeal way?

In the terms employed by Barrow and More, the question may be stated more generally. Did spirit and matter effect exchanges with each other or combine materially in some way in bodies? To that, Barrow would per-haps have answered in the affirmative, basing his judgment on the Hermetic philosophers who found a gross matter and a subtle spirit in all things. But More would have denied it, for to him it was a philosophical necessity to keep matter and spirit rigidly separated. For More, spirit only served to guide and direct the motions of bodies which could not be accounted for on the basis of mechanical impact.

Newton probably held the ideas of Barrow and More on matter–spirit relationships in unresolved tension in his mind for a long time. But even-tually he did effect a resolution of their conflicting views, and in doing so he effected a far greater transmutation than he ever managed in "the work in common gold." For it was in the resolution of that conflict that Newton's new concept of force was born, as will be seen in the next chap-ter, and the great intellectual spirits of Barrow and More could then rest easy, since Cartesian impact mechanism was then modified in just the right direction.

6 Newton's Integration of Alchemy and Mechanism

Introduction

If Newton really had such a remarkable success in making a philosophical mercury, and if he was really such a scientific alchemist as the present writer maintains, why then did he never publish a scientific paper on his process? Surely it would have been in order for him to make his knowledge public for the benefit of other natural philosophers, adhering to the new ideal of the open communication of secrets. Yet he did not, and his failure to do so might be taken to mean that the "Clavis" of Keynes MS 18 was not really his, or that he knew, good scientist that he was, that his experiments were really all failures. But another interpretation of his silence is possible and is indeed indicated by Newton's response in 1676 to a question from Boyle.

Newton's statement on the publication of alchemical secrets occurs in a letter he wrote to Henry Oldenburg, Secretary of the Royal Society, in April of 1676. The occasion for Newton's letter was ostensibly to thank Oldenburg for publishing another letter from Newton (on optics) and for having an optical experiment that Newton had suggested tried before the Royal Society. But the bulk of the letter concerned itself with an "uncommon expt" Boyle had just published in the *Philosophical Transactions* of that society.

Boyle had found a mercury which would grow hot when mixed with gold. The incalescence of mercury with gold being something that the alchemists claimed was a characteristic of philosophical mercury, Boyle said, he had been moved to make certain "trials" of the matter, and in 1652 he had succeeded in making such a mercury. He made many tests with it, mixing it in the palm of his hand with gold, and found the mixture to grow hot in about one minute. He had also had "the Learned Secretary of the Royal Society" (Oldenburg) and "the Noble and Judicious President of the Royal Society, the Lord Viscount Brouncker" try the experiment with their own hands, and they also had felt the incalescence. To Boyle, that meant that the mercury was an extraordinarily noble one and that probably mercurial medicines made from it would serve extraordinarily well in "Physick." But he was not certain whether the good which such medicines might provide would be greater than the "political inconveniences" that might ensue if the mercury should actually prove to be "of the best kind," i.e., a true philosophical mercury, and "fall into ill hands." So he was making the matter known to the learned world in order to solicit the opinions "of the wise and skilful" as to whether he should keep his silence.[1]

Newton certainly thought that Boyle should keep "high silence" and so informed Oldenburg. At that time Newton was probably not very far away from having prepared his own special mercury, so his remarks about Boyle's mercury must have had a great deal of personal meaning for himself. Newton presented his views on the matter to Oldenburg in the following manner, after saying that he did not think Boyle's mercury had any "great excellence."

> But yet because ye way by wch ☿ may be so impregnated, has been thought fit to be concealed by others that have known it, & therefore may possibly be an inlet to something more noble, not to be communicated wthout immense dammage to ye world if there should be any verity in ye Hermetick writers, therefore I question not but that ye great wisdom of ye noble Authour will sway him to high silence till he shall be resolved of what consequence ye thing may be either by his own experience, or ye judgmt of some other that throughly understands what he speakes about, that is of a true Hermetic Philosopher, whose judgmt (if there be any such) would be more to be regarded in this point then that of all ye world beside to ye contrary, there being other things besides ye transmutation of metalls (if those great pretenders bragg not) wch none but they understand.[2]

From those remarks it seems that Newton simply thought that it was not safe to make alchemical knowledge public. Why not, he did not know, but he took the alchemical writers at their word that there were "other things besides ye transmutation of metalls" involved. And until Boyle knew what is was "by his own experience" and could judge the safety of the world by that, or until he had obtained the necessary judgment from "a true Hermetic Philosopher," he had best not take the risk of bringing "immense dammage to ye world."

That being Newton's understanding of the matter, it is no wonder at all that he never made his own alchemical studies public knowledge, either the successes or the failures, if in fact he never was able to find out what the great secret was "by his own experience." The fact that he never published a work on alchemy cannot be taken to mean that he knew he had failed. On the contrary, it probably means that he had

---

[1] [Robert Boyle], "*Of the Incalescence* of Quicksilver *with* Gold, *generously imparted by* B. R.*,*" *Philosophical Transactions* **10** (1675–1675/76), 515–33, hereinafter referred to as Boyle, "Incalescence"; Boyle, *Works*, IV, 219–30, (3, n. 68).

[2] Isaac Newton to Henry Oldenburg, April 26, 1676, in Newton, *Correspondence*, II, 1–3, quotation from p. 2 (1, n. 52).

enough success to think that he might be on the track of something of fundamental importance and so had good reason for keeping his "high silence," even though there is nothing to indicate that he himself was searching for that mysterious "inlet to something more noble."

The argument might then be made that a study of Newton's alchemy is virtually useless, that at best it supplies only a curious footnote to the intellectual history of the seventeenth century, since Newton kept his thoughts on it so private that they had no influence on the world at all. But the exact opposite of that argument seems to be true, for a great deal of Newton's later writing reflects his long struggle to integrate alchemy with the rest of his science.

It is proposed in this final chapter to indicate some of the material in Newton's scientific writings which seems to be based in part on his alchemical ideas. By doing so, it is hoped that three things will be accomplished. First, it is hoped that the evidence that Newton included alchemical ideas in his scientific writings will round out and complete the argument of the present work that he was indeed a scientific alchemist. Second, it is hoped that a more theoretical dimension may be supplied for Newton's alchemical processes in some instances. And, finally, it is hoped that when alchemical modes of thought are seen to rest at the very heart of the great Newtonian system, this study of Newton's alchemy and all other studies of so-called "pseudo-sciences" will find some small measure of justification for their existence.

In order to do these things, it is necessary first to introduce a somewhat different, but really much more conventional, perspective from which Newton's work may be viewed. In previous chapters, this study has aimed almost exclusively at an elucidation of the alchemical background to Newton's alchemy and at an understanding of his early alchemical reading and experimentation. The time has now come, however, to broaden that picture and to remind the reader of Newton's other interests.

The most important aspect of Newton's career which has been deliberately neglected up to this point is the fact that he was of the second generation with respect to mechanical philosophies. Descartes had begun to publish before Newton was born and had died before Newton ever went up to Cambridge. Digby, Hobbes, and Gassendi had put out their most influential works while Newton was constructing toy mills and memorizing his declensions. Therefore Newton came upon the new mechanical philosophies virtually as a philosophical innocent, which no one before his generation had done. Even Newton's older contemporary, Henry More, for example, as was seen in Chapter 4, read widely in Platonism and Neoplatonism before encountering Descartes. But that was not the case with Newton, and it should never be forgotten that, although he was introduced to the standard outdated undergraduate curriculum, he was immersed in mechanical philosophies almost from the very beginning, and they formed the stable backdrop against which his alchemy was played out.

Newton's student notebook presents incontrovertible evidence of his early involvement with *philosophia mechanica*: it was she who seduced him almost immediately from Plato and Aristotle, as Westfall has demonstrated so definitively in his recent works. Not only did Newton read Descartes himself, that master "mechanician," but Newton also read most other important works of the period which presented variant mechanical systems. He read Charleton's epitome of Gassendi and perhaps Gassendi himself; he read Hobbes, Digby, and Boyle, and familiarized himself with at least a part of Galileo's work. And while yet an undergraduate, Newton also began to ponder and experiment on various problems of matter and motion.[3] He was in short a mechanical philosopher in his own right, although perhaps not yet fully fledged, for some years before he became an alchemist, and there is no reason whatsoever to suppose that *alchimia*, for all her charms, ever fully supplanted *philosophia mechanica* in his affections.

Almost everyone who cared at all about natural philosophy in the seventeenth century would have recognized that something new was afoot. There was a new world system in Copernicanism, there were new experiments on the magnet from Gilbert, there were whole new sciences from Galileo, there was a new "animal eoconomy" of circulation from Harvey, and even a new light on alchemy from Sendivogious.[4] But the new mechanical philosophies captured the general consciousness of natural philosophers as few other things had because of their apparent explanatory power. Their basic unifying concept – that all events could be explained by matter in motion, by bits of matter acting on each other by impact – was readily understandable, and many natural philosophers seem to have felt a sense of relief that the vague occult qualities and substantial forms of Aristotelianism were no longer required for explanation. In a very real sense, the early mechanical philosophers, such as Descartes, Digby, Hobbes (except for his supposed atheism), Gassendi, and Charleton, must have represented the best of the new natural philosophy to Newton's generation, and Newton was no exception.

Of those mechanical philosophers whom the young Newton read, two were very interested in alchemy, as has already been noted: Sir Kenelm Digby and the Honourable Robert Boyle. Digby had studied alchemy in

---

[3] Westfall, "Foundations" (4, n. 13); Richard S. Westfall, *Force in Newton's Physics. The Science of Dynamics in the Seventeenth Century* (London: Macdonald; New York: American Elsevier, 1971), esp. pp. 323–63, hereinafter referred to as Westfall, *Force*.

[4] *The Rise of Modern Science. External or Internal Factors?* ed., with introd., by George Basalla (Problems in European Civilization; Lexington, Mass.: D. C. Heath and Co., 1968), p. vii.

197

the 1630's and had to some extent incorporated alchemical ideas into the matter-theory of his 1644 *Two Treatises*, the book which Newton read. But Digby does not seem to have made much effort to integrate alchemy with his mechanical philosophy, and in none of his writings does he offer a mechanical explication of transmutation, not even in his book of *Secrets* which consists largely of alchemical recipes.[5] Boyle, on the other hand, although he generally confined himself in print to intriguing hints about alchemical possibilities, was quite sure that his mechanical philosophy offered a ready rationale for transmutation. Boyle's ideas along those lines seem to have influenced Newton rather heavily, and they provide a good way of approaching the theoretical ideas which Newton seems to have held but did not make very explicit.

It was Boyle's book *Of Formes* which Newton had read and taken notes on just before he began the experiments considered in the last chapter. In that book Boyle presented his ideas on the mechanical structure of matter and on two different corpuscularian mechanisms which might effect transmutation. One of these mechanisms may be designated as maturation by heat, the other as transmutation by tincture. If one may judge by Newton's later writings, he accepted the idea of one universal matter as advanced in Boyle's writings on mechanical philosophy, and also accepted the first of those Boylian mechanisms as one of the basic processes involved in transmutation, rejecting the second mechanism. In the present chapter, Boyle's basic mechanical underpinning for the concept of transmutation will be explored first, and then his two speculative mechanisms will be examined, along with the reasons for believing Newton to have accepted the first.

It will then be assumed that Newton adhered to a corpuscularian concept of matter throughout all his alchemical work, that he accepted the idea of maturation by heat, and that he built up certain other concepts on those fundamental ones. To make these assumptions no doubt has its dangers. But they are not inconsistent with the evidence at hand, and, until further studies of Newton's alchemy reveal them to be in error or are able to refine them with new subtleties, they will be offered as a reasonable way of entering into the maze.

The evidence of the last chapter suggests that the first subsidiary concept that Newton grafted onto those basic ones was the idea of "opening" metals to get out their "mercury." His first recorded chemical experiments were based on that idea, and he seems to have pursued it in one way or another until he finally obtained, he thought, a true philosophical mercury.

But the techniques he used to prepare the "philosophick" mercury

[5] Dobbs, "Digby. Part ɪ" (2, n. 28); Dobbs, "Digby. Part ɪɪ" (2, n. 58); Dobbs, "Digby. Part ɪɪɪ" (2, n. 59).

were derived from writers who were anything but mechanical philosophers. Sendivogious, d'Espagnet, and Eirenaeus Philalethes all relied on
a Neoplatonic world-view and spoke of "magnets" and of the radical
moisture or celestial virtue or vivifying spiritous "aire" which the
"magnet" might draw to itself. Since the process Newton derived from
that school of thought seemed to him in the 1670's to be a valid one, he
must have thought that their philosophy also had some truth in it. But
could it be reconciled in any way with the mechanical philosophy?

It could be and it was. In Newton's first published statement of a
natural philosophical position, "An Hypothesis Explaining the Properties
of Light," sent to the Royal Society in a letter to Oldenburg late in 1675,
one may see the Neoplatonic "aire" re-written as a mechanical system,
the first-fruits of Newton's integration of alchemy and mechanism. The
present chapter will examine that unusual hybrid creation.

Newton discarded his first synthetic effort within a few years, however,
and that for very good reasons. In its place he put a new concept, one
which was to prove of permanent scientific value. Much of Newton's
work in developing his new idea lies outside the scope of the present study
and can only be suggested here. But the new concept of force, as it grew
and developed in Newton's mind, undoubtedly had some roots in his
alchemical studies, and one may now at least point to those.

Although Newton never published a work fully devoted to alchemy,
after 1692 he did make available to the world his mature thoughts on
the sub-structure of matter and on the forces acting there. The influence
of the ideas he expressed in the published works of that later period on
chemistry as it actually developed in the eighteenth century has been
explored by Thackray[6] and will not be pursued here. Nevertheless, the
material has considerable intrinsic interest since it is Newton's and since
it came from a quarter of a century or more of alchemical experimentation.
In its turn, it throws a little new light on what sort of knowledge Newton
had really hoped to obtain from his alchemical processes, and in conclusion some of Newton's ideas on transmutation will be discussed in terms of
his mature matter theory.

"One Catholick Matter" and its transmutations

As has already been suggested, and as will be demonstrated in detail in
this and in succeeding sections, Newton's thoughts on the "one Catholick
Matter" of the mechanical philosophers closely paralleled Boyle's ideas.
Boyle expressed himself on the subject earlier and more fully than Newton
did, however, and so some of the Boylian writings offer a convenient
point of attack upon the problem of establishing the ideas Newton held

---

[6] Thackray, *Atoms and Powers* (2, n. 55).

about the corpuscles of bodies and the changes which might be wrought in them by the mechanical rearrangment of their parts.

In his book *Of Formes* of 1666 Boyle reported on two experiments which he thought might be important cases of fundamental transmutations, if they could be repeated and a few of his "scruples" allayed. The one was a transmutation of gold into silver – not "Lucriferous," Boyle said, but quite "Luciferous."[7] The other was a transmutation of one Peripatetic element into another, Water into Earth.[8]

Not waiting for the experiments to be repeated and verified, Boyle indulged himself freely in 1666 in corpuscularian explications of transmutation. Introducing the experiment in which gold was degraded, he said:

But to give you a plain and naked Account of this matter, that you may be able the better to judge of it, and if You please, to repeat it, I will freely tell You, That supposing all Metals, as well as other Bodies, to be made of one Catholick Matter common to them all, and to differ but in the shape, size, motion or rest, and texture of the small parts they consist of, from which Affections of Matter, the Qualities, that difference particular Bodies, result, I could not see any impossibility in the Nature of the Thing, that one kind of Metal should be transmuted into another; (that being in effect no more, then that one Parcel of the Universal Matter, wherein all Bodies agree, may have a Texture produc'd in it, like the Texture of some other Parcel of the Matter common to them both.[9]

In the other experiment, in which he thought he had obtained a white earth by repeated distillations of pure rain water, heat being the only active agent employed, Boyle interjected a speculation on the mechanism by which some parts of the water became a heavy, fixed substance.

Some bold Atomists...would perchance particularly tell you, how the continually, but slowly, agitated parts of the Water, by their innumerable occursions, may by degrees rub, and as it were grind themselves into such Surfaces, as *either* to stick very close to one another by immediate contact, (as *I* elsewhere observe polish'd pieces of Glass to do,) *or* implicate, and intangle themselves together so, as to make, as it were, little *knots*; which knots...or the newly mention'd *clusters* of coherent Particles, being then grown too great and heavy to be supported by the Water, must subside to the bottom in the form of a Powder....[10]

[7] Boyle, *Of Formes*, pp. 349–78 (Experiment vii) (4, n. 78); Boyle, *Works*, iii, 93–100 (3, n. 68).
[8] Boyle, *Of Formes*, pp. 387–420 (Experiment ix) (4, n. 78); Boyle, *Works*, iii, 102–09 (3, n. 68).
[9] Boyle, *Of Formes*, p. 350 (4, n. 78); Boyle, *Works*, iii, 93–94 (3, n. 68).
[10] Boyle, *Of Formes*, pp. 402–03 (4, n. 78); Boyle, *Works*, iii, 105 (3, n. 68).

The possibility of making alchemical gold was raised by Boyle and quite explicitly related to that mechanism he had introduced when speculating on the conversion of water into earth. Reiterating his caution that the event itself, the conversion of water into earth, had not really been proven, he nevertheless went on to say,

> if...this Powder, whether it be true Elementary Earth or not, be found to be really produc'd out of the Water it self, it may prove a *Magnale* in Nature, and of greater consequence then will be presently foreseen, and may make the Alchymists hopes of turning other Metals into Gold, appear less wild, since that by Experimentally evincing, that two such difficult Qualities to be introduc'd into a Body, as considerable degrees of Fixity & Weight, (whose requisitenesse to the making of Gold are two of the Principal things, that have kept me from easily expecting to find the Attempts of Alchymists successful,) may, without the mixture of the Homogeneous Matter, be generated in it, by varying the Texture of its parts.[11]

Newton did not take Boyle's doubts about the verity of the experiment very seriously, it seems, but treated the transmutation of water to earth as a proven fact, for much, much later he noted in the *Opticks* that "Water by frequent Distillations changes into fix'd Earth, as Mr. *Boyle* has try'd...."[12]

Newton also followed right along uncritically in Boyle's speculations about a universal matter and its gradual transformation by the action of heat, actually inserting in the *Principia* in 1687 remarks which expanded on Boyle's suggestions. In 1687 Newton said:

> The vapors which arise from the sun, the fixed stars, and the tails of the comets, may meet at last with, and fall into, the atmospheres of the planets by their gravity, and there be condensed and turned into water and humid spirits; and from thence, by a slow heat, pass gradually into the form of salts, and sulphurs, and tinctures, and mud, and clay, and sand, and stones, and coral, and other terrestrial substances.[13]

Certain parallel passages in papers written about the same time but not included in the published *Principia* express the same thoughts and make Newton's belief in a universal matter and its coagulation by heat a little clearer. The following appears in two manuscript versions of an unpublished "Conclusio" for the *Principia*.

[11] Boyle, *Of Formes*, p. 417 (4, n. 78); Boyle, *Works*, III, 108 (3, n. 68).

[12] Newton, *Opticks*, p. 374 (1, n. 9). This statement in Query 30 had first been introduced by Newton in Query 22 of the Latin edition of 1706, and it appeared in all editions subsequent to that.

[13] Newton, *Principia*, II, 542 (1, n. 4); Koyré and Cohen, *Newton's Principia*, II, 758 (1, n. 4). This passage appeared in the first edition and subsequent ones although another important passage on transmutation in the first edition was modified by Newton after 1687, as will be discussed below.

...[T]hat rare substance water can be transformed by continued fermentation into the more dense substances of animals, vegetables, salts, stones and various earths. And finally by the very long duration of the operations be coagulated into mineral and metallic substances. For the matter of all things is one and the same, which is transmuted into countless forms by the operations of nature, and more subtle and rare bodies are by fermentation and vegetation commonly made thicker and more condensed. By the same motion of fermentation bodies can expel certain particles....[14]

And in one of the variant manuscripts of the "Conclusio," Newton's words echo Boyle's suggestion of the mechanism by which coalescency occurs, except that Newton has added the concept of forces. That addition was, of course, very important, and will be discussed in its place below. For the present it suffices to note the close correspondence between Newton's "slow and continued motion of heat" and Boyle's "continually, but slowly, agitated parts," for Newton wrote:

Furthermore through the slow and continued motion of heat the particles of bodies can gradually change their arrangement and coalesce in new ways and by the attractive forces of contiguous particles (which are stronger than expulsive ones) come together more densely.[15]

Boyle offered a possible alternative mechanism for transmutation in his book *Of Formes*, in which scheme bodies differed because of the presence or absence of special "noble and subtle" particles rather than because the whole general texture of the body had been changed. This second mechanism does not seem to have appealed to Newton at all, perhaps because Boyle himself was equivocal about it. In the discussion of his degradation of a part of the body of the gold to silver, Boyle said that perhaps the old terms used by "Chymists" – the "*Tinctura Auri*" or the "*Anima Auri*," the tincture or the soul of gold – might be explained mechanically as

[14] Isaac Newton, "Conclusio," in Newton, *Unpublished Papers*, pp. 328 and 341 (3, n. 157). See Chapter 5 above for a suggested explanation of Newton's concept of "fermentation," a notion with both mechanical and alchemical overtones as he used it. In this passage from the "Conclusio" the present writer has modified the Halls' translation slightly, their rendering of "the process of growth" being changed to the more literal "vegetation" to bring it into accord with Newton's usage of the word "vegetation" in Burndy MS 16, "Of natures obvious laws & processes in vegetation." Cf. Rattansi, "Newton's Alchemical Studies" (1, n. 49), where that manuscript, which was Sotheby Lot 516, is discussed. Although Burndy MS 16 was not included in the alchemical papers at the Sotheby sale, it does treat the "vegetation of metals," and it would seem that Newton probably meant to include the concept of metallic growth, as well as that of animals and vegetables, in his usage in the "Conclusio."

[15] Isaac Newton, draft "Conclusio," in Newton, *Unpublished Papers*, pp. 332–33 and 346 (3, n. 157). The same idea was expressed in similar language by Newton in a partial draft of the "Praefatio" to the first edition of the *Principia*: Newton, *Unpublished Papers*, pp. 303–04 and 306–07.

some noble and subtle Corpuscles, [which] being duely conjoyn'd with the rest of the Matter, whereof Gold consists, may qualifie that Matter to look Yellow, to resist *Aqua fortis*, and to exhibit those other peculiar *Phaenomena*, that discriminate Gold from Silver....[16]

Boyle was not at all sure about the matter. After all, he had treated his gold with a certain menstruum which conceivably had extracted just that kind of "noble and subtle Corpuscles" and left his matter in the form of silver. Also a "Mineralist of great Veracity" had told him of "Extracting a Blew Tincture out of Copper, so as to leave the Body White." But on the whole he thought it more likely that the color of a body depended on the general texture of the body rather than on some special tingeing corpuscles.[17]

There is no evidence that Newton ever took up Boyle's suggestions on tinctures in a serious way. On the contrary, Newton was really quite explicit later on in saying that the colors of bodies depended solely on the size of the particles which composed them. In the "Conclusio" of 1687, for example, he noted that "the various colours of bodies arise from various sizes of reflecting particles...."[18] And in Book II of the *Opticks* he engaged in an elaborate consideration of the colors of bodies and related the various colors to certain experiments of his own which he thought would enable him to ascertain the actual size of the corpuscles which produced the colors by reflecting light. That study of colored bodies by Newton will be treated in more detail below, but here it may be observed that the particles involved in producing colors were not conceived by Newton as special tingeing ones, for he spoke in the *Opticks* of the "Particles on which the Operations in Chymistry, and the Colours of natural Bodies depend, and which by cohering compose Bodies of a sensible Magnitude."[19] In other words, the particles that produced color were the ordinary particles of the general texture of the body and not "some noble and subtle " ones.

Newton read Boyle's *Of Formes* in 1667 or 1668. Besides that, he had already been deeply immersed for some years before that in mechanical ideas of particles of matter in motion. He pretty clearly adhered to Boyle's basic mechanical position about the universal matter and was following one of Boyle's mechanical explications of transmutation twenty years later when he wrote the *Principia*. The various editions of the *Opticks* of his old age are all packed with corpuscularian concepts. All of those facts are now well known and generally recognized, but it seems well to reiterate and emphasize them at this point in order to provide a rationale

[16] Boyle, *Of Formes*, p. 359 (4, n. 78); Boyle, *Works*, III, 95–96 (3, n. 68).

[17] Boyle, *Of Formes*, pp. 359–63 (4, n. 78); Boyle, *Works*, III, 96 (3, n. 68).

[18] Isaac Newton, "Conclusio," in Newton, *Unpublished Papers*, pp. 329 and 342 (3, n. 157).

[19] Newton, *Opticks*, p. 394 (1, n. 9).

for the following assumption: a mechanical, particulate concept of the changes in the "forms" of "one Catholick Matter" was fundamental to Newton's thought throughout all the intervening period when he was studying alchemy so intensively even though neither the alchemical manuscripts nor the laboratory notes reflect it.

If Newton believed in one universal matter and in a mechanical maturation by heat, why then had he precipitously involved himself in looking for the "mercury" of metals? Certainly, the search for that elusive "chymical" principle looks anachronistic, suddenly appearing as it does in Newton's experimental notes, against the sensible, rational background of Boyle's mechanical philosophy. Something of the nature of the answer to that question – which seems to lie in Newton's concept of the hierarchical structure of matter – has been indicated in Chapter 5, and an attempt to answer it in more detail will be given later in the present chapter after Newton's mature chemical thought has been considered. But although he may have done so, there is no direct evidence that he had worked out all his thoughts on complex particles by 1675. He published nothing on that subject until later, and certainly his mature formulation had to have been effected after he elaborated his concept of force.

Newton's material aether and more about the concept of mediation

But by the end of 1675 he had worked out his integration of Neoplatonic "aire" with mechanism. He called the product the "aether," and he thought it to be material rather than spiritual and thought it explained all sorts of things. In the concluding remarks to Chapter 5, it was observed that Newton gave no indication of having decided one very important question at the time he wrote Keynes MS 18, the "Clavis." That question was, whether what his "magnet" had drawn in was present in a material way or in a spiritual way. By 1675, however, *philosophia mechanica* had asserted herself, and he had opted for the first choice, at least temporarily.

In his letter to Oldenburg in 1675 which contained the "Hypothesis" on light, Newton said he himself would not "assume" any hypothesis to explain the properties of light, not deeming it necessary. Nevertheless, he presented rather a full one, one which he must have been thinking seriously about for some time. In the hypothesis the first supposition was to be "that there is an aethereall Medium much of the same constitution with air, but far rarer, subtiler & more strongly Elastic."[20]

[20] Isaac Newton to Henry Oldenburg, Cambridge, Dec. 7, 1675, in Newton, *Correspondence*, I, 362–89, esp. 363–64 (1, n. 52).

The aether was not uniform, according to Newton, and his statements on its constitution are reminiscent of d'Espagnet's "universal spirit," that vast repository of celestial virtues. Newton said,

But it is not to be supposed, that this Medium is one uniforme matter, but compounded partly of the maine flegmatic body of aether partly of other various aethereall Spirits. . . . For the Electric & Magnetic effluvia and gravitating principle seem to argue such variety.[21]

One may compare also d'Espagnet's concept, that all things were generated from the "universal spirit," with Newton's statements about the aether.

Perhaps the whole frame of Nature {may be nothing but aether condensed by a fermental principle} [may be nothing but various Contextures of some certaine aethereall Spirits or vapours condens'd as it were by praecipitation, much after the manner that vapours are condensed into water or exhalations into grosser Substances, though not so easily condensible; and after condensation wrought into various formes, at first by the immediate hand of the Creator, and ever since by the power of Nature, wch by vertue of the command Increase & Multiply, became a complete Imitator of the copies sett her by the Protoplast]. Thus perhaps may all things be originated from aether.[22]

But yet Newton's aether was not really quite the same as d'Espagnet's "universal spirit," for it was material – much like air, Newton had said, and therefore not a true spirit at all. It also had the frictional qualities associated with matter in Newton's mind, for he added,

Of the existence of this Medium the [loss of] motion of a Pendulum in a glasse exhausted of Air almost as quickly as in the open Air, is no inconsiderable argument.[23]

Since the aether was material, in some respects it was much more like the "one Catholick Matter" of Robert Boyle, a "universal matter" rather than a "universal spirit."[24] Even though Newton frequently called it a "Spirit," it is apparent that he thought that its condensation produced material bodies. He went on in the letter to Oldenburg of 1675 to say the following.

For if such an aethereall Spirit may be condensed in fermenting or

---

[21] Ibid., I, 364.

[22] Ibid. The material in the curly brackets was in the original letter sent by Newton to Oldenburg, but Newton later wrote and asked Oldenburg to change that to what is in the square brackets. See ibid., I, 386–87, n. 1, and also Isaac Newton to Henry Oldenburg, Cambridge, Jan. 25, 1675/76, in ibid., I, 413–15.

[23] Ibid., I, 364. The bracketed material in this quotation was added for clarity by the present writer.

[24] It is instructive to compare Newton's ideas on the aether in 1675 with the ideas Boyle expressed on Effluviums (1673) and on Hidden Qualities in the Air (1674): Fulton, Bibliography, pp. 73–78, 83–86 (3, n. 99); Boyle, Works, III, 659–730, and IV, 79–144 (3, n. 68).

burning bodies, {or otherwise inspissated in ye pores of ye earth to a
tender matter wch may be as it were ye succus nutritious of ye earth
or primary substance out of wch things generable grow} [or otherwise
coagulated, in the pores of the earth and water, into some kind of
humid active matter for the continuall uses of nature, adhereing to
the sides of those pores after the manner that vapours condense on the
sides of a Vessell subtily set]; the vast body of the Earth, wch may be
every where to the very center in perpetuall working, may continually
condense so much of this Spirit as to cause it from above to descend
with great celerity for a supply.[25]

But Newton's aether does not remain in a condensed form, and one is
reminded of the theory of the generation of all things of Sendivogius.
The Newtonian aether seems to go into the center of the earth, much like
the sperm of the Sendivogian elements, what Sendivogius called their
"thinness." The aether cannot stay there, however, for, as Newton puts it,
the earth is in "perpetuall working" to the very center. In the center of
the earth Newton's aether seems to meet the Sendivogian "Archeus,"
for the aether is again cast forth, as Newton said, "nature makeing a
circulation by the slow ascent of as much matter out of the bowells of
the Earth in an aereall forme...."[26] When the matter is sent out again,
Newton continues, it

> for a time constitutes the Atmosphere, but being continually boyed
> up by the new Air, Exhalations, & Vapours riseing underneath, at
> length, (Some part of the vapours wch returne in rain excepted)
> vanishes againe into the aethereall Spaces, & there perhaps in time
> relents, & is attenuated into its first principle. For nature is a perpet-
> uall circulatory worker, generating fluids out of solids, and solids out
> of fluids, fixed things out of volatile, & volatile out of fixed, subtile
> out of gross, & gross out of subtile....[27]

One can hardly miss the alchemical overtones in that last passage, and
again Newton appears to echo Sendivogious, who saw all natural processes as parallel to alchemical ones.

With his aether Newton undertook the explanation of a large variety
of natural phenomena, one of which was animal motion, commonly
thought in the seventeenth century to depend upon something called
"animal spirits." The "Animall Spirits" are "Aethereall" in nature,
Newton said, and operate by way of "Mediation" between the "muscular
juices" and the "common externall aether," the latter of which he
thought was required to fill up the muscles and make them swell and
move. Newton's speculative discussion on animal motion need not be

[25] Newton, *Correspondence*, i, 365–66 (1, n. 52). For an explanation of the bracketed
material, see n. 22 above.

[26] *Ibid.*, i, 366. [27] *Ibid.*

fully analyzed here, interesting though it is, since most of it does not derive from alchemical concepts. But his idea that the "animal spirits" act by mediation does deserve to be pursued, for he presents the notion of mediation by the use of several examples of chemical specificity, and the whole concept probably derived from the mediation or "melioration" of Philalethes' "doves."

"[F]or knowing how the Spirit may be used for Animal motion," Newton said,

you may consider, how some things unsociable are made Sociable by the Mediation of a Third. Water, wch will not dissolve copper, will do it if the copper be melted with Sulphur: Aqua fortis, wch will not pervade Gold will do it by addition of a little Sal Armoniac, or Spirit of Salt; Lead will not mix in melting with copper, but if a little Tin or Antimony be added, they mix readily, & part againe of their own accord, if the Antimony be wasted by throwing Saltpeter or otherwise.[28]

One more chemical example followed, that lead melted with silver allows the silver to be fused in a lesser heat than usual. Then Newton concluded that his speculative "aethereal" mechanism might work the same way.

And in like manner the aethereal Animal Spirit in a man may be a mediator between the common aether & the muscular juices to make them mix more freely; & so by sending a little of this Spirit into any muscle, though so little as to cause no sensible tension of the muscle by its owne force, yet by rendering the juices more Sociable to the common external aether, it may cause that aether to pervade the muscle of its owne accord in a moment more freely & copiously then it would otherwise do & to recede againe as freely so soon as this Mediator of Sociablenes is retracted.[29]

Whatever one may think of the merits of Newton's venture into animal physiology, his discussion of "Mediation" in this paper of 1675 is certainly related to the alchemical mediation in the "Clavis" of Keynes MS 18. There, it will be recalled, Newton thought "mediating" agents to be active in two stages of his process for making the philosophical mercury. The first such agent was pure silver (the virgin Diana) which enabled the operator to join common mercury with metallic antimony. That regulus, the metallic antimony, however, already had in it a "spiritual semen" which carried with it a "fermental virtue" drawn in from the "aire," and, once the regulus and the common mercury were joined, that "fermental virtue" intervened or mediated so that the "spiritual semen" might effect the purgation of the common mercury and make it an "actuated" mercury.

[28] *Ibid.*, I, 369.　　　　[29] *Ibid.*

Newton's use of the concept of mediation in the "Hypothesis" of 1675 and in the "Clavis" of about the same time makes possible the explication of an obscure passage in another letter from Newton to Oldenburg a few months later, the letter mentioned earlier in the present chapter in which Newton was responding to Boyle's publication on his incalescent mercury.

Boyle had not actually said how he had made his mercury which grew hot with gold, but he hinted that it might

> with skill and pains be at length obtained from common Mercury
> skilfully freed from its recrementitious and heterogeneous parts, and
> richly impregnated with the subtle and active ones of congruous
> Metals or Minerals.[30]

Newton took Boyle at his word that it was an "impregnated" mercury. Then he went on to assume that the impregnating particles were "grosser" than the particles of the mercury. That assumption was merely fairly standard mechanical philosophy, for in such philosophies mercury was frequently assumed to have smaller particles than other metals because of its volatility. That being the case, however, a problem in formulating a mechanical explication of Boyle's experimental results arose. For Boyle's mercury had mixed and amalgamated more readily with the gold than ordinary mercury would, and, according to standard mechanical theories on particle size, that should have meant that the particles of Boyle's special mercury were *smaller* than those of ordinary mercury: it was usually thought that smallness or "subtility" in the particles of the dissolving medium produced easier dissolutions since the small particles could more readily enter into the pores of the solute and break it apart. But Newton saw a way around the problem by applying his ideas on mediation, and he had the following to say.

> [F]or it seems to me yt ye metalline particles with wch yt ☿ is im-
> pregnated may be grosser yn ye particles of ye ☿ and be disposed
> to mix more readily wth ye ☉ upon some other account then their
> subtility....[31]

Recalling what Newton had recently had to say about "sociability" and the "mediation of sociability" in his "Hypothesis of Light," it would seem that he was thinking in 1676 that the impregnating particles in Boyle's mercury were acting in a process of mediation. And when one recalls that Newton spoke of a similarly "actuated" mercury in his "Clavis," it becomes a virtual certainty that he thought of his own mercury the same way, even though he did not discuss particles in that alchemical manuscript.

In the letter to Oldenburg in 1676, Newton went on to expand the

[30] Boyle, "Incalescence," pp. 525–26 (6, n. 1); Boyle, *Works*, IV, 225–26 (3, n. 68).
[31] Newton, *Correspondence*, II, 1 (1, n. 52).

idea of mediation and give it a mechanical rationale in particle size. Because the impregnating particles are bigger than those of the mercury, Newton said,

> their grossnes may enable them to give ye parts of ye gold ye greater shock, & so put ym into a brisker motion then smaller particles could do: much after ye manner that ye saline particles wherewith corrosive liquors are impregnated heate many things wch they are put to dissolve, whilst ye finer parts of common water scarce heat anything dissolved therein be ye dissolution never so quick.... I would compare therefore this impregnated ☿ to some corrosive liquor (as Aqua fortis) the ☿ial part of ye one to ye watry or flegmatic part of ye other, & ye metallick particles wth wch ye one is impregnated to ye saline particles wth wch ye other is impregnated, both wch I suppose may be of a middle nature between ye liquor wch they impregnate & ye bodies they dissolve & so enter those bodies more freely & by their grossness shake ye dissolved particles more strongly then a subtiler agent would do.[32]

Newton relied on a kinetic theory of heat in that passage and assumed that the motion of larger particles would produce more heat. Apparently he thought that the momentum of the particle was the crucial factor, since he said the larger ones would give "ye greater shock." But the really intriguing idea he presents is concerned with the "middle nature" of the impregnating particles.

Since he is discussing what he thinks of as solution processes, one is inevitably reminded of the special cases of "solution" which he had mentioned a short time before in his "Hypothesis." In all of those, he had said, some things "unsociable" had been "made Sociable by the Mediation of a Third." Evidently, he now considers the impregnating substance in Boyle's mercury to be a new example of such a "Third." But he has extended his ideas somewhat by applying a common chemical dictum and then translating the whole into corpuscularian terms.

That dictum, a truly venerable idea in chemical circles, was that "Like dissolves like." As a matter of fact, that saying still lives as a chemical rule of thumb, for a solvent is more likely to be effective if its molecular constitution resembles that of the solute. But Newton applied the idea not to the solvent but to the mediating agent, for he says he supposes that the impregnating particles "may be of a middle nature between ye liquor wch they impregnate & ye bodies they dissolve." Thus the impregnating particles being like *both* the solvent and the solute – having a "middle nature" between them – act as the necessary "Third" to increase "sociability." Translating that into corpuscular terminology, he then came up with the idea that their "middle nature" meant that the impregnating particles were of an intermediate size.

[32] *Ibid.*

The resulting conceptualization was not exactly adequate by modern standards. But Newton appears to have thought that it was a satisfactory way of rendering chemical specificity of reaction in generalized mechanical terms. In any case, his concept of "middle natures" of intermediate size was one he held onto, for it appears again in a letter to Boyle in 1678/79, complete with diagrams,[33] and it also appears in *De natura acidorum* in 1692 applied to acids, as will be discussed below.

The creation of the new concept of force

A material aether such as Newton offered in his "Hypothesis" of 1675 carries within it the seeds of its own undoing, and Newton soon realized that he had to discard his. That dramatic story has recently been reconstructed by R. S. Westfall, and his full account is recommended.[34] Although the following brief sketch adheres closely to Westfall's arguments, the full picture of Newton's maturing ideas on dynamics is omitted here as lying too far afield from the alchemy and chemistry which are the central concerns of the present study.

In both the "Hypothesis" of 1675 and the student notebook of 1661–65, Newton tended to attribute gravity to the pressure of a descending aetherial shower. Since all mechanical philosophies required that motions be caused by impact phenomena, the descent of heavy bodies to the surface of the earth had presented something of a problem in them. What was pushing the bodies down? The mechanical philosophers all worked out their individual answers to the question, utilizing invisible mechanisms of one sort or another. "Attraction" and "repulsion" are only appearances, they said, and the phenomena attributed to attraction and repulsions in the common language are really caused by the impacts of innumerable particles of a subtle matter.

Newton had thought in 1675 that the aether – his invisible subtle matter needed for a mechanical explanation of gravity – was indeed material, saying, it will be recalled, that the damping of the swing of the pendulum in a vessel from which the air had been exhausted argued in favor of its existence. But sometime around 1679 his thoughts underwent a complete inversion.

One problem with material aethers acting mechanically was that of infinite regression in the aetherial systems, which may be briefly summarized as follows. If it be particles of aether pushing bodies towards the earth in the action of gravity, then what pushes the particles of aether? Must there then be postulated particles of a yet more subtle aether, *ad*

---

[33] Isaac Newton to Robert Boyle, Cambridge, Dec. 28, 1678/79, in *ibid.*, II, 288–96.
[34] Westfall, *Force*, pp. 323–423 (6, n. 3).

infinitum? Westfall has cogently argued that Newton confronted that problem about 1679 when he attempted to compose a tract which he never published, entitled *De aere et aethere*.[35]

Then there was another problem with the aether which seems to have arisen in Newton's mind about the same time. It was the result of a pendulum experiment more refined than the earlier one which he thought argued for the existence of the aether. In the new experiment, Newton found that his hypothetical aether, which should have offered more resistance to the internal parts of bodies than to the surfaces, actually offered no resistance at all to the internal parts. He was forced to the conclusion then that the aether he had postulated actually did not exist.[36]

With the disappearance of his invisible mechanism, if one may so speak, Newton was thrown back upon the concepts of attraction and repulsion, actions which take place at a distance rather than by impact. From that time on Newton spoke of forces of attraction and repulsion, of active principles, of "Powers," and of "Virtues" by which particles of matter act on each other at a distance. Strictly speaking, his philosophy of nature was no longer orthodox mechanism.[37]

When the *Principia* was published, Newton's opponents were quick to cry out that Newton's forces were occult qualities. Newton claimed they were derived from phenomena and so were not really occult, even though their causes were not yet known, but in a very real sense his critics were right: Newton's forces were very much like the hidden sympathies and antipathies found in much of the occult literature of the Renaissance period.[38]

But Newton had given forces an ontological status equivalent to that of matter and motion. By so doing, and by quantifying the forces, he enabled the mechanical philosophies to rise above the level of imaginary impact mechanisms. Westfall argues that it was thus the *wedding* of the Hermetic tradition with the mechanical philosophy which produced modern science as an offspring.[39]

Westfall has also pointed out that Newton's concept of forces between particles derived initially from terrestrial phenomena, especially chemical ones. Most importantly, it was the concept of attraction, that fundamental tenet of Newton's law of universal gravitation, which was so derived. It was only in 1679–80, Westfall says, when Newton first began to work seriously on the dynamics of orbital motion, that he applied his chemical idea of attraction to the cosmos.[40]

[35] *Ibid.*, pp. 373–75. *De aere et aethere* is now available in Newton, *Unpublished Papers*, pp. 214–28 (3, n. 157).

[36] Westfall, *Force*, pp. 375–77 (6, n. 3).   [37] *Ibid.*, pp. 377–78.   [38] *Ibid.*, pp. 386–91.

[39] *Ibid.*, pp. 377–91; Westfall, "Newton and the Hermetic Tradition" (1, n. 50).

[40] Westfall, "Newton and the Hermetic Tradition," pp. 193–94 (1, n. 50).

If that was indeed the case, then the results of the present study point rather plainly to a specific episode which might well have pre-disposed Newton to think in terms of attractive forces: the alchemical success he reported in the "Clavis" of Keynes MS 18. When his aether failed him, he perhaps remembered that very special mercury which the attractive powers of the "magnet" had given him. He might not have been very clear about just what it was that the "magnet" had drawn in, but he certainly thought it had worked, and he also certainly knew that the alchemical theory behind it claimed that it worked by attractive powers. In any event, it is quite clear that about 1679–80 he came to accept the notion that active forces were operating generally.

The universe lived again as Newton's thoughts swung on towards the *Principia* in the 1680's, for forces and active principles were everywhere. Not only was there the attractive force of gravity binding the planets and the stars into a vibrant whole, there was also activity in the substructure of matter. Gone, to Newton's mind, were the inert particles of Cartesian matter resting quiescently together between impacts. In their place were structured corpuscles of increasing complexity, held together upon occasion by attractive forces of their own, but also capable upon other occasions of repelling each other. Change was the order of the day in the little world and matter matured and decayed and was constantly replenished by active principles.

Activity increased in the laboratory in the front garden at Trinity also, moving towards a crescendo as Newton sought to elucidate those forces acting below the surfaces of everyday matter in time to include them in the *Principia*. Humphrey Newton was there and they took turns sitting up by night to tend the furnace while the composition on celestial matters moved on apace in the daytime. R. S. Westfall, who has done a quantitative assessment of all Newton's chemical papers, including the laboratory notes, has found that Newton produced over twice as many pages with chemical content in the period just before publication of the *Principia* as in the period before 1674.[41] Though any definitive statement on the nature of Newton's experimental program must await an analysis of the "middle" period alchemical manuscripts and laboratory notes, it seems only reasonable to assume that the work he did during those years was the natural outgrowth of his early alchemical experimentation discussed in Chapter 5.

It seems reasonable to assume also that Newton expected during the 1680's to be able to build up a broad system of natural philosophy which would include not only celestial phenomena but terrestrial chemical ones as well. If the concept of alchemical attraction had allowed him to accept the ontological reality of celestial forces through analogy with

[41] Private communication, Feb. 7, 1974.

the alchemical "magnets" he had studied, why should he not then expect that the concept of attractive force would apply to earthly chemical phenomena as well? The *Principia* was probably intended to be a more thoroughly universal system than anyone suspected at the time, but that full comprehensiveness was not quite realized. Chemical matters were first relegated to the "Conclusio" and the "Praefatio" and then finally suppressed altogether except for a few odd bits. But in the "Praefatio" which was published in the first edition, Newton betrayed his intent in an uncharacteristically wistful passage. "The whole burden of philosophy," he said,

> seems to consist in this – from the phenomena of motions to investigate the forces of nature, and then from these forces to demonstrate the other phenomena....[42]

I have done that, he modestly points out, by way of the forces of gravity, for the sun, the planets, the comets, the moon, and the sea. Then, undoubtedly referring to his Herculean alchemical efforts which the public knew not of, he reluctantly omitted the rest.

> I wish we could derive the rest of the phenomena of Nature by the same kind of reasoning from mechanical principles, for I am induced by many reasons to suspect that they may all depend upon certain forces by which the particles of bodies, by some causes hitherto unknown, are either mutually impelled towards one another, and cohere in regular figures, or are repelled and recede from one another. These forces being unknown, philosophers have hitherto attempted the search of Nature in vain; but I hope the principles here laid down will afford some light either to this or some truer method of philosophy.[43]

Newton's mature chemical thought

Once Newton had accepted the notion of forces acting between particles, his thoughts on the sub-structure of matter did not undergo as many fundamental changes as one might think. So although there are undoubted advantages to a strict chronological treatment in most instances, in the case of Newton's mature chemical thought it seems unnecessary for the most part. Therefore the procedure to be followed in the present section to elucidate the ideas Newton held most consistently will be to choose for examination his fullest and most complete statement on them, without regard to chronological order.

[42] Newton, *Principia*, I, xvII–xvIII (1, n. 4); Koyré and Cohen, *Newton's Principia*, I, 16 (1, n. 4).
[43] Newton, *Principia*, I, xvIII (1, n. 4); Koyré and Cohen, *Newton's Principia*, I, 16 (1, n. 4).

The most primitive particles in Newton's mature matter theory were hard and solid, without pores, and indivisible. He may or may not have considered them as all alike in our particular part of the universe, but probably he did. In the most famous and oft-quoted passage from Query 31 of the final version of the *Opticks*, Newton seems to say that they were in fact different in some properties.

> ...[I]t seems probable to me, that God in the Beginning form'd Matter in solid, massy, hard, impenetrable, moveable Particles, of such Sizes and Figures, and with such other Properties, and in such Proportion to Space, as most conduced to the End for which he formed them....[44]

Those words imply differing geometric shapes and even possibly differing densities and consequently different species of matter in the ultimate particles. They might even suggest that Newton here anticipated Dalton's atomic theory except in the final step of identifying the different particle densities with the observably different densities of chemical substances, but that seems to be stretching a point. For Newton did not consider the most primitive particles to engage directly in chemical structure or processes, as will be seen. The primitives were, on the contrary, merely the smallest building blocks of complex particle systems.

A different interpretation of the above passage is possible also when it is read in conjunction with a comment later in Query 31.

> And since Space is divisible *in infinitum*, and Matter is not necessarily in all places, it may be also allow'd that God is able to create Particles of Matter of several Sizes and Figures, and in several Proportions to Space, and perhaps of different Densities and Forces, and thereby to vary the Laws of Nature, and make Worlds of several sorts in several Parts of the Universe.[45]

The implication of the latter passage is of course that as long as one finds the "Laws of Nature" to be invariable in one's own area, the properties of matter are likewise constant and all the primitive particles of matter are alike. This interpretation seems to be more consistent with the whole tenor of Newton's thought, for he frequently spoke of "Nature" as being "consonant" and "conformable to her self." Furthermore, there are many indications elsewhere that Newton thought the matter of all things to be one and the same. And, as McGuire has correctly observed, Newton did not rely on the geometric qualities of primordials to explain the qualities of bodies as Descartes, Gassendi, and Boyle did. Rather he based his explanations on the operation of forces.[46] Newton said in the *Opticks*:

[44] Newton, *Opticks*, p. 400 (1, n. 9).   [45] *Ibid.*, pp. 403–04.

[46] J. E. McGuire, "Transmutation and immutability: Newton's doctrine of physical qualities," *Ambix* **14** (1967), 69–95, esp. p. 84, hereinafter referred to as McGuire, "Transmutation."

It seems to me...that these Particles have not only a *Vis inertiae*, accompanied with such passive Laws of Motion as naturally result from that Force, but also that they are moved by certain active Principles, such as is that of Gravity, and that which causes Fermentation, and the Cohesion of Bodies.[47]

Newton's "passive Laws" were those of impact phenomena and were all Descartes would ever have allowed in his mechanical system. But Newton, having added his "active Principles" to impact mechanism, was no longer bound to the explanatory geometric devices of his predecessors. His "active Principles," which liberated him from the strange hooked atoms and screw-shaped pores and springy textures of the others, were the forces of attraction and repulsion between particles. Those "active Principles" caused the primordials to cohere in larger corpuscles and also caused "Fermentation." What Newton meant by "Fermentation" was suggested in the last chapter. Here the discussion will follow the chain of thought suggested by "Cohesion."

Newton apparently came to view the primitive particles as surrounded by two spheres of forces. At very close range the attractive force predominated, but its sphere of influence reached only to an extremely short distance. When two particles were close enough for the attractive force to control, they were bound together by it. If, however, they were agitated by heat or other means so as to move away from each other, the repulsive force came to control. The repulsive sphere of influence extended much farther and was dominant at any point past the very short range attractive force.

And as in Algebra, where affirmative Quantities vanish and cease, there negative ones begin; so in Mechanicks, where Attraction ceases, there a repulsive Virtue ought to succeed.[48]

When the primordials cohere due to the attractive force or forces, they form various complex arrangements and so form the multitudinous bodies of the sensory world.

Depending on the force and manner of the coming together and cohering of the particles, they form bodies which are hard, soft, fluid, elastic, malleable, dense, rare, volatile, fixt, [capable of] emitting, refracting, reflecting or stopping light.[49]

Their "manner of coming together and cohering" was explored further by Newton in the *Opticks*, where he inserted two major statements on the complicated hierarchical arrangements he attributed to micro-

[47] Newton, *Opticks*, p. 401 (1, n. 9).
[48] Isaac Newton, draft "Praefatio," in Newton, *Unpublished Papers*, pp. 302–04 and 305–07 (3, n. 157); Newton, *Opticks*, Query 31, *passim*, quotation from *Opticks*, p. 395 (1, n. 9).
[49] Isaac Newton, draft "Praefatio," in Newton, *Unpublished Papers*, pp. 303 and 306 (3, n. 157).

matter. In both he may be seen developing speculations that matter in the gross form comprehended by human sensory apparatus is comprised of both parts and pores.

In the one he approaches his formulation from the very smallest particles of the "universal matter" and builds up to the complex structure of "Bodies of a sensible Magnitude." It may be seen that the very short-range attractive forces are at their maximum strength between primordials and decrease in "Virtue" at each successive level of the ladder.

> Now the smallest Particles of Matter may cohere by the strongest Attraction, and may compose bigger Particles of weaker Virtue [or attractive force]; and many of these may cohere and compose bigger Particles whose Virtue is still weaker, and so on for divers Successions, until the Progression end in the biggest Particles on which the Operations in Chymistry, and the Colours of natural Bodies depend, and which by cohering compose Bodies of a sensible Magnitude.[50]

In the other passage, which is more extended, Newton is attempting to explain the experimental fact that "Light is transmitted through pellucid solid Bodies in right Lines to very great distances." Since he held that a ray of light would be "stifled and lost" in the event of hitting a solid particle of matter inside the bodies, he thought that such bodies must be "exceeding porous." Also Newton pointed out that even very dense and opaque bodies allow magnetic and gravitational forces to pass through "without any diminution"; hence they also must have a "quantity of Pores." How bodies might have enough pores to account for these effects and yet in their resulting rarity "not be capable of compression by force," was "difficult to conceive, but perhaps not altogether impossible." He offered an arrangement of the parts which might possibly give bodies "a sufficient quantity of Pores." In this particular statement Newton was mentally analyzing bodies from the largest particles down to the smallest ones, which to his mind were solid and had no pores.

> ...[T]he Colours of Bodies arise from the Magnitudes of the Particles which reflect them.... Now if we conceive these [largest] Particles of Bodies to be so disposed amongst themselves, that the Intervals or empty Spaces between them may be equal in magnitude to them all; and that these Particles may be composed of other Particles much smaller, which have as much empty Space between them as equals all the Magnitudes of these smaller Particles: And that in like manner these smaller Particles are again composed of others much smaller, all which together are equal to all the Pores or empty Spaces between them; and so on perpetually till you come to solid Particles such as

[50] Newton, *Opticks*, p. 394, bracketed material supplied (1, n. 9).

have no Pores or empty Spaces within them [then the gross body might be given any desired quantity of pores]....[51]

Continuing his speculations in that same passage in the *Opticks*, Newton offered figures on the proportion of pores to parts which would arise in his basic scheme when the particles of greatest magnitude had various degrees of internal complexity.

> And if in any gross Body there be, for instance, three such degrees of Particles, the least of which are solid; this Body will have seven times more Pores than solid Parts. But if there be four such degrees of Particles, the least of which are solid, the Body will have fifteen times more Pores than solid Parts. If there be five degrees, the Body will have one and thirty times more Pores than solid Parts. If six degrees, the Body will have sixty and three times more Pores than solid Parts. And so on perpetually.[52]

Newton ended this particular statement in the *Opticks* with the observation that other arrangements might be conceived and that the "inward Frame" of bodies "is not yet known to us."[53] But he seems to have meant that other arrangements or proportions of pores and parts might be envisioned and not to have been seriously questioning the internal complexity of the largest particles.

About 1691 or early in 1691/92 Newton wrote out a short tract in which he made much more explicit the relationship of his hierarchical system to certain chemical processes and substances. The tract was entitled *De natura acidorum* in the Latin versions, in which it seems to have circulated for several years, and "Some Thoughts about the Nature of Acids" in the English version published by John Harris in 1710.[54] Harris, who also included one of the Latin versions, said that what he published had been "supervised and approved by the Illustrious Author." For that reason, and also because Harris' versions were the only ones readily available for two and a half centuries, the Harris versions have been the only ones studied extensively. Most studies of the material have been oriented towards eighteenth-century developments in any case, and in those studies it was more important to focus on what had been published and so could have influenced later developments. For the present study, however, which is focused more on determining Newton's own thoughts

---

[51] *Ibid.*, pp. 266–69, quotation from p. 268, bracketed material supplied.

[52] *Ibid.*, pp. 268–69.

[53] *Ibid.*, p. 269. Although they are not germane to present discussion, the interested reader may find more extended descriptions of Newton's hierarchical system of particles in (1) Thackray, *Atoms and Powers* (2, n. 55); (2) Arnold Thackray, "'Matter in a nut-shell'; Newton's *Opticks* and eighteenth-century chemistry," *Ambix* **15** (1968), 29–53; (3) Vavilov, "Newton and the Atomic Theory" (1, n. 41); and (4) Gregory, "Hierarchic system" (1, n. 41).

[54] Isaac Newton, "Some Thoughts about the Nature of Acids," in Harris, *Lexicon Technicum*, II, sig. b 4r,v (1, n. 56).

on certain subjects, it is important to use the fullest version available. Consequently, a recent publication of *De natura acidorum* in which all known manuscript variants are combined has been utilized for the discussion of it given here.[55]

De natura acidorum makes clear, in a way which the statements in the *Opticks* did not, that Newton had speculated on the internal arrangements of the biggest particles in chemically different types of gross bodies, and also it makes clear that he considered those largest particles to be *chemically* complex. But before considering those finer details, it will be well to establish that he was thinking in terms of hierarchical structure in 1691/92 as he was in 1727.

In *De natura acidorum* Newton set out that by now familiar idea of hierarchies in discussing the specific example of gold.

Gold has particles which are mutually in contact: their sums are to be called sums of the first composition and their sums of sums, of the second composition, and so on [to particles of the last order or composition][56]

Continuing with his discussion of gold, Newton took up the question of how that metal, or any other body, might be reduced into the common universal matter and then transmuted.

Mercury can pass, and so can Aqua Regia, through the pores that lie between particles of last order, but not others. If a menstruum could pass through those others or if parts of gold of the first and second composition could be separated, it would be liquid gold. If gold could ferment, it could be transformed into any other substance.

And so of tin, or any other bodies, as common nourishment is turned into the bodies of animals and vegetables.[57]

Now it is quite apparent from the preceding passage that Newton thought in terms of a universal matter in the smallest particles, those which were "mutually in contact," the same sort of universal matter he had mentioned at the time of the writing of the *Principia* and a concept which he probably originally got from Boyle. Rearrangement of the basic particles of that matter would effect transmutation, but in the matter of rearranging them Newton's approach was much more sophisticated than Boyle's. Boyle, it will be recalled, had only spoken of changes in texture effected by the slow agitation of heat without speaking of the *analysis* which complex bodies might have to undergo to get back to simpler particles which could then be rearranged to give a new substance. But Newton, with his particles constantly increasing in complexity up a hierarchical ladder, recognized that some sort of reduction to simpler forms would be a necessary first step in transmutation. For

[55] Isaac Newton, *De natura acidorum*, in Newton, *Correspondence*, III, 205–14 (1, n. 52).
[56] *Ibid.*, III, 207 and 211, bracketed material supplied.   [57] *Ibid.*

Newton it was not the particles of greatest magnitude which could be transformed. Rather he thought that the biggest particles had to be broken down. Mercury or *aqua regia* applied to gold only broke down the coherence among the largest particles, the coherence which held them together in a massy body, but did not attack the largest particles themselves: those two menstruums only passed through the pores *between* particles of the last order. To get down the scale to simpler particles would require a really powerful analytical agent.

A question may now be raised concerning Newton's conception of the internal structure of the largest particles of different substances, and *De natura acidorum* answers that query rather explicitly. Newton spoke first of *sal alkali* (probably sodium or potassium hydroxide).

Sal alkali he thought to consist of acid and earthy parts and he also thought it probable that the acid particles had surrounded each earthy particle so as to "adhere to them very closely on all sides."[58] Thus the particles of *sal alkali* of the "last order" would be complex both structurally and chemically, each consisting of a single earthy particle entirely covered over with acid ones.

The largest particles of certain "fatty bodies" seemed to Newton to be somewhat similar, but the constituent acid and earthy particles were present in a proportion different from that of the *sal alkali* particles and also were present in a different arrangement. The substances which Newton used for examples of "fatty bodies" were *Mercurius dulcis* (mercurous chloride, Hg_2Cl_2), ordinary sulfur (S), *luna cornea* (silver chloride, AgCl), and "copper which has been corroded by mercury sublimate" (cupric chloride, $CuCl_2$). In those bodies Newton thought the acidic particles to be proportionately less than in *sal alkali* and, because the "fatty" substances did not exhibit the properties he associated with acids, he said that the acid parts "are so closely held" by the earthy ones "that they are, as it were suppressed and hidden by them."[59] In other words, the particles of "last order" in a "fatty body" were conceived as having at least one central particle of an earth which had some acid particles embedded in it deeply enough that the acid could not exert its influence towards the exterior of the large particle.

Unfortunately Newton did not spell out his conception of the internal structure of metallic particles of the "last order" in *De natura acidorum*, but he did append to his discussion of the particles of *sal alkali* and of "fatty bodies" a perfectly general statement which in that context means that he thought that the largest particles of everything consisted of acid and earthy parts. The statement is in its generality applicable to metals and is especially interesting in view of the link it provides with the pre-

[58] *Ibid.*, III, 205 and 209.
[59] *Ibid.*, III, 205–06 and 209.

mechanical theory, derived from Arabic "chymistry," of the sulfur–mercury composition of metals. Here is Newton's statement on the relationship of the older terminology to his conception.

Note that what is said by chemists, that everything is made from sulphur and mercury is true, because by sulphur they mean acid, and by mercury they mean earth.[60]

So from Newton's statements in *De natura acidorum*, it appears that he considered the largest particles of all substances whatever to be compounded of earthy particles and acid particles in arrangements which varied from one substance to another, and also that he identified the older terms "mercury" and "sulphur" with his earthy and acidic particles.

There are also statements in *De natura acidorum* which strongly reflect Newton's acceptance of a kinetic theory of heat and also the restless activity he sensed in micromatter. For example, he said:

Heat is the agitation of particles in every direction.

Nothing is absolutely at rest as regards its particles and therefore nothing is absolutely cold except atoms destitute of vacuum.[61]

Those cold atoms "destitute of vacuum" were his ultimate uncuttables, devoid of pores and parts, his smallest particles which constituted the "universal matter" of all things. Only they lacked internal motion. Between them, as between all more complex corpuscles made of them, forces came into play, and motion, heat, and activity ensued.

In the *Opticks* Newton conveyed his sense of movement *within* complex corpuscles even more graphically, by way of an analogy between the great "Globe of the Earth" and the particles of salt. There, not only do large corpuscles have their own internal motion, the parts of them even have their own individualistic "endeavours."

As Gravity makes the Sea flow round the denser and weightier Parts of the Globe of the Earth, so the Attraction may make the watry Acid flow round the denser and compacter Particles of Earth for composing the Particles of Salt.... Now, as in the great Globe of the Earth and Sea, the densest Bodies by their Gravity sink down in Water, and always endeavour to go towards the Center of the Globe; so in Particles of Salt, the densest Matter may always endeavour to approach the Center of the Particle: So that a Particle of Salt may be compared to a Chaos; being dense, hard, dry, and earthy in the Center; and rare, soft, moist, and watry in the Circumference.[62]

Now in Newton's "biggest Particles," the ones "in which the Progression end" and upon which "the Operations in Chymistry...depend," one may see his integration of alchemy and mechanism reflected in two

---

[60] *Ibid.*, III, 206 and 210. [61] *Ibid.*
[62] Newton, *Opticks*, p. 386 (1, n. 9).

ways. The one is his inclusion of attractive forces, the "active Principles" of the Hermetic tradition which Westfall has also seen incorporated into the attractive force of gravity.

The other is the synthesis of the new corpuscularianism of particles and the old "chymistry" of substances. For just below the surfaces of Newton's "biggest Particles" there exist in all cases an "earth" and an "acid." Albeit the "earth" and "acid" (or "mercury" and "sulphur," if one wishes to use the older terminology) are themselves particulate, nevertheless they are also the bearers of chemical properties as were the old "chymic" principles. And it is with the varying particulate complexities internal to his "biggest Particles," that Newton struggles to explain the various chemical properties of ordinary substances. On that point it is instructive to consider in more detail his speculations on the composition of "fatty bodies" in *De natura acidorum*.

> If the acid particles are joined with the earthy ones in a lesser proportion they are so closely held by the latter that they are, as it were, suppressed and hidden by them. For they no longer excite the organs of sense, nor do they attract water, but they compose bodies which are sweet and which do not readily mix with water; that is, they compose fatty bodies....[63]

In that passage it seems that the "acid" is the substance responsible for various sensory and chemical effects, just as in the "chymic" system, but the diminution of those effects has been given a particulate, structural explication which a "chymist" would never have considered.

Newton's scheme thus allowed the similarities and differences within a class of chemical substances to be explained structurally on the basis of internal arrangements of the "type" particles of "earth" and "acid," an explanation which would have suited a mechanical philosopher. But also it allowed an explanation at a more chemical level since Newton's "earth" and "acid" were the bearers of chemical properties like the old "chymic" principles.

Newton made another elaborate attempt to relate sensory data to the properties of particulate matter, in this case the color of bodies to the size of the largest particles. Apparently rejecting Boyle's concept of special "noble and subtle" tingeing particles, Newton said that it was the "biggest Particles" on which "the Colours of natural Bodies depend," and he devised an ingenious approach based on color to determine the sizes of his "biggest Particles." Factors of atomic and molecular structure beyond his ken render his method worthless by modern standards, but, since it is another place in which his alchemical studies influenced his thinking, it is perhaps worth exploring.

The usual mechanical theories about particle size had some conflicts

[63] Newton, *Correspondence*, III, 205 and 209 (1, n. 52).

built into them which left them unable to deal with real material bodies. For example, dense substances were supposed to be composed of larger particles. An approach to particle size based on that idea would suggest that if all substances were ranked according to density, then they would likewise be ranked according to relative particle size. But on the other hand, volatile substances should have smaller particles according to the mechanical philosophers. How is one to rank mercury then, which is both dense and volatile?

Newton recognized that the density of substances must be inversely related to the total volume of pores. But he also thought that pore size at any given level of composition was independent of density and dependent rather on the corpuscle size at that level.[64] From that recognition it follows that he had no method based on his theory of matter for relating the size of his "biggest Particles" directly to their observable density, as differences in "packing" and size at the different levels of composition, and also different numbers of levels of composition, might obviate the obvious correlation.

The method he worked out must have been suggested to him by his alchemical readings. In the progression of the Great Work of alchemy, matter in its most dissolute form was always black, and a definite sequence of colors arose as the matter was urged towards perfection. As will be seen, Newton, translating this process into particulate terms, thought of the black material as comprised of the smallest particles and of the other colors as belonging to more complex particles of some definite "bigness."

Newton's method was based on his careful measurements and observations upon the "*Reflexions, Refractions, and Colours of thin transparent Bodies*" in Book II of the *Opticks*. "It has been observed by others," he said in introducing these studies,

> that transparent Substances, as Glass, Water, Air, &c. when made very thin by being blown into Bubbles, or otherwise formed into Plates, do exhibit various Colours according to their various thinness, altho' at a greater thickness they appear very clear and colourless. In the former Book I forbore to treat of these Colours, because they seemed of a more difficult Consideration, and were not necessary for establishing the Properties of Light there discoursed of. But because they may conduce to farther Discoveries for compleating the Theory of Light, especially as to the constitution of the parts of natural Bodies, on which their Colours or Transparency depend; I have here set down an account of them.[65]

There follows in Book II, Part I, a series of observations on the colored

[64] Kuhn, "Density and pore-size" (1, n. 41).
[65] Newton, *Opticks*, pp. 193–94 (1, n. 9).

arcs and rings of color or of black and white produced when two convex glasses or two prisms are pressed together and illuminated in various ways, also observations on the colors in soap bubbles. Book II, Part II, contains some elegant geometry erected upon the experimental measurements. Then a table is derived in which "the thickness of Air, Water, and Glass, at which each Colour is most intense and specifick, is expressed in parts of an Inch divided into ten hundred thousand equal parts." The observed rings start with "Very black," followed by colors in definite sequence; these repeat themselves as one proceeds out from the center, but with some variations specific to each set. Newton presents these measurements through seven sets or "Orders" of the colored rings.[66]

In Part III of Book II Newton comes finally to his design.

> For since the parts of these Bodies...do most probably exhibit the same Colours with a Plate of equal thickness, provided they have the same refractive density; and since their parts seem for the most part to have much the same density with Water or Glass, as by many circumstances is obvious to collect; to determine the sizes of those parts, you need only have recourse to the precedent Tables, in which the thickness of Water or Glass exhibiting any Colour is expressed. Thus if it be desired to know the diameter of a Corpuscle, which being of equal density with Glass shall reflect green of the third Order; the Number $16\frac{1}{4}$ shews it to be $\dfrac{16\frac{1}{4}}{10000}$ parts of an Inch.

> The greatest difficulty is here to know of what Order the Colour of any Body is.[67]

There follow various particular observations which help in choosing the "Order" to which the color of a given body belongs. His reasoning on the green of "Vegetables" illustrates his approach.

> There may be good *Greens* of the fourth Order, but the purest are of the third. And of this Order the green of all Vegetables seems to be, partly by reason of the Intenseness of their Colours, and partly because when they wither some of them turn to a greenish yellow, and others to a more perfect yellow or orange, or perhaps to red, passing first through all the aforesaid intermediate Colours. Which Changes seem to be effected by the exhaling of the Moisture which may leave the tinging Corpuscles more dense, and somewhat augmented by the Accretion of the oily and earthy Part of that Moisture. Now the green, without doubt, is of the same Order with those Colours into which it changeth, because the Changes are gradual, and those Colours, though usually not very full, yet are often too full and lively to be of the fourth Order.[68]

[66] *Ibid.*, pp. 232–33.   [67] *Ibid.*, p. 255.   [68] *Ibid.*, p. 256.

If one refers to the table after having reached this decision that vegetable green is of the third order, one may readily see that the green particles themselves are, assuming equal density with water, $18\frac{9}{10}/10\,000$ parts of an inch in diameter, and when they change to yellow or red in the fall, they become larger – up to $21\frac{3}{4}/10\,000$ parts of an inch.

The size of metallic particles is a more complicated question, however, because of their greater density. From various considerations Newton deduces that the colors of gold and copper belong to the second and third orders and that the particles of white metals are of such a size as to reflect the white of the first order. Then he goes on to say:

> The first and only Colour which white Metals take by grinding their Particles smaller, is black, and therefore their white ought to be that which borders upon the black Spot in the Center of the Rings of Colours, that is, the white of the first order. But, if you would hence gather the Bigness of metallick Particles, you must allow for their Density. For were Mercury transparent, its Density is such that the Sine of Incidence upon it (by my Computation) would be to the Sine of its Refraction, as 71 to 20, or 7 to 2. And therefore the Thickness of its Particles, that they may exhibit the same Colours with those of Bubbles of Water, ought to be less than the Thickness of the Skin of those Bubbles in the Proportion of 2 to 7. Whence it's possible, that the Particles of Mercury may be as little as the Particles of some transparent and volatile Fluids, and yet reflect the white of the first order.[69]

Changes in color are definitely related to changes in particle size and vice versa, in Newton's thinking, but there is no direct correlation between size and type of matter, such as volatile or fixed, because of the density factor which must enter into the calculations if the density differs much from those given in the table.

"[F]or the production of *black*," Newton goes on to say, "the Corpuscles must be less than any of those which exhibit Colours." He justified that statement on the basis of his observations and then elaborated a corpuscularian explanation of the black color produced by burning or by the alchemical process of putrefaction: "And from hence may be understood why Fire, and the more subtile dissolver Putrefaction, by dividing the Particles of Substances, turn them to black...."[70]

In these last statements one may see something of the complex interaction of Newton's observations of optical phenomena, his thoughts on the dissolution of matter in the alchemical process, and his mechanical philosophy. He found black to be produced when the thickness of the plate or film was least; putrefaction to his mind produced decomposed, black matter; such matter was particulate to him and because decomposed was in small particles. It seems to be from these three strands of

[69] *Ibid.*, pp. 259–60.   [70] *Ibid.*, p. 260.

thought that he had synthesized his method for determining the sizes of the particles of bodies.

The method depended upon two assumptions, one of which he recognized and one of which he apparently did not. The one he recognized was the questionable validity of considering "That transparent Corpuscles of the same thickness and density with a Plate, do exhibit the same Colour."[71] In that assumption lies the whole problem of transduction (or transdiction, as it is sometimes called), the epistemological problem of making valid inferences from observations in the realm of macro-matter to what is in principle unobservable in the realm of micromatter. Newton was perfectly willing to make the assumption that valid inferences about micromatter could be drawn from observations on larger bodies, because he accepted the idea of continuity in nature, as McGuire has shown.[72] There are no sudden breaks in the properties of matter, Newton thought, for nature operates with slight gradations. Thus by what he called the "analogy of Nature," Newton thought it was quite reasonable to assume that bodies in nature's invisible realm acted in a way similar to those in the visible world.

The other assumption upon which Newton's method of determining corpuscular sizes rested apparently never was consciously analyzed by him. That assumption was drawn directly from the alchemical doctrine that the black matter of putrefaction was in a relatively unformed condition, or, in mechanical terms, that it was composed of matter in particles smaller than those produced later in the alchemical process as the matter "matured" or was shaped into various complex substances. There was really no justification for equating black with the smallest particles, except that unquestioned assumption from alchemy.

Transmutation

This survey of the residue of Newtonian alchemy in his later science may now be concluded with an examination or re-examination of some of his ideas on the transmutation of bodies in the light of his mature chemical thought. Although some of his ideas may have changed, they did not change very much, and to the very end they continued to reflect their origins in the alchemical studies of 1668–75.

As has been seen, Newton believed massy bodies were composed of large complex corpuscles of the "last order," each of some definite "bigness" which he thought he had determined with considerable exactness by way of the "analogy of Nature" from the greater world. Those largest particles were composed of "earthy" particles and "acid" ones at the

[71] *Ibid.*, p. 261.
[72] J. E. McGuire, "Atoms and the 'analogy of nature': Newton's third rule of philosophizing," *Studies in History and Philosophy of Science* **1** (1970), 3–58.

level of composition just one step down the ladder of complexity from the largest ones. Those secondary particles in their turn were composed of pores and parts, and so on down the scale of complexity to the smallest particles which had no pores and were the same for all substances. At every level or "order" forces came into play, forces whose powers grew increasingly strong as one moved towards the ultimate primitives. Ordinary chemical actions took place at the highest level of complexity where the forces were weakest. The largest particles had to be divided into their lesser components before any transmutation could be accomplished, although it was not clear how far down into the hierarchical structure that sub-division needed to reach. Ordinary solution could not effect it. Were there any other processes which might?

One answer Newton had made to that question had been drawn directly from Robert Boyle in 1667–68, as already noted. Newton's conceptualization of that process of "opening" metals may now be re-examined in terms of his mature matter theory, for the *Principia's* unpublished "Conclusio" of 1687 carries a discussion which links those early laboratory notes with his corpuscularian formulations. Newton's discussion in 1687 also offers an explanatory context for his ideas on the division of large particles, and his statements will here be treated sentence by sentence so that their full import may be realized.

> And because all the particles of a compound are greater than the component particles, and larger particles are agitated with greater difficulty, so the particles of sal ammoniac are less volatile than the smaller particles of the spirits of which they are composed.[73]

This first sentence on the complex particles of the "last order" has quite a modern ring, as "sal ammoniac" is ammonium chloride, NH_4Cl, which is made by the direct union of ammonia, NH_3, and hydrogen chloride, HCl. The particles of ammonium chloride undoubtedly are "greater than the component particles" and undoubtedly are "agitated with greater difficulty." But "sal ammoniac" is only an example in this sentence, and Newton's real point is that large particles are hard to agitate, as is more strongly suggested by the next sentence.

> So gold, which is the most fixed of all bodies, seems to consist of compound particles, all which can not be carried up by the agitation of heat on account of their massiveness, and whose component parts cohere with one another too strongly to be separated by that agitation alone.[74]

[73] Isaac Newton, "Conclusio," in Newton, *Unpublished Papers*, pp. 322 and 335 (3, n. 157).

[74] *Ibid.*, pp. 322–23 and 335. The Halls' translation has been modified slightly in the interest of a more accurate description of the physical event: Newton's meaning must have been that *none* of the large compound particles of gold could be carried up by the agitation of heat.

The complex particles of gold are too massive to be volatilized by heat; also they are too tightly bound internally for their component parts to be disjoined by heat. But why was Newton concerned at all with the "agitation" of these large particles? The answer must surely lie in his concept of maturation by heat in which transmutation is brought about by "the slow and continued motion of heat" which causes the particles of bodies gradually to "change their arrangement and coalesce in new ways." Newton is saying in effect that transmutation cannot take place when the large complex particles of the "last order" are agitated by heat because the large particles are so hard to agitate, and in the next sentence his discourse leaps to small particles which are easily volatilized and agitated: "That the particles of water and spirits, however, are most subtle and small of all, and for that reason exceedingly volatile, is consonant with reason."[75]

In the final sentence of the paragraph Newton offers a practical solution to the difficulty of dividing the complex particles of metals and rendering them smaller and more volatile.

And the acid spirit of salts subtilized in a sublimation of mercury divides Antimony and metals more in a new sublimation of the sublimate, and, by dividing, volatilizes [them] more than does the same spirit distilled by itself from the same salts.[76]

Although alone the last sentence makes very little sense to the modern reader, when it is read in conjunction with Newton's belief that only small particles were capable of being transformed into new substances, it seems that he means to divide the particles of antimony ore and metals by the method offered and so put them in readiness for transmutation. First the "acid spirit" is itself to be made more subtle by subliming it with mercury. Then the product is to be sublimed with antimony ore or with metals in order to make the metallic substances more volatile. That method works better, Newton says, because the "acid spirit" which has been treated with mercury is more subtle than the same "acid spirit" obtained by itself, by a simple distillation from the salts it was in. Presumably, Newton means that the process divides the metallic particles into smaller ones since the process of division makes them volatile and he has just associated smallness of particles with the property of volatility in the preceding sentence. He seems to suggest that the division might be done by treating antimony ore or the metals with "acid spirit of salts" alone, but has found that substance to be more effective when used in conjunction with mercury sublimate.

For the process of distilling or subliming the salts with mercury would

---

[75] *Ibid.*, pp. 323 and 335.
[76] *Ibid.*, The Halls' translation has been modified somewhat in an effort to achieve an accurate representation of the chemical events.

probably have yielded mercuric chloride (or mercury sublimate) plus an admixture of various other volatile components. And it is the odd mixture of mercury sublimate and some sort of acid spirit, the whole being used as an agent for the "opening" and volatilizing of antimony ore and metals, which links this entire passage from the "Conclusio" so definitely to Newton's laboratory notes. Those were precisely the reagents he used time and again in his experiments. He evidently intended to break down the complex particles of the "last order" of the antimony ore and of the metals, and so obtain smaller particles which would be fit for transmutation. If one may judge from the passages in *De natura acidorum*, where he talks about earthy particles and acid particles comprising the most complex particles of all things, he expected to obtain particles of "earths" and "acids" by his process. And if one may judge from his identification of "earth" with "mercury" and of "acid" with "sulphur" in 1691/92, he considered the terms "earth" and "mercury" to be interchangeable. Thus his early attempt to obtain a "mercury" of metals takes on a theoretical dimension altogether lacking in the laboratory notes. He had attempted to obtain the "mercury" of metals by volatilizing the metals because the mechanical philosophy had led him to believe that volatile particles of matter were smaller than "fixed" ones and therefore probably less complex. Actually, what he would have obtained in most cases would have been metallic chlorides – more volatile than the metals certainly, but hardly less complex.

One final insight into the theoretical bases of Newton's earliest efforts at transmutation is to be gained from his mature treatment of the concept of mediation. Newton begins *De natura acidorum* with a statement on the intermediate corpuscular size of acids in which he lays the groundwork for demonstrating that acids act as "Mediators."

> The particles of acids are coarser than those of water and therefore less volatile; but they are much finer than those of earth, and therefore much less fixed than they.[77]

No experimental evidence is offered for those designated particle sizes. Newton only says that the acids "are of a middle nature between water and [terrestrial] bodies," which indicates that he has assigned the sizes on the basis of his own corpuscularian formulation of mediation worked out in the 1670's. He has of course added the concept of forces by the 1690's, and the mediating function of the acids is now attributed to their attractive force.

> They are endowed with a great attractive force and in this force their activity consists by which they dissolve bodies and affect and stimulate the organs of the senses. They are of a middle nature between water and [terrestrial] bodies and they attract both.

[77] Newton, *Correspondence*, III, 205 and 209 (1, n. 52).

228

By their attractive force they surround the particles of bodies be they stony or metallic, and they adhere to them very closely on all sides, so that they can scarcely be separated from them by distillation or sublimation. When they are attracted and gathered together on all sides they raise, disjoin and shake the particles of bodies one from another, that is, they dissolve the bodies, and by their force of attraction by which they rush to the [particles of] bodies, they move the fluid and excite heat and shake asunder some particles to such a degree as to turn them into air and generate bubbles: and this is the reason of dissolution and violent fermentation.[78]

Now not only has mediation been set out in terms of forces (in addition to corpuscular size), but the burden of solution in the alchemical process has been shifted to the acid from the philosophical mercury of the early period. Remembering that Newton identified "mercury" with "earth" and "sulphur" with "acid" in this same tract, the following passage makes it seem that his search for the philosophical mercury has been converted into a search for a philosophical sulfur, but it also perhaps indicates something of how he thought his early philosophical mercury had worked in dissolving gold.

But the acid, suppressed in sulphureous bodies, by attracting the particles of other bodies (for example, earthy ones) more strongly than its own, causes a gentle and natural fermentation and promotes it even to the stage of putrefaction in the compound. This putrefaction arises from this, that the acid particles which have for some time kept up the fermentation do at length insinuate themselves into the minutest interstices, even those which lie between the parts of the first composition, and so, uniting closely with those particles, give rise to a new mixture which may not be done away with or changed back into its earlier form.[79]

In that passage Newton was surely considering the "acid" or "sulphur" as the agent capable of working its way down into the pores of the hierarchic particles of ordinary matter, "even those which lie between the parts of the first composition," the ultimate uncuttables. He seems to think he has probably gotten some matter divided that finely by using an acidic agent to "ferment" and "putrefy" the matter, but concedes that the acid itself combined with the putrefied matter so as to "give rise to a new mixture" and did not in fact free the smallest particles for unlimited transmutation, as he may have hoped it would do. In any case, it is the "sulphur" and not the "mercury" of Keynes MS 18 which is the important agent of dissolution in 1692. But remembering that Newton considered the "mercury" of Keynes MS 18 to be "actuated" with some sort of seminal sulfureous particles, it may be that he

[78] *Ibid.*   [79] *Ibid.*, III, 206 and 209–10.

thought the "mercury," which he had then, effected the dissolution of gold with the same mechanism he suggested for the action of acids in 1692.

De natura acidorum seems to bear evidence that in 1692 Newton no longer considered the philosophical mercury of Keynes MS 18 as capable of dissolving gold, for it will be recalled that it was in *De natura acidorum* that Newton spoke of mercury and *aqua regia* as only passing through the pores "that lie between particles of last order" in gold and seemed doubtful that any other menstruum might do better.

That is probably a safe interpretation, but one cannot surely say that that was what he meant and that he had changed his mind about the effectiveness of the process he described in Keynes MS 18. For he almost certainly never had any intention of making his successful alchemical processes public until he was sure what all was involved in the Great Work, and the changing details of his thoughts on transmutation can never be reconstructed without a thorough analysis of the "middle" and "late" alchemical manuscripts and laboratory notes.

Conclusion

It may safely be said nevertheless that Newton's alchemical thoughts were so securely established on their basic foundations that he never came to deny their general validity, and in a sense the whole of his career after 1675 may be seen as one long attempt to integrate alchemy and the mechanical philosophy.

It has been seen in this study that although he probably began his work on transmutation with Boylian mechanical concepts in mind, he soon plunged into alchemical studies that did not have a built-in mechanical rationale to guide his thoughts. From those alchemical studies came his first recorded success, a "mercury" which he believed to be a true "philosophick" one. There were three crucial alchemical concepts behind that particular experimental program which resulted in the "Clavis" of Keynes MS 18. One was the Neoplatonic "universal spirit" or spiritous "aire," the source of all specific forms of matter. Another was the concept of the "Mediator," the "middle nature" required for the joining of disparate substances. The third was the concept of active principles, exemplified in the process by the "attraction" of the antimonial "magnet" and by the "fermental virtue" drawn in by the "magnet."

Newton managed very soon to restate all three of those concepts in his own way. At first he conceived the Neoplatonic "universal spirit" as a particulate mechanical system and the "Mediators" as particles of intermediate corpuscular size. The original concept of the aether failing to stand further experimental testing, Newton abandoned it for a new concept of force which in fact incorporated the alchemical active principles into mechanism in a radically new way.

But he did not abandon, with the mechanical aether, his idea of 1675 that "the whole frame of Nature" might "be nothing but aether condensed by a fermental principle." Newton never really changed his conception that all matter was generated by fermentation and condensation from some common material. During the *Principia* period around 1687 he substituted vapors from the sun, stars, and comets for the aetherial substance, suggesting, as has been noted, that such vapors might be condensed into "water and humid spirits" first, then "by continued fermentation" into all the more dense substances. Those passages were never removed from later editions of the *Principia*, but the concept was given a new formulation in the *Opticks*, in which the "vapors" seem to have become "Light." However, since Newton held that light was corpuscular in nature, and since those celestial bodies do emit light, that change is not really as great as it appears at first. Newton's final formulation in the *Opticks* follows.

> Are not gross Bodies and Light convertible into one another, and may not Bodies receive much of their Activity from the Particles of Light which enter their Composition? For all fix'd Bodies being heated emit Light so long as they continue sufficiently hot, and Light mutually stops in Bodies as often as its Rays strike upon their Parts....
> I know no Body less apt to shine than Water; and yet Water by frequent Distillations changes into fix'd Earth, as Mr. *Boyle* has try'd; and then this Earth being enabled to endure a sufficient Heat, shines by Heat like other Bodies.
> The changing of Bodies into Light, and Light into Bodies, is very conformable to the Course of Nature, which seems delighted with Transmutations.[80]

In this context, the very mention of Boyle's supposed conversion of water into earth, studied by Newton sixty years earlier, is evidence that Newton still thought just before his death in terms of "one Catholick Matter" and its transmutations. Only the details have changed in all that time.

Similar evidence appears in the fate of Hypothesis III of the first edition of the *Principia*. In the first edition, Book III of the *Principia*, which is entitled *De systemate mundi*, *On the System of the World*, was headed by nine "hypotheses." In later editions those "hypotheses" were changed. The last six of them were distributed elsewhere as "Phaenomena." The first two, which were methodological in nature, were kept as "Regulae philosophandi," "Rules of Reasoning in Philosophy," and other meth-

---

[80] Newton, *Opticks*, p. 374 (1, n. 9).

odological, logical, and epistemological items were added to them. The third of the original "hypotheses" actually disappeared, according to Koyré, who was the first to study them in detail.[81]

That third "hypothesis" had affirmed the unity of matter and the total possibility of transformation or transmutation. In Koyré's translation, it reads as follows. "Any body can be transformed into another, of whatever kind, and all the intermediate degrees of qualities can be induced in it."[82]

Koyré pointed out the heterogeneous and illogical nature of the original list, supposing it to be explained by the haste with which the *Principia* was composed. Although there may be a great deal of truth in Koyré's explanation, it is hard to see why such a "chymical" hypothesis would have been there at all had not Newton at one time intended chemistry to form an integral part of the whole work. Newton had meant to demonstrate the two or three forces controlling cohesion, fermentation, and repulsion, and from the forces to derive the "Properties and Actions of all corporeal Things" and their ease of convertibility, one to another. But since that planned program was never realized, Hypothesis III does look a bit out of place in the *Principia*, and its "disappearance" as Newton revised the work for the second edition might be taken to mean that he had abandoned not only the attempted elucidation of chemical forces but also the concept of the unity of matter. Koyré suggested in fact that Newton had suppressed the hypothesis because he came to believe in the non-transformability of one kind of matter into another.[83]

In a more recent and fuller study of the matter, however, McGuire has reached the opposite conclusion. Newton's manuscripts in which he revised Hypothesis III and eventually converted it into Rule III of the second and third editions show no such thing, McGuire says. Rather they indicate that Newton had clarified his thinking on the distinction between the primary and secondary qualities of bodies, finding certain primary ones to be immutable and essential to matter and not subject to change in degree. But the secondary qualities of bodies suffered no such restriction in his view, and, if anything, Newton came to see transmutation as even more widespread than he had thought before. McGuire suggests that had Newton formulated a new statement on transmutation for inclusion in the *Principia* in his later years, he might even have characterized it as an "Axiom" rather than as an "Hypothesis" because he thought by then that transmutation could be deduced from phenomena.[84]

[81] Alexandre Koyré, "Newton's 'Regulae Philosophandi'" in Alexandre Koyré, *Newtonian Studies* (Cambridge, Mass.: Harvard University Press, 1965), pp. 261–72.
[82] *Ibid.*, p. 263.   [83] *Ibid.*
[84] McGuire, "Transmutation" (6, n. 46).

Epilogue

Newton concerned himself with the transmutations of the forms of the universal matter for almost all of his long and fruitful life, and he tried to use the techniques of alchemy to probe the internal structure of its particles, to find those "certain forces by which the particles of bodies... are either mutually impelled towards one another... or are repelled and recede from one another," for that, he said, was the "burden of philosophy." And if his efforts met with something less than the success he had once hoped for, still he did succeed in modifying Descartes in just the right direction.

Chemistry had to go rather a long way around in order to do it, but it has achieved in the present century the total integration of alchemy and mechanism that Newton hoped for, with quite real metallic transmutations, a return to the position of "one Catholick Matter," and some considerable elucidation of the forces acting in its substructure. So perhaps Newton's attempted synthesis should be compared to the green lion he sought. The lion was there all right; it was after all no chimera. But it wanted a maturity which the seventeenth century could not provide.

> All haile to the noble Companie
> Of true Students in holy *Alchimie*,
> Whose noble practise doth hem teach
> To vaile their secrets wyth mistie speach;
> Mought yt please your worshipfulnes
> To heare my fitly soothfastnes,
> Of that practise which I have seene,
> In hunting of the *Lyon Greene*:
>
> . . . . . .
>
> Whose collour doubtles ys not soe,
> And that your wisdomes well doe know;
> For no man lives that ever hath seene
> Upon foure feete a *Lyon* colloured *greene*:
> But our *Lyon* wanting maturity,
> Is called *greene* for unripeness trust me,
> And yet full quickly can he run,
> And soone can overtake the Sun:
>
> . . . . . .
>
> He bringeth hym to more perfection,
> Than ever he had by Natures direccion.[85]

[85] *The Hunting of the Greene Lyon, Written by the Viccar of Malden* in Ashmole, *TCB*, pp. 278–79 (2, n. 6); Keynes MS 20, f. 1r,v (1, n. 1).

The Alchemical Papers of Sir Isaac Newton

Most of Newton's alchemical manuscripts that are available for study are in the Keynes Collection, King's College, Cambridge, as noted in the text. However, the Keynes Collection has been microfilmed in order better to preserve the manuscripts themselves and also in order to make the material more readily available to scholars. The microfilms have been deposited at University Library, Cambridge, and the numbers following each of the Keynes MS numbers in this appendix are the University Library call numbers for the microfilms.

The following descriptions of the manuscripts were drawn up at the time of their sale by Sotheby and Company in 1936. Although subsequent research has indicated that some of the Sotheby descriptions need to be modified somewhat, the Sotheby Catalogue is, and probably will continue to be, the standard scholarly reference work for Newton's alchemical papers.

Sotheby Lot 1. Keynes MS 12. U.L.C. Microfilm 660.	Alchemical Propositions [in Latin] about 400 words, 3 pp.; Miscellaneous Notes on Multiplication by Solution and Coagulation, etc. [in Latin], about 2000 words, 7 pp. All autograph. Sm. 4to.
Sotheby Lot 2. Keynes MS 13. U.L.C. Microfilm 660.	Alchemical Writers. A long classified list of writers on alchemy, on 3 pp. folio; on p. 4: "An account of Gold and Silver moneys coyned since Christmas" (draft); Another List of Writers, signed at foot "Jeova sanctus unus" [an anagram used by Newton] 2 pp. folio; Another list of "Authores optimi," 1 p. folio, on reverse, an Extract from "An Act for encouraging Coynage." All autograph. Folio.
Sotheby Lot 3. Babson MS 418.	Alchemical Writers. A list of 113 Writers on Alchemy, arranged under nationalities, 2 pp.; List of Books, with shelf marks (perhaps in Trinity Library), 2 pp. Both autograph. Sm. 4to.
Sotheby Lot 4.	Alchemical Writers. "Of Chemical Authors and their Writings," [an Alphabetical List of about 120 writers on Alchemy, with comments on the works of the more important], 7 pp. Autograph. Sm. 4to.
Sotheby Lot 5.	Alchemical Writers. List of Authors and Books on Alchemy, including a list of 27 items headed; "Desiderantur opera Lullii sequentia," on 8 pp. 4to and folio, enclosed in a folded sheet containing geometrical diagrams. Autograph.
Sotheby Lot 6. Stanford University, Frederick E. Brasch Collection of Sir Isaac Newton and the History of Scientific Thought.	Alchemical Writers. "De Scriptoribus Chemicis" [a Bibliography containing titles and particulars of over 80 Printed Books, and several MSS.], 5 pp. Autograph. Sm. 4to.

Sotheby Lot 7. Keynes MS 14. U.L.C. Micro- film 660.	Artephius his secret Book, [in English] about 9000 words, 20 pp., followed by "The Epistle of John Pontanus, wherein he beareth witness of ye book of Artephius," 2¼ pp. Both autograph. Sm. 4to.
Sotheby Lot 8. Keynes MS 15. U.L.C. Micro- film 660.	Bloomfield. "Out of Bloomfield's Blossoms," in verse, beginning: "Father Time set me at ye gate," 212 lines, followed by "A short work that beareth the name of Sr. George Ripley," beginning: "Take heavy, soft, cold, and dry" in verse, 92 lines, and "Fragments," 19 lines, 8 pp. All autograph. Sm. 4to.
Sotheby Lot 9. Keynes MS 16. U.L.C. Micro- film 660.	"Causae et initia naturalium" [Notes on an un-named book] about 3000 words, 7 pp. Autograph. Sm. 4to.
Sotheby Lot 10. Keynes MS 17. U.L.C. Micro- film 660.	Clavis Aureae Portae [Notes and Abstracts]; Notes on "Medulla Alchemiae"; "Pupilla Alchemiae," etc. about 4000 words, 8 pp. Autograph. Sm. 4to.
Sotheby Lot 11. Keynes MS 18. U.L.C. Micro- film 660.	"Clavis" [Directions for a lengthy operation beginning with the digestion of Antimony, Iron and Sulphur] in Latin, about 1200 words, 3 pp. Autograph. Sm. 4to.
Sotheby Lot 12. Keynes MS 19. U.L.C. Micro- film 660.	Collectiones ex Novo Lumine Chymico quae ad Praxin spectant [and] Collectionum Explicationes, written in parallel columns, over 3000 words, 8 pp. Autograph. Sm 4to.
Sotheby Lot 13. Keynes MS 20. U.L.C. Micro- film 660.	Copper. "The Hunting of ye Green Lyon," in verse [English] about 180 lines; The Standing of ye Glass for ye time of Putrefaction & Congela- tion of ye medicine, in verse [English], 14 lines, 6 pp.; Notes upon ye hunting of ye green Lyon [in Latin], about 500 words, 2 pp. All auto- graph. Sm. 4to.
Sotheby Lot 14.	Copper. "The Hunting of ye Green Lyon," [in English], about 800 words, 2 pp.; Miscellaneous Notes [in Latin] about 400 words, 1 p. Both autograph. Folio.
Sotheby Lot 15.	Dickenson (Edmund) Ex Epist. Edmundi Dickenson and Theodorum Mundanum. Dat Londini prid. Cal. Aug. 1683 edit 1686, about 2500 words, 8 pp. Autograph. Sm. 4to.
Sotheby Lot 16. Bodleian Lib- rary, Oxford, MS Don. b. 15.	Dictionary of Terms, Materials & Apparatus used in Alchemy, with Directions for performing various Operations, etc., about 7000 words, 16 pp. Autograph. Sm. 4to.
Sotheby Lot 17. Keynes MS 21. U.L.C. Micro- film 660.	[Didier's "Six Keys"; a Commentary, mostly in English] about 8500 words, 35 pp. Autograph. Sm. 4to.
Sotheby Lot 18.	Earths. Notes on "Terra Lemnia & Terra Sigillata...sold in but one shop in London...for 2s. 6d. per pound," and its distillation products, about 230 words, with note at foot: "Mr. Leibnitz is Counsellor to ye Elector of Brunswick. Mr. Fatio has no correspondence with him. This

direction will find him out. Cl. viro. D. G. G. Leibnitio Hannoverae," 1 p. 8vo; "De Peste" [notes from Van Helmont] about 650 words, 2 pp., sm. 4to; Notes on "Roth Mallor's work," about 250 words, 1 p. folio. All autograph.

Sotheby Lot 19.
Keynes MS 22.
U.L.C. Micro-film 661.

Edwardus Generosus. "The Epitome of the Treasure of Health written by Edwardus Generosus Anglicus innominatus who lived Anno Domini 1562" [a Treatise on the Philosopher's Stone; the Animal or Angelical Stone; the Perspective Stone, or the Magical Stone of Moses, etc. with an Alchemic Poem at the end, all in English] about 14 000 words, 28 pp. Autograph. Sm. 4to.

Sotheby Lot 20.
Keynes MS 23.
U.L.C. Micro-film 661.

Epistola ad veros Hermetis discipulos continens claves seu principales Philosophiae secretae, about 4750 words, 19 pp. Autograph. Sm. 4to.

Sotheby Lot 21.
Keynes MS 24.
U.L.C. Micro-film 661.

Epistola. "Anno 1656. Serenissimi Principis Frederici Ducis Holsatiae et Sleswici, etc. communicatione sequens epistola me sibi vendicat, inaudila memorans. Veni et vide," about 1300 words, 7 pp. Autograph. Sm. 4to.

Sotheby Lot 22.

Faber. Extracts from the Works of Faber & other Alchemical Writers [in Latin and English] about 7000 words, 2 pp. 4to and 20 pp. folio. Autograph.

Sotheby Lot 23.

"Fabri Hydrographo Spagyrico" [Notes and Abstracts], about 1500 words, 4 pp. Autograph. Folio.

Sotheby Lot 24.

Fermentation. Notes on Fermentation, etc. [in Latin] about 800 words, 3 pp., the first page containing the beginning of the draft of a letter; Miscellaneous Notes [in Latin] about 1200 words, 4 pp.; Miscellaneous Notes, headed "Opus primum...quintum," about 1200 words, the first page containing the beginning of the draft of a letter to "Mr. Proctor, an Attorney," about succeeding his father in the management of Newton's "concerns." All autograph. Sm. 4to.

Sotheby Lot 25.

Flammel (N.) The Book of Nicholas Flamel conteining the explication of the Hieroglyphical Figures wch he caused to be put in the Church of the SS. Innocents at Paris, about 15 000 words, 61 pp. Autograph transcript. Sm. 4to.

Sotheby Lot 26.
Keynes MS 25.
U.L.C. Micro-film 661.

Flammel (N.) Notes out of Flammel [concerning "The first Angel painted in Abraham ye Jew's Hieroglyphics"]; Notes "ex libro de Metallorum Metamorphosi"; "ex Epistola Com Trevisani ad Thom. Bonon."; etc. about 3000 words, 7 pp. Autograph. Sm. 4to.

Sotheby Lot 27.

Furnaces. [Notes on the Making of Portable Furnaces] about 300 words on 2 pp. sm. 4to, with rough sketches, on 1 p. folio. Autograph.

Sotheby Lot 28, part.
Babson MS 725.

Gold. "Experimts. of refining Gold wth Antimony made by Dr. Jonathan Goddard," about 1500 words, 4 pp.; Draft of a Note asking for "of Antimony about sixty pounds...of Copper oare 12 or 16 pounds," and other metallic ores, about 200 words, 1 p.; Notes on the amalgamation of Tin & Mercury; Preparation of Stannic Chloride, etc., about 700 words, 3 pp. All autograph.

Sotheby Lot 28, part.
Keynes MS 91.
U.L.C. Micro-film 661.

Experiment the 4th. Being the coralery of all the former, containing a true Process of the whole Worke (ff. 3–6b). On f. 1a part of another work beginning "The Pondus in Dissecting the Subject Matter..." In all 8 pp. Sm. 4to. Not autograph except for 3 lines added by Newton at the foot of f. 1a.

237

Sotheby Lot 29.	"Hercula Piochymico" [by Faber. Notes & Abstracts], about 1300 words, 4 pp. Autograph. Folio.
Sotheby Lot 30. Keynes MS 27. U.L.C. Micro-film 661.	[Hermes Trismegistus] "The Seven Chapters" [a Treatise on Transmutation] about 4750 words, 19 pp. Autograph, with many alterations and re-writings. Sm 4to. At the end is the Autograph Draft of a Letter (unsigned) regarding the use of the circle in constructing equations with 3 real roots, 9 lines.
Sotheby Lot 31. Keynes MS 28. U.L.C. Micro-film 661.	Hermes Trismegistus. Tabula Smaragdina [et Commentarium] and a Translation into English of the Tabula, 3½ pp. Autograph. Sm. 4to.
Sotheby Lot 32. Keynes MS 29. U.L.C. Micro-film 661.	Hermes Trismegistus. Notes on Hermes Trismegistus [partly in English]; Notes on Flammel and other authors, about 5000 words, 9 pp. Autograph. Sm. 4to.
Sotheby Lot 33. Keynes MS 30. U.L.C. Micro-film 661.	Index Chemicus [an elaborate Subject-index to the literature of Alchemy, giving page references to a very great number of different works] over 20 000 words, 113 pp.; Three other similar Indexes, over 5000 words, 49 pp. All autograph, Sm. 4to.
Sotheby Lot 34.	Index Chemicus ["Ablutio" to "Aqua Foetida" only] about 2000 words, 5 pp. Autograph. Folio.
Sotheby Lot 35. Jewish National and University Library, MS Var. 259.	[Jodochus a Rhe] "Le Procede Universelle pour faire la Pierre Philosophiale laquelle l'auteur dit davoir faict quatre fois" [in French] about 1500 words, 6 pp. Autograph. Sm. 4to.
Sotheby Lot 35, cont. MS Var. 259, cont.	Artephius. "De Arte Occulta et lap. Philos. Liber secretus," about 3500 words, 8 pp. Autograph. Sm. 4to.
Sotheby Lot 35, cont. MS Var. 259, cont.	Flammel (N.) "The Hieroglyphicall figures of Nicholas Flammel explained," anno 1399, about 3000 words, with elaborate pen-and-ink drawing of the figures, 7 pp. Autograph. Sm. 4to.
Sotheby Lot 35, cont. MS Var. 259, cont.	Novum Lumen Chymicum [Notes and Abstracts] about 4000 words, 8 pp. Autograph. Sm. 4to.
Sotheby Lot 35, cont. MS Var. 259, cont.	Spagnetus. "Enchiridion Physicae Joh. Spagneti," about 3500 words, 8 pp. Autograph. Sm. 4to.
Sotheby Lot 35, cont. MS Var. 259, cont.	Norton's Ordinall [Notes and Abstracts]; Notes on Ripley's Twelve Gates, about 4000 words, 8 pp. Autograph. Sm. 4to.
Sotheby Lot 35, cont. MS Var. 259, cont.	"Ex Augurelli Chrysopoeia" [Notes and Abstracts], 2 pp., followed by Notes on "The Marrow of Alchymy" [in English] 6 pp., in all about 3000 words. Autograph. Sm. 4to.

Sotheby Lot 35, cont. MS Var. 259, cont.	Notes on Ripley's Third Tract; on Tabula Smaragdina; on the Tincture, etc., about 1800 words, 5 pp. Autograph. Sm. 4to.
Sotheby Lot 35, cont. MS Var. 259, cont.	"Observanda [Notes and Abstracts relating to "Instructio de Arbori solari"; "Lucerna Salis Philosophorum," etc.; Newton has crossed out one paragraph and written in the margin: "Credo hic nihil adeptus"], about 4500 words, 12 pp. Autograph. Sm. 4to.
Sotheby Lot 35, cont. MS Var. 259, cont.	Snyder (John de Monte) "Commentatio de Pharmaco Catholico... donata per Authorem Chymicae Vannus" [Notes and Abstracts] about 5000 words, 13 pp. Autograph. Sm. 4to.
Sotheby Lot 35, cont. MS Var. 259, cont.	Valentine (B.) "B. Valentine's Process described in his 12 Keys and other writings," a fine set of notes [in English] about 5000 words, 13 pp.; Miscellaneous Notes [in Latin], 8 pp. All autograph. Sm. 4to.

These eleven items of Sotheby Lot 35 were enclosed in a wrapper bearing a list of contents in Newton's hand.

Sotheby Lot 36.	Lead. Notes on Saturn [Lead], the mining, preparation, and properties of the metal, etc. [mostly in English, with verse extracts from "Bloomfeild's Blossoms," etc.] about 3500 words, 8 pp. Autograph. Sm. 4to.
Sotheby Lot 37. Keynes MS 31. U.L.C. Microfilm 661.	Liber Mercuriorum Corporum [in English] with table of Alchemical Symbols, about 3000 words, 8 pp. Autograph. Sm. 4to.
Sotheby Lot 38. Stanford University, Frederick E. Brasch Collection of Sir Isaac Newton and the History of Scientific Thought	Lucatello's Balsam: "To make Lucatello's Balsome" [a Recipe], 1 p. Autograph. Sm. 4to.
Sotheby Lot 39.	Lully (R.) Ex Codicillo R. Lullii, impress. Coloniae 1563 [Notes and Abstracts], about 3000 words, 6 pp. Autograph. Sm. 4to.
Sotheby Lot 40. Jewish National and University Library, Yahuda MS Var. 1, Newton MS 30.	Lumina de Tenebris. "Out of La Lumiere sortant des Tenebres" [Abstracts, in English], about 1700 words, 4 pp. Autograph. Folio.
Sotheby Lot 41.	Lumina de tenebris [Notes and Abstracts, in English], about 2500 words, 5 pp. Autograph. Folio.
Sotheby Lot 42.	Maier (M.) Emblemata Michaelis Maieri Comitis Germani [Notes and Abstracts]; Notes on the "Speculum Alchymiae" of Arnoldus, etc., about 5000 words, 8 pp. Autograph. Sm. 4to.

Sotheby Lot 43. Keynes MS 32. U.L.C. Micro- film 661.	Maier (M.) Symbola aurea mensae duodecim nationum, Authore Michaele Majero. Dat. Francofurti mense Decemb. 1616 Edit Francofurti 1617. Anno aetatis Majeri 49, 24 pp.; Atalanta Fugiens, hoc est Emblemata nova, etc. Dat Francofurti mens. Aug. 1617. Edit 1618, 24 pp.; Viatorium, hoc est De Montibus Planetarum septem, Datum Francofurti ad Moen. 1618, mense Septembri. Edit Rothomagi 1651, 16 pp.; Septimana Philisophica [Days 1–6 only] Dat. Magdeburgi Anno 1620, Jan. 11, styl. vet., 24 pp., totalling about 50 000 words on 88 pp. Autograph transcripts. Sm. 4to.
Sotheby Lot 44. Keynes MS 33. U.L.C. Micro- film 661.	"Manna" [a Tract on Alchemy] not in Newton's hand, but containing at the end Autograph Additions and Notes of about 1500 words, 11 pp. Folio.
Sotheby Lot 45. Keynes MS 26. U.L.C. Micro- film 661.	Memorandum made by Newton, begins: "On Munday March 2d or Tuesday March 3 1695/6, A Londoner acquainted wth Mr. Boyle and Dr. Dickinson making me a visit, affirmed that the work of Jodochus a Rhe wth [vitriol] twas not necessary that the [vitriol] should be purified but the oyle or spirit might be taken as sold in shops." Followed by an interesting account of an alchemical operation, about 700 words, 3 pp. Autograph. Sm. 4to.
Sotheby Lot 46.	Memorandum made by Newton: another version, with differences, of Lot 45, 1 p. folio; on reverse, the draft of a short letter: "...Last autumn when I was in London Mr. Pepys asking me about the possibilities of finding ye Longitude at Sea and desiring my leave that he might say that I thought it possible, I desired him not to mention me about it least it might be a means to engage me in it and reflect upon me if I did not compass it." Autograph. Folio.
Sotheby Lot 47.	Mercury. "De Virga [Mercurii]" Notes on 1 p., with Five Queries: "What Lute for distilling O. Vitr."; "Whether ye Spt. in ye first digestion stink, & how soon, & with what odor"; "How he contrives his Lamp," etc.; other Notes, fragmentary, on 11 pp. All autograph. Sm. 4to.
Sotheby Lot 48. Keynes MS 34. U.L.C. Micro- film 661.	Mercury [Notes on the preparation of Philosophical Mercury by fermentation and "ye mediation of Diana's Doves"; Preparation of Menstrues, etc., in English], about 2500 words, 6 pp. Autograph. Sm. 4to.
Sotheby Lot 49.	Mercury. [Notes on Mercury and its Purification], about 350 words, 1 p. Autograph. Sm. 4to.
Sotheby Lot 50. Keynes MS 35. U.L.C. Micro- film 661.	Metals. "Quomodo metalla generantur," etc., about 1200 words, 3 pp. Autograph. Sm. 4to.
Sotheby Lot 50, cont. Keynes MS 35, cont.	Minerals. "De radice semine spermate et corpore mineralium," about 1200 words, 3 pp. Autograph. Sm. 4to.
Sotheby Lot 50, cont. Keynes MS 35, cont.	Minerals. "De Mineralibus ex quibus desumitur," about 3500 words, 11 pp. (some edges burnt, affecting text). Autograph. Sm. 4to.

Sotheby Lot 50, cont. Keynes MS 35, cont.	Mercury. [Notes on Mercury, Lead, Tin, Sulphur & Iron, partly in English], about 4500 words, 14 pp. Autograph. Sm. 4to.
Sotheby Lot 50, cont. Keynes MS 35, cont.	Mercury. "De Mercurio duplato, ex Turba," about 1600 words, 5 pp. Autograph. Sm. 4to., etc.
Sotheby Lot 50, cont. Keynes MS 35, cont.	Conjunction. "De Conjunctione in hora nativitatis" [partly in English], about 2500 words, 4 pp., edges slightly burnt, affecting a few words. Autograph. Sm. 4to.

These six items of Sotheby Lot 50 were enclosed in a wrapper bearing a list of contents in Newton's hand.

Sotheby Lot 51. Keynes MS 36. U.L.C. Microfilm 661.	Metals. De Metallorum Metamorphosi [3 Tracts: Prefatio; Brevis Manuductio ad rubinum coelestem; Fons Chemicae Philosophiae], over 4000 words, 8 pp. Autograph. Sm. 4to.
Sotheby Lot 52.	Metals. "De Mineralibus" [Notes on Geber, Basil Valentine, etc.] about 800 words, 3 pp. Autograph. Sm. 4to.
Sotheby Lot 53. Trinity College, Cambridge, MS R. 16. 38, ff. 439–440A, excluding underlined section.	Method how to use the tincture of sol [in English], about 600 words, 3 pp.; A medecine to clear the eye sight: To Extract the tincture of Coral; To extract fixed salt of Amber; about 600 words, 2 pp.; all autograph; other Recipes (not in Newton's hand). Folio.
Sotheby Lot 54.	Mynsicht (Hadrian) Aureum saeculum redivivum [and] Testamentum Hadrianeum de aureo Philosophorem lapide, diagram, about 300 lines [verse], 8 pp. Autograph. Sm. 4to.
Sotheby Lot 55. Keynes MS 37. U.L.C. Microfilm 661.	Norton. "Out of Norton's Ordinal," in verse, beginning: "Good Mr. Tensile, then teach me truly," 200 lines, followed by "Out of Chaucer's Tale of ye Chanon's Yeoman," beginning: "Tell me ye name of ye privy stone," 12 lines; and a Poem by Richard Carpenter, 14 lines, 7 pp. All autograph. Sm. 4to.
Sotheby Lot 56.	"Notanda Chymica" [Notes, mostly on Maier, ending; "Populi Americani in Peru mollificare norant ut instar cerae digitis tractetur"], about 1500 words, 4 pp. Autograph. Sm. 4to.
Sotheby Lot 57.	"Notanda Chemica" [Notes on Philalethes, Ripley, Maier, etc.], about 1500 words, 4 pp. Autograph. Sm. 4to.
Sotheby Lot 58. Keynes MS 38. U.L.C. Microfilm 661.	Note-Book: Containing "Notanda Chemica," on 3 pp. and "Sententiae notabiles" [in English and Latin] on 22 pp., about 7500 words, plus 76 blank ll., marbled paper wrappers. Autograph. Sm. 4to.
Sotheby Lot 59. Jewish National and University Library, MS Var. 260.	Note-Book: Arranged for entries under "De Sale"; "Solutio"; "Conjunctio et Liquefactio," etc., but containing little else than headings and analytical sub-headings. 15 ll., unstitched. Autograph. Sm. 4to.

Sotheby Lot 60.	Notes "ex Turba Philosophorum," over 2000 words, 6 pp. Autograph. Sm. 4to.
Sotheby Lot 61.	Nova Lumina Chemica. Loca difficilia in Novo Lumine Chymico explicata, etc., 2 pp. Autograph. Sm. 4to.
Sotheby Lot 62. Keynes MS 39. U.L.C. Microfilm 661.	Observations of ye matter in ye Glass. Authore Anonymo [in English], about 2500 words, 6 pp. Autograph. Sm. 4to.
Sotheby Lot 63. Babson MS 417, excluding underlined sections.	Operations. An Important Collection of Papers [in Latin] headed: "Operationum Ordo"; "Materiae Mineralis praeparatio primo et conversio in aquam"; "Extractio et Rectificatio Spiritus"; "Extractio et Rectificatio Animae et calcinatio Corporis"; "Reductio"; "Reductio et Sublimatio"; "Elementorum Qualitates"; "Separatio Elementorum," etc., about 20 000 words, 63 pp., some pp. containing numerous marginal references, and many pp. much corrected. Autograph. Folio.
Sotheby Lot 64. Keynes MS 40. U.L.C. Microfilm 661.	Operations. Opus Primum; Opus Quintum, Two sets of Notes; Opus Sextum, Two sets of Notes; Opus Octavum; Extractio auri vivi et conjunctio ejus; in all, about 12000 words, 42 pp. Autograph. Sm. 4to.
Sotheby Lot 65. Keynes MS 41. U.L.C. Microfilm 661.	[Operations: 1–6, beginning with "Extractio et Rectificatio Spiritus"; ending with "Solutio sicca et humida metallorum vulgi eorumque purgatio et multiplicatio infinita mercurii sophici et extractio auri vivi," in Latin but with verse extracts in English from Ripley], about 12 000 words, with alterations and re-writings. Autograph. Sm. 4to.
Sotheby Lot 66.	[Operations: Notes on the Operations, 1, 2, and 6–9, some fragmentary, in Latin], about 13 000 words, 43 pp. Autograph, with very many alterations and re-writings. Sm. 4to.
Sotheby Lot 67.	"Opus Galli Anonymi," with note: "Simile est hoc opus operi Fabri ...", Chapters headed: "Practica"; "De Igne"; "Multiplicatio"; "Facere aurum potabile," etc., about 2750 words, 8 pp. and wrapper. Autograph. Folio.
Sotheby Lot 68. Keynes MS 42. U.L.C. Microfilm 661.	"Pearce the black Monck upon ye Elixir," in verse [English], 226 lines, 7 pp. Autograph. Sm. 4to.
Sotheby Lot 69. Stanford University, Frederick E. Brasch Collection of Sir Isaac Newton and the History of Scientific Thought.	Philalethes. Notes out of Philalethes, beginning: "Mercury vulgar is prepared for conjunction with sol vulgar by frequent cohobations of Reg of [Iron], 2 pp. Autograph. Sm. 4to.
Sotheby Lot 70.	Philosophers Stone. "Lapis Philosophicus cum suis rotis elementaribus"; a Diagram, consisting of a central circle ("Prima Materia") with 7 other circles (3 feminine and 4 masculine) round it, with MS. descriptions of the colours, 1 p. 8vo; Miscellaneous Notes, etc. on 25 pp. (one leaf containing on the reverse the beginning of the draft of a

letter: "Mr. Aubrey: I understand you have a letter from Mr. Lucas for me. Pray forbear to send me anything more of that nature"). All autograph. Sm. 4to., etc.

Sotheby Lot 71. Keynes MS 43. U.L.C. Microfilm 661.	Philosopher's Stone. "Regulae seu canones aliquot Philosophici de Lapie Philosophico. Authore docto quodam Anonymo. Impress in fine Curationum Paracelsi," followed by "Mayer's Figures praefixed to Basil Valentine's Keys," about 3000 words, 8 pp. Autograph. Sm. 4to.
Sotheby Lot 72. Keynes MS 44. U.L.C. Microfilm 661.	Philosopher's Stone. Several Questions concerning the Philosopher's Stone [in English], about 3000 words, 8 pp. Autograph. Sm. 4to.
Sotheby Lot 73. Keynes MS 45. U.L.C. Microfilm 661.	Practica Mariae Prophetissae in Artem Alchemicam, about 1500 words, 4 pp. Autograph. Sm. 4to.
Sotheby Lot 74. Babson MS 420.	Praxis [A Treatise in Five Chapters with the following headings: "Chap. 1. De Materie Spermaticis"; "Chap. 2. De materia prima"; "Chap. 3. De Sulphure Philosophorum"; "Chap. 4. De agente prima"; "Chap. 5. Praxis", in English], with very numerous alterations and re-writings, about 5500 words, 26 pp., in a wrapper of which 2 pp. contain notes on the derivation of the names and symbols of the metals from the Egyptian Gods, the Planets, etc. All autograph. Sm. 4to.
Sotheby Lot 75.	Preparatio Mercurii ad Lapidem per Regulam et Lunam, ex MSS. Phi. Americani, about 1000 words, 6 pp. Autograph. Sm. 4to.
Sotheby Lot 76. Jewish National and University Library, Yahuda MS Var. 1, Newton MS 38.	Pyrotechny. De Igne sophorem et materia quam cale facit [with passages in English on the Exaltation of Tinctures, etc.], about 2500 words, 6 pp. Autograph. Sm. 4to.
Sotheby Lot 77. Keynes MS 46. U.L.C. Microfilm 661.	Pyrotechny. The Three Fires [in English] about 1000 words, 3 pp. Autograph. Sm. 4to.
Sotheby Lot 78.	Pyrotechny. ["The Three Mysterious Fires," mostly in English], etc. about 1200 words, 3 pp. Autograph.
Sotheby Lot 79.	Pyrotechny. A Summary of the Operations and Technique of Pyrotechny, analysed and arranged in tabular form, with details of all the various pieces of Apparatus used; Directions for the construction and working of Furnaces, illustrated by six diagrams, over 3000 words, written on 2 sides of a folio sheet folded in 4to. Autograph.
Sotheby Lot 80. Keynes MS 47. U.L.C. Microfilm 661.	Raymundus. "Experimenta Raymundi," [24 Experiments], about 2500 words, 6 pp. Autograph. Folio.

Sotheby Lot 81.	Raymundus. [Notes on Experiments out of Raymundus, the "Theatrum Chymicum," etc., headed "Miscellanea," about 1200 words, 4 pp.; other Notes, on 6 pp.] All autograph. Folio.
Sotheby Lot 82. Stanford University, Frederick E. Brasch Collection of Sir Isaac Newton and the History of Scientific Thought.	Raymundus. Ex Raymundi libro secretorum, seu de Quintessentiis, about 2000 words, 5 pp. and Notes [in Latin] on the separation of elements under Saturn and Jupiter, about 3000 words, 7 pp. All autograph. Sm. 4to.
Sotheby Lot 83.	Recipes. "A Medicine to transmute Copper"; "To make artificial pearle"; and other Recipes and Notes, not in Newton's hand, on 7 pp. sm. 4to and 3 pp. folio.
Sotheby Lot 84. Keynes MS 48. U.L.C. Microfilm 661.	[Regimens. A Treatise on the Regimens, etc.] mostly in English, under the following headings: "Lapidis Compositio"; "Elementorum Conversio Conjunctio et Decoctio"; "Regimen Ignis"; "Materia" [from Philalethes, Maier, Snyder, Basil Valentine, etc.]; "Decoctio Regimen Mercurii"; "Regimen Jovis"; "Regimen Lunae"; "Regimen Veneris, Martis et Solis," with an Early Draft, much altered, of the "Decoctio Regimen Mercurii," in all about 23 000 words, 76 pp. enclosed in a wrapper covered with notes and rough drawings of stills, retorts, etc. All autograph. Sm. 4to.
Sotheby Lot 85.	Regimen. Notes on the Regimen, three sets, about 5300 words, 18 pp., 8 pp. badly discoloured. Autograph. Sm. 4to.
Sotheby Lot 86. Keynes MS 49. U.L.C. Microfilm 661.	Regimen [a Series of Seven Aphorisms, in English; followed by Annotations, mostly in Latin], about 3000 words, 7 pp. Autograph. Folio.
Sotheby Lot 87.	Regimens. The Regimens described, wth ye time & signes [in English], about 2000 words, 6 pp. Autograph. Sm. 4to.
Sotheby Lot 87, cont.	Proportions. "Of Proportions" [in Latin], about 1200 words, 4 pp. Autograph. Sm. 4to.
Sotheby Lot 87, cont.	Elements. "De Primis materialibus principiis [et eorum praeparatione], about 1000 words. Autograph. Sm. 4to.
Sotheby Lot 87, cont.	Axioms. "The Series of the Work in Axioms [43] Clueply of ye work in common gold," about 800 words, 4 pp. Autograph. Sm. 4to.
Sotheby Lot 87, cont.	Ripley. "Conclusione ex Riplei Operibus deductae"; "Praeparatio Salis ex Riplei operibus," etc., about 2000 words, 11 pp. Autograph. Sm. 4to.
Sotheby Lot 87, cont.	Alcahest. "Of the Alcahest" [in English, except for quotations], about 3000 words, 8 pp. Autograph. Sm. 4to.
Sotheby Lot 87, cont.	Philosopher's Stone. "De medendi Arte, et usu Lapidis" [partly in English], about 2000 words, 6 pp. Autograph. Sm. 4to.

These seven items of Sotheby Lot 87 were enclosed in a wrapper bearing a list of contents in Newton's hand.

Sotheby Lot 88.	Regulus Martis. [A Recipe for making an alloy of Iron and Antimony, in English], about 300 words, 1 p., folio, autograph; Recipes in the hand of John Conduitt for making Aqua Fortis, and for Refining Silver, 1 p., folio,; "Experimentum Bellini," 1 p. 4to.
Sotheby Lot 89. Keynes MS 50. U.L.C. Micro-film 661.	Rehe (Jodocus a) Opera Chymica, with Transcripts of Letters to Dr. John Twysden from A. C. Faber, and Notes on Faber's Works [in Latin], about 7500 words, 22 pp. Autograph. Sm. 4to.
Sotheby Lot 90. Trinity Col-lege, Cam-bridge, MS R. 16. 38, f. 441A, underlined section only.	Remedies. A Collection of Remedies "Against ye Plague"; "For ye Scurvy & Goute"; "For a Tetter Worme on ye face"; "for ye Tooth-ach"; "For ye Spleene"; etc., about 850 words, on 5 pp.; "To make Ink." etc., about 600 words. on 4 pp. sm. 4to. Autograph.
Sotheby Lot 91. Keynes MS 51. U.L.C. Micro-film 661.	Ripley (Sir George) Notes on Sir George Ripley: Ripley's Vision; Preface to his Gates; and Opening of the Six Gates (Calcination, Solution, Separation, Conjunction, Putrefaction, Congelation), about 8000 words, 14 pp. Autograph. Sm. 4to.
Sotheby Lot 92. Keynes MS 52. U.L.C. Micro-film 661.	Ripley (Sir George) Epistle to K[ing] Edward unfolded, about 10 000 words, 17 pp. Autograph. Sm. 4to.
Sotheby Lot 93. Keynes MS 53. U.L.C. Micro-film 661.	[Ripley (Sir George)] "Of ye first Gate," [in English], about 4000 words, 9 pp. Autograph. Sm. 4to.
Sotheby Lot 94. Keynes MS 54. U.L.C. Micro-film 661.	Ripley (Sir George) "Ripley expounded" [the Twelve Gates], about 4500 words, 12 pp.; "Notes upon Ripley," about 1500 words, 8 pp. All autograph. Sm. 4to.
Sotheby Lot 95.	"Rosario Magno" [Notes and Abstracts], about 750 words, 1½ pp. Autograph. Sm. 4to.
Sotheby Lot 96.	Schroeder. [Notes] Out of Schroderus Pharmacopoeia; and other Notes (fragmentary) on 10 pp. Autograph. Sm. 4to.
Sotheby Lot 97. Keynes MS 55. U.L.C. Micro-film 661.	Sendivogius explained [Notes on the Treatises 1–7, and 9–12 of Sendi-vogius; Notes on "Ye Philosophick Riddle," etc., mostly in English], about 12 000 words, 37 pp. Autograph. Sm. 4to.
Sotheby Lot 98. Keynes MS 56. U.L.C. Micro-film 661.	Sententiae luciferae et Conclusiones notabiles [with passages in English], etc., about 6000 words, 18 pp. An autograph note on p. 1 reads: "King's Receivers take all but brass money till next May 24. None clipt passes after Feb. 10 except 6 pences. Broad money punched. 4 Mints in several parts of the Nation." All autograph. 4to.

Sotheby Lot 99. Keynes MS 57. U.L.C. Microfilm 661.	Separations. Notes on Separations, Processions, Sublimations, Distillations, etc. [mostly in Latin], with references to Raymundus, Maria Ferrar, Avicenna, Sendivogius, Flammel, and others, about 9000 words, 17 pp. Autograph. Sm. 4to.
Sotheby Lot 100. Keynes MS 58. U.L.C. Microfilm 661.	Silver. Recipe for Multiplying Silver (by adding silver, antimony and lead to Cinnabar) [in English], about 160 words, ½p.; Notes on "Aqua Sicca," etc. [in English and Latin], about 3000 words (much altered), with rough sketches of furnaces, and calculations, 12 pp. All autograph. Sm. 4to.
Sotheby Lot 101. Keynes MS 59. U.L.C. Microfilm 661.	Solution. "De Secreto Sol[utions]" in English, [on Diana's doves, etc.], part of heading and several words missing, about 1000 words, 3 pp. Autograph. Sm. 4to.
Sotheby Lot 102. Yale University, Medical Library.	Snyders (John de Monte) The Metamorphosis of the Planets, That is A Wonderfull Transmutation of the Planets and Metallique formes into their first Essence (with an annexed Process), being a discovery of the three keys pertinent to the obteining of ye three Principles. Likewise in what manner the most generall universall is to be obteined is in many places of this treatise described by John de Monte Snyders, about 22 000 words, 62 pp., with beautifully written title containing a pen-and-ink drawing on reverse, autograph transcript; Notes by Newton on the book, 4 pp. Autograph. Sm. 4to.
Sotheby Lot 103.	Snyders. A Key to Snyders [in Latin], about 1000 words, on 3 pp.; Sententiae notabilii expositiae, about 1400 words, on 4 pp. All autograph. Sm. 4to.
Sotheby Lot 104.	Sublimation. Notes on Sublimation; The Stone, The Tincture, etc. [in Latin], about 1500 words, 6 pp. Autograph. Sm. 4to.
Sotheby Lot 105. Keynes MS 60. U.L.C. Microfilm 661.	"Tabula Smaragdyna" and "Hieroglyphica Planetarum", about 1750 words, 4 pp. Autograph. Sm. 4to.
Sotheby Lot 106. Stanford University, Frederick E. Brasch Collection of Sir Isaac Newton and the History of Scientific Thought.	"Theatrum Astronomiae Terrestris" [and ye conversion of one element into another in ye composition of ye stone," etc., in English and Latin], about 1000 words, with 2 pen-and-ink diagrams. Autograph. Folio.
Sotheby Lot 107. Keynes MS 61. U.L.C. Microfilm 661.	Thesaurus Thesaurorum sive Medicina Aurea [in English], about 1200 words, 5 pp. Autograph. Sm. 4to.
Sotheby Lot 108. Keynes MS 62. U.L.C. Microfilm 661.	Transmutation. "The Work of an old Priest, viz: B," etc. [containing directions "To turn ☽ into ☉ (i.e., silver into gold)," "☽ out of ☿ and ♂" (silver out of mercury and iron); directions to make Aurum Potabile, etc.], with drawings of retorts, receivers, etc., about 3500 words, 8 pp. Autograph. Sm. 4to.

Sotheby Lot 109.

Transmutation. Fragment only, of a Treatise [in Latin] containing several curious pen-and-ink drawings of processes, one occupying a whole page, described [in English] in mystical terms, about 2200 words, 8 pp. Autograph. Folio.

Sotheby Lot 110.

Valentine (B.) Abstracts from the Works of Basil Valentine on Minerals; Transmutation of Metals; Vitriol, etc. [in English], about 6000 words, 14 pp. Autograph. Sm. 4to.

Sotheby Lot 111.
Keynes MS 63.
U.L.C. Micro-film 661.

Valentine (B.) Verses at the end of B. Valentine's Mystery of the Microcosm, about 580 lines, with a small pen-and-ink drawing. Autograph Sm. 8vo.

Sotheby Lot 112.
Keynes MS 64.
U.L.C. Micro-film 661.

Valentine (B.) Currus Triumphalis Antimonii [Notes and Abstracts], about 4500 words, 8 pp. Autograph. Sm. 4to.

Sotheby Lot 113.

[Vegetation of Metals: draft of a Short Treatise, incomplete (in English)], about 4500 words, 12 pp. Autograph, with many corrections and alterations. Sm. 4to.

Sotheby Lot 114.
Keynes MS 221 (photostat).
U.L.C. Micro-film 661.

[Villa nova (Arnoldus de)] "Ex Rosarii abbreviati tract. quinq." – "Ex Petri Boni Lombardi Ferrariensis Margarita Pretiosa" – "Ex Dionysii Zachariis Opusculo" – "Out of Philalethe's Work concerning the extraction of [sulphur] out of [mercury]"; Commentary on Ripley, etc. (partly in English) – "Ex Clangore Buccinae"; about 24 000 words, on 61 pp., preceded by 5 pp. prepared as an index, the remainder of the volume blank, except for a note on the fly-leaf: "Sep. 25 1727. Not fit to be printed. Tho. Pellett." Contemporary calf. Autograph. Folio.

Sotheby Lot 115.

Vitriol. "Notes upon ye working of ——" [An account of an actual purification of Vitriol, which took 6 months to accomplish], about 350 words, 1 p.; on the reverse are seven "Quares," and also notes of various sums owing to Newton (dated 1673–77). Autograph. Sm. 4to.

Sotheby Lot 116.

Yarworth (W.) Processus Mysterii Magni Philosophicus, 1701, a transcript up to Experiment 9 only (incomplete), not in Newton's hand, on 107 pp., contemporary calf, joints broken, with note on fly-leaf: "Sep. 25 1727. Not fit to be printed. T. Pellet." 8vo.

Sotheby Lot 117.

Yarworth (W.) Processus Mysterii Magni Philosophicus, 1702, a transcript of Chapters 1–6 (incomplete), not in Newton's hand, on 91 pp., contemporary calf, with clasp, g. e., with note on fly-leaf: "Sep. 25. 1727. Not fit to be printed. Tho. Pellet." Oblong sm. 8vo.

Sotheby Lot 118.
Keynes MS 65.
U.L.C. Micro-film 661.

Yarworth (W.) Processus Mysterii Magni Philosophicus, a neatly written transcript of Chapters 1–10 (incomplete), not in Newton's hand, on 75 pp. Folio.

The texts of the foregoing transcripts of the Processus are not identical.

Sotheby Lot 119.
Keynes MS 66.
U.L.C. Micro-film 661.

[Yarworth (W.) Processus Mysterii Magni Philosophicus: an Abstract of Chapters 1–5 (incomplete) [in English], about 3300 words, on 12 pp. Autograph. Folio.

Sotheby Lot 120. Keynes MS 67. U.L.C. Microfilm 661.	Alchemical MS. A Collection of Treatises on Alchemy: "The Apocalypse or Revelacon of the Secret Spirit"; "The true order in ye worke of ye Philosophers stone"; "De Alkymiae veritate"; Extracts from the Alchemical Writers, and a large collection of Recipes and Experiments, mostly in English, 112 ll., contemporary limp vellum, wrapper (lower part missing) apparently from Newton's Collection but not containing anything in his hand. [16th–17th Cent.] Sm. 4to.
Sotheby Lot 121. University of Wisconsin, Duveen Collection.	*Secrets Reveal'd: or, An Open Entrance to the Shut-Palace of the King: containing the greatest Treasure in Chemistry...composed By a most famous English-man styling himself Anonymus, or Eyraeneus Philaletha Cosmopolita... Published for the Benefit of all English-men, by W. C. Esq.* [i.e., William Cooper], containing corrections and additions in the hand of Newton on almost every page, some completely filling the margins, with Newton's press mark "A/n⊙" on inner cover, and autograph inscription on fly-leaf; "cost dg Mr. Story," contemporary sheepskin, sm. 8vo., *W. Godbid for W. Cooper,* 1669.

APPENDIX B

Newton's Essay on the Preparation of Star Reguluses

University Library, Cambridge, Portsmouth Collection
MS Add. 3975, f. 42r,v (Newton's pp. 81–82)

/ f. 42r / To make Regulus of ♁, ♂, ♄, ♀ &c. Take of ♁ 12 ℥ of ♂ 4½ ↗ or 5¼ ↙ or ♀ 6 ↗ or 6¼ ↙, or of ♄ 8½ or proportionably more to ye ♁ if it will beare it. When they are melted pour them of ⟨*sic*⟩ & you will have a Reg. You may when they are molten throw 2 or 3 ℥ of ⦶ on them which having done working pour them of ⟨*sic*⟩. Jf ye scoria of ♄ bee full of small eaven rays there is two ⟨*sic*⟩ little ♄ in proportion. Jf any reg swell much in the midst of the upper surface it argues two ⟨*sic*⟩ much ♁ if it bee flat it argues two ⟨*sic*⟩ little. The better yoe proportions are the brighter and britler will ye Reg bee & ye darker ye scoria & the easier will they part: And also ye more perfect the starr, unlesse the salts on ye top worke & bubble in the cooling to disturb ye sd superficies. The work succeeds best in least quantitys. Jf there bee stuff like pitch long in cooling tis noe good signe & often argues too much Antimony. Twelve ounces of ♁ gives 4⅓ of Reg of ♂ 3⅓ of Reg of ♀ or ♄ when refined. To refine it, so soone as it is molten throw in ¼ or ⅕ pt as much salt peeter as there is reg in weight; then blow to give a good heate till ye mettall & salt boyle ↗ well ↙ together, & ↗ also till ↙ they have done ↗ boyling & ↙ working, yn poure them of ⟨*sic*⟩. This you may repeate till the salt come of ⟨*sic*⟩ white, wch will bee at the second or 3d refining. Mix noe charcoale wth the peeter least ye peeters force be otherwise spent then upon the mettall. Tin may bee 5⅓ to 12 of ♁ or 4 to 9. Jf ye quantity bee but small as 2 ℥ of ↗ tin ↙ then take 4¼ of ♁ but if bigger take 4 to 9. Note yt in Tin & Lead if ye scoria bee full of very small striiae like hairs or rays tending from ye center of ye metall it ↗ argues ↙ too much ♁. Jf it bee branched wth grosser graines (wch in tin ↗ especially ↙ will appeare continuous to ye ↗ centrall ↙ metall) it argues two ⟨*sic*⟩ little ♁. Tis best when ye scoria is / f. 42v / haire-grained inwards towards ye center of ye metall but not quite to ye outside, unlesse it happen yt ye scoria look black.

These rules in generall should bee observed. 1st yt ye fire bee quick. 2dly yt ye crucible bee throughly heated before any thing bee put in; 3dly yt metalls bee put in successively according to their degree of fusibility ♂ . ♀ . ♁ . ♃ . ♄ . 4tly That ↗ they ↙ stand ↗ some time ↙ after fusion before they bee poured of ⟨*sic*⟩ accordingly to ye quantity of regulus they yield, ♂, ♁ ♃ ♄. 5tly That at ye first time noe salt bee thrown on, unlesse upon ♂ to keep it from hardeing ⟨*sic*⟩ on ye top & then let it bee poured of ⟨*sic*⟩ when ye fury of ye salt is over before it have quite done working. 6 That if you would have ye saltp ↗ eter ↙ flow wthout two ⟨*sic*⟩ great a heat, you may quicken it by throwing in a little more saltpeter mixed wth ⅛ or 1/16 of charcoal finely poudered.

Also these signes may bee observed in generall. That if ye scoria & Regulus part not well there is two ⟨*sic*⟩ much metall; that if they doe part well & yet yeild not a dew quantity of Regulus there is too little metall (unlesse ye fire hath not been quick enough or the regulus not had time to sattle) That if the reg bee tough it argues too much metall unlesse in tin wch is therby made ye brittler. That possibly the proportions of ye metalls may alter in the refining Thus ☿ of ♂ being more volatile yn that of ♄, if there bee two ⟨*sic*⟩ much ♂ at first, it may in 3 or 4 times refining come to a good proportion. That the degrees of fire may cause some variations in the proportions. Thus wth a good quick & smart fire 4 of ♂ to 9 of ♄ gave a most black & filthy scoria & ye Reg after a purgation or two starred very well. But in a lesse heat a greater proportion of ♄ gave ye blackest scoria.

Jf ye Regulus be poudered & mixed wth ↗ ½ or ↙ ⅓ of niter & so thrown gradually into a crucible, the better half of ye regulus will be lost in ye Salts, but if a little charcoal be mixed wth ye salt (suppose an ⅛ or 1/12 part to make ye salt deflagrate, it will not consume so much of ye metall.

APPENDIX C

"The Key": Keynes MS 18, Latin Text and English Translation

King's College, Cambridge

Clavis.

/ f. 1r / Imprimis scito minerale crudum et immaturum esse ♁ habens in se quid metallici proprium materialiter, utut ⟨*sic*⟩ aliter crudum et indigestum sit minerale. Digeritur autem veré per 🜍 quod in ♂te et alibi nuspiam reperitur.

Partes ♁ij duae cum j, ♂ dant ♉ qui in quarta sui fusione stellam exhibet, hoc signo cognosces ♂tis animam virtute ♁ij factam totaliter volatilem. Hic stellatus ♉ si fundatur cum † ☉ aut ☽ totus in testa et cineritio evaporatur, quod mysterium est. Hic item ♉ si cum ☿io vulgari a̅a̅a̅ et ad tempus breve 2 vel 3 horarum spatium digeratur in vase clauso leni calore ac postmodum per 1/8 horae teratur in mortario mediocriter calenti absque humiditate donec nigredinem suam expuat, dein ad nigredinis suae maxima parte depositionem lavetur donec vix amplius nigredine inficiatura aqua, quae admodum in principio nigrescit, quod multiplici affusione aquae et effusione obtinetur, siccetur hoc a̅a̅a̅, igni denuò apponatur atque in calore supradicto detineatur per tres horas, postea iterum ut prius in mortario cicco ⟨*sic*⟩ calidoque teratur, novam excruit nigredinem quae iterum abluenda, hoc tamdiu iterandum donec a̅a̅a̅ totum fiat instar ☽ae splendidae et cupellatae. Existit autem in principio colore plumbeo et obscuro.

Dein ☿ hunc sic ablutam distilla et denuò septies aut novies * iterum a̅a̅a̅, & in singulis a̅a̅a̅ calefactionem triturationem ablutionem toties observa & toties ipsum distilla ut prius. habebis vice septima ☿um omnia metalla solventem, imprimis solem. Sciens scribo, nam habeo varia in igne vitra cum ☉ et hoc ☿io in quibus arboris forma crescunt et circulatione continua solvuntur arbores iterum cum opere in ☿um novum. Tale vas cum ☉ sic soluto in igne habeo ubi ☉ non est solutum per corrosivum in attomos ad oculum sed ex- et intrinsicè in ☿um tam vivum ac mobilem ac ⟨*sic*⟩ ullus in mundo reperitur ☿us. Tumescere enim facit ☉ turgere et putrescere ac crescere in surculos et ramos, mutando indies colores quorum aspectus me quotidiè tenent. quod magnum arcanum puto in Alchymia, idque non quaeri ↗ censeo ↙ recta via ab artistis quorum nimia est sapientia ut existiment ☿um vulgarem per reiteratam cohobationem a regulo ♈ is [i.e. ♂tis seu ♁ij] peti debere, quod tamen unicum corpus ☿io familiare utpote proxime cum isto cog- et agnoveris in toto minerali regno, huicque proximum / f. 1v / ☉em. Atque haec est methodus philosophica meliorandi naturam in natura consanguinitatem in consanguinitate.

⟨Newton's own footnotes⟩ † forte sit 🜍☉ ita
* ⟨No note appears in the MS relative to this asterisk.⟩

Quoad hanc operationem evolve Epistolam responsoriam ad Thomas de Bononia, quaestionem hanc planè solutam invenies.

Alterum secretum est quod opus sit tibi mediatione dianae virginis [q̄. ē. purissima ☽a] aliter non uniuntur ☿ et ♋︎.

♋︎ fit ex ♁ ℥ iv ↗ p. ix ↙ ♂ ℥ ij ↗ p. iv ↙, sic est bona proportio. Jmpar ♁ numerus super ⟨Blank in MS.⟩ non est negligendus. Nam si hic error frustraberis. ffac ♋︎ inijciendo ⟨sic⟩ per vices ☉, ut materia fluat injice ad superiorem proportionem ☉ ℥ iv minimum ℥ iij.

Majorem quantitatem in uno +lo preparare non est consultum ♁ tritum ⅄lo inde simul cum ♂te quicquid alij dicant scribantve.

Possunt sumi minores clavi et imprimis eorum extremitates ab equińis soleis confracti. Jgnis sit fortis ut fluere poterit materia [instar ∇ae] quod facile fit, ubi fluit cochleare nitri injice, ubi deflagraverit alium injice idque continuo donec tres quatuorve injeceris ℥. Tum auge carbones circa +lum cavendo tamen ne in ⅄lum cadant. Jgnem auge qualem requirit lunae vulgaris fusio, atque in hoc statu relinque per 1/8 horae. [Materia instar aquae tenuis esse debet si recte operatus fueris.] Tum in conum effunde, subsidebit ♋︎ a quo cineritiam separa scoriam. Refrigeratum in vase sicco serva.

Bonae fusionis signum est si ferrum totaliter sit fusam scoriaeque per se concidant in pulverem.

♋︎ pulsa et ipsi ☉i ℥ ij, ad summum ℥ ijss, adde et exactè tere simul, iterum funde scorias arsenicales et inutiles adjice.

♋︎ tertio et quarto cum ☉ ℥ j ad summum tere et in novo ⅄ funde et habebis quarta vice scorias aureo colore tinctas et ♋︎ stellatum.

NB   tribus ultimis vicibus scoriae tanquam arsenicales sunt abjiciendae, tamen in chirurgia utiles.

NB   in tribus ultimis fusionibus ♋︎ est pulsandus et cum ☉ terendus ac miscendus, contra id, quod nitrum aliqui frustulatim ⅄ injiciunt quod non laudandum nam jmò sic prolongatur fusio ♋︎ non sine aliquali ejusdem exhalatione. 2do ☉ sic injectum tantum accupat superficiem, et ad tempus ♋︎ refrigerat, et cum ☉ facilè fluat, ipsum prius / f. 2r / fluit et nitrum incrustat ut citra magnum ignem non fluat, sicque ♋︎ pars optima per conflagrationem perit, unde sit quod nonnunquam stella perit quod falso adscribitur constellationi. Hâc autem ratione videbis ♋︎ cum ☉ mistum facile cum ipso fluere, nec obdurescere ullo modo, praeter differentiam depurationis quae longe major in mixto quam si ☉ injiciatur.

℞ hujus * ♋︎ pt j. ☽ae pt ij, funde sumul ⟨sic⟩ donec funde simul donec ⟨sic⟩ existant instar metalli fusi, effunde et habebis massam friabilem coloris plumbi

NB   ♋︎ junctus cum ☽a faciliús fluunt quàm utrumque seorsim, [manebuntque tamdiu fusa et ♄ licet sic partes ij ☽ae sint quae tum mutatur in naturam ♂ij friabilem et saturneam.

Massam hanc friabilem ♄ eam pulsa, eamque cum ☿io vulgi ☉ et ♁ loti (puta decuplo) et iterum resiccati (duplo) conjice in mortarium marmoreum eo usque calefactum ut ejus calorem digitis perferre possis. ☿ per ¼ horae ferreo pistillo tere, atque sic mediantibus Dianae columbis ipsum cum fratre suo philosophico ☉e conjunge ex quo spirituale concipiet semen, quod ignis est quod omnes ipsius purgabit superfluitates interveniente virtute fermentali. Tunc ℞ parum ✳ pulsati et cum ipso tere, ubi planè erit a̅a̅a̅ adde tantillum humiditatis quae sufficit ad humectandum et occurret tibi hoc unum philosophicum signum quod in

* [♋︎ ℥ β, ☽ ℥ j.]

252

conficiend⟨o⟩ suo ☿io est magnus foetor. Deinde lava tuum ☿um aquam affundendo terendo decantando iterumque novam affundendo toties donec faeces paucae appareant.

Key.

/ f. lr /   First of all know antimony to be a crude and immature mineral having in itself materially what is uniquely metallic, even though otherwise it is a crude and indigested mineral. Moreover, it is truly digested by the sulfur that is found in iron and never elsewhere.

Two parts of antimony with iron give a regulus which in its fourth fusion exhibits a star; by this sign you may know that the soul of the iron has been made totally volatile by the virtue of the antimony. If this stellate regulus is melted with † gold or silver by an ash heat in an earthen pot, the whole regulus is evaporated, which is a mystery. Also, if this regulus is amalgamated with common mercury and is digested in a sealed vessel on a slow fire for a short time – two or three hours – and then ground for ¼ of an hour in a mortar without moisture while being warmed moderately, until it spits out its blackness, then it may be washed to deposit the greatest part of its blackness, until the water, which in the beginning becomes quite black, is scarcely more tinged by the blackness. This can be done by flushing it with water many times. Let the amalgam be dried, again placed near the fire, and kept in the above-mentioned heat for three hours. Afterwards let it be ground again as before in a dry and warm mortar. It pushes out new blackness, which must be washed away again; this must be repeated continually until the whole amalgam becomes like shining and cupellated silver, whereas at first it had a dark leaden color.

Then distill this mercury which has been so washed and * amalgamate over again seven or nine times, and in each amalgamation see to the heating, grinding, and washing as many times as before, Distill the whole as before. On the seventh time you will have a mercury dissolving all metals, particularly gold. I know whereof I write, for I have in the fire manifold glasses with gold and this mercury. They grow in these glasses in the form of a tree, and by a continued circulation the trees are dissolved again with the work into new mercury. I have such a vessel in the fire with gold thus dissolved, where the gold was visibly not dissolved by a corrosive into atoms, but extrinsically and intrinsically into a mercury as living and mobile as any mercury found in the world. For it makes gold begin to swell, to be swollen, and to putrefy, and to spring forth into sprouts and branches, changing colors daily, the appearances of which fascinate me every day. I reckon this a great secret in Alchemy, and ↗ I judge ↙ it is not rightly to be sought from artists who have too much wisdom to decide that common mercury ought to be attacked through reiterated cohobation by the regulus of leo [that is, of iron or antimony]. That unique body, that regulus, however, is familial with mercury seeing that it is closest to that mercury you have known and recognized in the whole mineral kingdom, and hence most closely related to / f. lv / gold. And this is the philosophical method of meliorating nature in nature, consanguinity in consanguinity.

With regard to this operation, look at the Letter responding to Thomas of Bologna, and you will find this question fully solved.

Another secret is that you need the mediation of the virgin Diana [quintessence, most pure silver]; otherwise the mercury and the regulus are not united.

The regulus is made from antimony four ounces ↗ nine parts ↙, iron two

---

† Thus perhaps it is the sulfur of gold
* ⟨No note appears in the MS relative to this asterisk.⟩

ounces ⟋ four parts ⟍; this is a good proportion. Do not neglect to have a mass of antimony greater than that of iron, for if an error is made here you will be disappointed. Make the regulus by casting in nitre bit by bit; cast in between three and four ounces of nitre so that the matter may flow.

It is not a good idea to prepare in one crucible a greater quantity than the above measure of antimony. The antimony is ground, then cupelled together with iron, whatever others may say or write.

Little nails may be used and especially the ends of those broken from horse-shoes. Let the fire be strong so that the matter may flow [like water], which is easily done. When it flows, cast in a spoonful of nitre; and when that nitre has been destroyed by the fire, cast in another. Continue that process until you have cast in three or four ounces. Then pile up the charcoals about the crucible, taking care that they do not fall into it. Increase the fire as much as the fusion of common silver requires, and keep it in that state for ⅛ of an hour. [The matter ought to be like a subtle water if you have labored correctly.] Then pour the matter out into a cone. The regulus will subside. Separate the ashy scoria from it. Keep the cooled material in a dry vessel.

It is a sign of a good fusion if the iron is completely fused and if the scoriae break up by themselves into powder.

Beat the regulus and add to it two, or at most 2½, ounces of nitre. Grind the regulus and the nitre together completely and again melt. Throw away the arsenical and useless scoriae.

Grind the regulus a third and fourth time with at most one ounce of nitre and melt in a new crucible, and on the fourth time you will have scoriae tinged with a golden color and a stellate regulus.

NB   In the last three times the scoriae must be thrown away because they are arsenical; however, they are useful in surgery.

NB   In the last three fusions the regulus must be beaten, and ground and mixed with nitre. Some cast the nitre into the crucible, but this is not recommended, for, firstly, the fusion is as a result prolonged and the regulus is not without some loss of itself by exhalation. Secondly, nitre thrown in in this way stays on the surface and in time it cools the regulus. And since nitre flows easily, / f. 2r / it may flow at first and encrust so that it will not flow again without a large fire. If that happens, the best part of the regulus perishes in the conflagration, whence it is that sometimes a star perishes because it is falsely ascribed to a constellation. You will see that the regulus mixed with nitre in this way flows easily with it; and you will not see it become hard in any manner, except for the difference in the depuration, which is far greater if it is mixed than if the nitre is just tossed in.

Take of this * regulus one part, of silver two parts, and melt them together until they are like fused metal. Pour out, and you will have a friable mass of the color of lead.

NB   If the regulus is joined with the silver, they flow more easily than either one separately [and they remain fused as long as lead even though there are thus two parts of silver, which is then changed into the nature of antimony, friable and leaden.

Beat this friable mass, this lead, and cast it together with the mercury of the vulgar into a marble mortar. The mercury should be washed (say ten times) with nitre and distilled vinegar and likewise dried (twice), and the mortar should

* [regulus½ ounce, silver one ounce.]

be constantly heated just so much as you are able to bear the heat of with your fingers. Grind the mercury ¼ of an hour with an iron pestle and thus join the mercury, the doves of Diana mediating, with its brother, philosophical gold, from which it will receive spiritual semen. The spiritual semen is a fire which will purge all the superfluities of the mercury, the fermental virtue intervening. Then take a little beaten sal ammoniac and grind with the mercury. When it is fully amalgamated, add just enough humidity to moisten it, and this one philosophical sign will appear to you: that in the very making of the mercury there is a great stink. Finally, wash your mercury by pouring on water, grinding, decanting, and again pouring on fresh water, until few feces appear.

A Suggested Periodization for Newton's Alchemical Studies

On the basis of Newton's handwriting and a few other odd bits of information, the following periodization for Newton's alchemical studies may be suggested: (1) "very early": 1667–69; (2) "middle early": 1670–75; (3) "late early": 1675–80; (4) "bad ink": 1680–81; (5) "middle confident": 1682–92; (6) "late": 1693–1727.

Newton's "very early" writing is carefully formed, lines straight, *extremely* small – so tiny as to require the use of a hand lens for interlineations. About 1670 a gradual increase in size sets in and, although Newton's formation of individual letters seems to the present investigator to stay about the same, the overall effect of a page of script from "middle early" or "late early" is that the writing is fuller and more rounded. The lines remain quite straight and regular. It must be admitted that separation of the decade of the 1670's into two periods is problematical and depends almost entirely on relative sizes of the writing. But two changes in Newton's writing habits appear during the 1670's and sometimes assist in determining whether a manuscript was written early or late in the decade. One of these is that Newton begins to provide a crossbar for the symbol for lead: it becomes the conventional " ♄ " whereas before he had always written it without the bar, as " ♄ ". The other is that he ceases to include accent marks in his Latin writings.

The so-called "bad ink" period of 1680–81 is also somewhat problematical. In the laboratory notes material dated during that time is virtually illegible in places because the ink ran through the paper and heavily discolored the reverse side. Some of the alchemical manuscripts show similar difficulties, but of course there is no assurance that Newton necessarily dipped his pen in that same batch of bad ink for everything he wrote for two years or that he never used ink of poor quality at any other period.

The "middle confident" period of the 1680's and early 1690's, the period in which he wrote the *Principia*, exhibits a handwriting which is altogether bolder and more expansive. It is fairly easy to distinguish from the other periods but no quirks have been discovered which permit a more specific periodization. About 1692 Newton is thought to have undergone some sort of psychic crisis for a few months and that episode has been chosen as a convenient cut-off date for the " middle confident" period.

It is certain that sometime in the 1690's his handwriting became somewhat smaller again though never quite as small as in the "very early" time. There are also some other characteristics of the "late" period which allow it to be readily distinguished from the "very early," as from the others. The letters are formed in a somewhat sharper manner, giving the writing a slightly crabbed appearance, and in addition the lines tend to slope upward instead of preserving the regular formation of "very early" script. No useful distinctions have appeared within the "late" period. Indeed Sir John Craig, who has studied the Mint-related manuscripts more than anyone else, all of which fail in the "late" period, remarked that Newton's handwriting did not vary appreciably in size or steadiness during the thirty years he spent at the Mint and also that it was so small as to verge on the microscopic in corrections between the lines.[1]

Craig attributed the smallness of the writing to Newton's nearsightedness, which Newton himself mentioned in a letter to Robert Hooke in 1679 in explanation of his hesitation to undertake some astronomical observations.[2] And Stukeley's eye witness account of Newton's eyes and vision likewise seems to provide some external verification of the periodization offered here, based on the handwriting.

Sir Isaac's eyes were very full and protuberant, which rendered him near-sighted in youth and manhood: and was the reason of his seeing so well in age: the eye being betterd by growing somewhat flatter: whereby the visual rays unite at a convenient distance, neither too near, or too far off the eyes. He never used spectacles. In the year 1725 I saw him cast up the accounts of the Treasurer Pitfield, for the Royal Society, being a whole sheet of paper, without spectacles, without pen and ink, an indication of the strength of his memory, as well as of his eyesight.[3]

Some additional support for the dating of the manuscripts offered in this study has come from Professor R. S. Westfall, who has in recent months undertaken his own dating survey of the alchemical portion of the Keynes Collection. Professor Westfall reports himself to be in basic agreement with the present writer's results, although inclined to a slightly later dating for some of the early items than is here indicated.[4]

[1] Craig, *Mint*, p. 124 (1, n. 5).
[2] Isaac Newton to Robert Hooke, Nov. 28, 1679, in Newton, *Correspondence*, II, 300–04 (1, n. 52). The letter is also printed in More, *Newton*, pp. 223–26 (1, n. 2).
[3] Stukeley, *Memoirs*, p. 66 (1, n. 16).
[4] Private communication, July 28, 1973.

Selected Bibliography

A. NEWTONIAN MATERIALS

NEWTON'S OWN WRITINGS

Newton, Isaac. "Alchemical Papers." See Appendix A.
  "Chemical Laboratory Notes." Portsmouth Collection, MS Add. 3973.
  University Library, Cambridge.
  "Chemical Laboratory Notes." Portsmouth Collection, MS Add. 3975.
  University Library, Cambridge.
  The Chronology of Ancient Kingdoms Amended. To which is Prefix'd, A Short Chronicle from the First Memory of Things in Europe, to the Conquest of Persia by Alexander the Great. London: Printed for J. Tonson in the *Strand*, and J. Osborn and T. Longman in *Paternoster Row*, 1728.
  The Correspondence of Isaac Newton. Edited by H. W. Turnbull, J. P. Scott, A. R. Hall and Laura Tilling. 5 Volumes, continuing. Cambridge: Published for the Royal Society at the University Press, 1959– .
  Isaac Newton's Papers and Letters on Natural Philosophy and Related Documents. Edited, with general introduction, by I. Bernard Cohen assisted by Robert E. Schofield. Cambridge, Massachusetts: Harvard University Press, 1958.
  Isaac Newton's Philosophiae naturalis principia mathematica. The Third Edition (1726) with Variant Readings. Assembled and Edited by Alexandre Koyré and I. Bernard Cohen with the Assistance of Anne Whitman. 2 Volumes. Cambridge, Massachusetts: Harvard University Press, 1972.
  Isaac Newtoni opera quae exstant omnia. Commentariis illustrabat Samuel Horsley, LL.D., R.S.S., Reverendo admodum in Christo patri Roberto espiscopo Londinensi a sacris. 5 Volumes. Londini: Excudebat Joannes Nichols, 1779–85.
  "List of Expenses." Yahuda MS Var. 1, Newton MS 34. Jewish National and University Library, Jerusalem.
  The Mathematical Papers of Isaac Newton. Edited by Derek T. Whiteside with the assistance in publication of M. A. Hoskin. 6 Volumes, continuing. Cambridge: At the University Press, 1967– .
  "Memorial on the Gold and Silver Coin." In *Observations on the Subjects Treated Of in Dr Smith's Inquiry into the Nature and Causes of the Wealth of Nations by David Buchanan* [*1817*] *with Appendices: Sir Isaac Newton's Memorial on the Gold and Silver Coins 1717 And Other Documents.* The Adam Smith Library. Reprints of Economic Classics. New York: Augustus M. Kelley, 1966.
  "Of natures obvious laws & processes in vegetation." Burndy MS 16. Burndy Library, Norwalk, Connecticut.
  Opticks, or A Treatise of the Reflections, Refractions, Inflections & Colours of Light. Foreword by Albert Einstein, introduction by Sir Edmund Whittaker, preface by I. Bernard Cohen, analytical table of contents by Duane H. D. Roller. Based on the 4th London edition of 1730. New York: Dover Publications, 1952.

Opticks: or, a Treatise of the Reflexions, Refractions, Inflexions and Colours of Light. Also Two Treatises of the Species and Magnitude of Curvilinear Figures. London: Printed for Sam. Smith, and Benj. Walford, Printers to the Royal Society, at the *Prince's Arms* in St. *Paul*'s Church-yard, 1704.

Philosophiae naturalis principia mathematica. Londini: Jussu *Societatis Regiae* ac Typis *Josephi Streater.* Prostat apud plures Bibliopolas, 1687.

Sir Isaac Newton: Theological Manuscripts. Selected and edited by Herbert McLachlan. Liverpool: At the University Press, 1950.

Sir Isaac Newton's Chronology, Abridged by Himself. To which are Added, Some Observations on the Chronology of Sir Isaac Newton. Done from the French, by a Gentleman. London: Printed for J. Peele, at *Locke's Head*, in *Pater-Noster-Row*, 1728.

Sir Isaac Newton's Mathematical Principles of Natural Philosophy and His System of the World. Translated into English by Andrew Motte in 1729. The translations revised, and supplied with an historical and explanatory appendix, by Florian Cajori. 2 Volumes. Berkeley and Los Angeles: University of California Press, 1966.

Unpublished Scientific Papers of Isaac Newton. A Selection from the Portsmouth Collection in the University Library, Cambridge. Chosen, edited, and translated by A. Rupert Hall and Marie Boas Hall. Cambridge: At the University Press, 1962.

BIOGRAPHICAL MATERIALS ON NEWTON

Andrade, E. N. da C. *Isaac Newton, with eight plates in photogravure, five illustrations in line.* Personal Portraits. Edited by Patric Dickinson and Sheila Shannon. New York: Chanticleer Press, 1950.

Brewster, Sir David. *The Life of Sir Isaac Newton.* Harper's Family Library. New York: J. &. J. Harper, 1831.

Memoirs of the Life, Writings, and Discoveries of Sir Isaac Newton. 2 Volumes. Edinburgh: Thomas Constable and Co.; Boston: Little, Brown, and Co., 1855.

Broad, C. D. "Annual Lecture on a Master Mind, Henriette Hertz Trust: Sir Isaac Newton." *Proceedings of the British Academy* **13** (1927), 173–202.

Brodetsky, S. *Sir Isaac Newton. A Brief Account of his Life and Work.* 2nd Edition. London: Methuen & Co., 1929.

Conduitt, John. "Memoirs of Sir Isaac Newton, sent by Mr. Conduitt to Monsieur Fontenelle, in 1727." In Edmund Turnor, *Collections for the History of the Town and Soke of Grantham. Containing Authentic Memoirs of Sir Isaac Newton, Now First Published From the Original MSS. in the Possession of the Earl of Portsmouth.* London: Printed for William Miller, Albemarle Street, by W. Bulmer and Co. Cleveland-Row, St. James's, 1806.

Craig, John. *Newton at the Mint.* Cambridge: At the University Press, 1946.

de Morgan, Augustus. *Essays on the Life and Work of Newton.* Edited, with notes and appendices, by Philip E. B. Jourdain. Chicago and London: The Open Court Publishing Co., 1914.

de Villamil, Richard. *Newton: the Man.* London: G. D. Knox, n.d. [1931].

Manuel, Frank E. *A Portrait of Isaac Newton.* Cambridge, Massachusetts: The Belknap Press of Harvard University Press, 1968.

Isaac Newton, Historian. Cambridge, Massachusetts: The Belknap Press of Harvard University Press, 1963.

More, Louis Trenchard. *Isaac Newton. A Biography.* London: Constable and Co., 1934; New York: Charles Schribner's Sons, 1934; New York: Dover Publications, 1962.

North, J. D. *Isaac Newton*. The Clarendon Biographies. General editors, C. L. Mowat and M. R. Price. N.p.: Oxford University Press, 1967.

Stukeley, William. *Memoirs of Sir Isaac Newton's Life by William Stukeley, M.D., F.R.S. 1752 Being some Account of his Family and Chiefly of the Junior Part of his Life*. Edited by A. Hastings White. London: Taylor and Francis, 1936.

Turnbull, H. W. *The Mathematical Discoveries of Newton*. London and Glasgow: Blackie & Son, 1947.

ADDITIONAL MATERIALS ON NEWTON

"The *Annus Mirabilis* of Sir Isaac Newton Tricentennial Celebration." *The Texas Quarterly* **10**, no. 3 (Autumn, 1967).

The Annus Mirabilis of Sir Isaac Newton 1666–1966. Edited by Robert Palter. Cambridge, Massachusetts, and London: The M.I.T. Press, 1970.

Boas (Hall), Marie. "Newton and the theory of chemical solution." *Isis* **43** (1952), 123.

Boas (Hall), Marie and Hall, A. Rupert. "Newton's chemical experiments." *Archives internationales d'histoire des sciences* **11** (1958), 113–52.

"Newton's theory of matter." *Isis* **51** (1960), 131–44.

Buchdahl, Gerd. *The Image of Newton and Locke in the Age of Reason*. Newman History and Philosophy of Science Series, no. 6. General editor, M. A. Hoskin. London and New York: Sheed and Ward, 1961.

Churchill, Mary S. "*The Seven Chapters*, with explanatory notes." *Chymia* **12** (1967), 29–57.

Cohen, I. Bernard. *Introduction to Newton's 'Principia'*. Cambridge, Massachusetts: Harvard University Press, 1971.

Feisenberger, H. A. "The libraries of Newton, Hooke, and Boyle." *Notes and Records of the Royal Society of London* **21** (1966), 42–55.

Forbes, R. J. "Was Newton an alchemist?" *Chymia* **2** (1949), 27–36.

Geoghegan, D. "Some indications of Newton's attitude toward alchemy." *Ambix* **6** (1957), 102–06.

Gray, George J. *A Bibliography of the Works of Sir Isaac Newton. Together with a List of Books illustrating his Works, with Notes*. 2nd Edition. Cambridge: Bowes and Bowes, 1907.

Gregory, Joshua C. "The Newtonian hierarchic system of particles." *Archives internationales d'histoire des sciences* **7** (1954), 243–47.

Guerlac, Henry. *Newton et Epicure. Conférence donnée au Palais de la Découverte, Université de Paris, le 2 Mars 1963, Histoire des Sciences*. Paris: Sur les presses de l'imprimerie Alençonnaise, 1963.

Hall, A. Rupert. "Sir Isaac Newton's Note-book, 1661–1665." *Cambridge Historical Journal* **9** (1948), 239–50.

Isaac Newton 1642–1727. A Memorial Volume Edited for the Mathematical Association by W. J. Greenstreet. London: G. Bell and Sons, 1927.

Kargon, Robert Hugh. *Atomism in England from Hariot to Newton*. Oxford: Clarendon Press, 1966.

Koyré, Alexandre. *Newtonian Studies*. Cambridge, Massachusetts: Harvard University Press, 1965.

Kubrin, David. "Newton and the cyclical cosmos: providence and the mechanical philosophy." *Journal of the History of Ideas* **28** (1967), 325–46.

Kuhn, Thomas S. "Newton's '31st Query' and the degradation of gold." *Isis* **42** (1951), 296–98.

"Reply to Marie Boas (Hall)." *Isis* **43** (1952), 123–24.

"The independence of density and pore-size in Newton's theory of matter." *Isis* **43** (1952), 364–65.

Law, William. *An Appeal To all that Doubt, or Disbelieve The Truths of the Gospel, Whether they be Deists, Arians, Socinians, or Nominal Christians. In which, the true Grounds and Reasons of the whole Christian Faith and Life are plainly and fully demonstrated. To which are added, Some Animadversions upon Dr. Trap's late Reply.* London: Printed for W. Innys, at the West-End of St. *Paul's*, 1742.

Selected Mystical Writings of William Law edited with Notes and Twenty-four Studies in the Mystical Theology of Willaim Law and Jacob Boehme and an Enquiry into the Influence of Jacob Boehme on Isaac Newton by Stephen Hobhouse. Foreword by Aldous Huxley. 2nd Edition revised. New York and London: Harper and Brothers, 1948.

Library of Sir Isaac Newton. Presentation by the Pilgrim Trust to Trinity College, Cambridge, 30 October 1943. Address of Presentation by The Right Hon. Lord Macmillan, G.C.V.O., Chairman of the Pilgrim Trust, and of Acceptance by George Macauley Trevelyan, O.M., Master of Trinity College. Cambridge: Printed at the University Press, 1944.

McDonald, John F. "Properties and causes: an approach to the problem of hypothesis in the scientific methodology of Sir Isaac Newton." *Annals of Science* **28** (1972), 217–33.

McGuire, J. E. "Atoms and the 'analogy of nature': Newton's third rule of philosophizing." *Studies in History and Philosophy of Science* **1** (1970), 3–58.

"Force, active principles, and Newton's invisible realm." *Ambix* **15** (1968), 154–208.

"The origin of Newton's doctrine of essential qualities." *Centaurus* **12** (1968), 233–60.

"Transmutation and immutability: Newton's doctrine of physical qualities." *Ambix* **14** (1967), 69–95.

McGuire, J. E. and Rattansi, P. M. "Newton and the 'Pipes of Pan.'" *Notes and Records of the Royal Society of London* **21** (1966), 108–43.

McKie, Douglas. "Somes notes on Newton's chemical philosophy written upon the occasion of the tercentenary of his birth." *Philosophical Magazine* **33** (1942), 847–70.

McLachlan, Herbert. *Religious Opinions of Milton, Locke, and Newton.* Publications of the University of Manchester, no. 276; Theological Series, no. 6. Manchester: Manchester University Press, 1941.

Munby, A. N. L. "The Keynes Collection of the works of Sir Isaac Newton at King's College, Cambridge." *Notes and Records of the Royal Society of London* **10** (1952), 40–50.

Sir Isaac Newton, 1727–1927. A Bicentenary Evaluation of His Work. A Series of Papers Prepared under the Auspices of The History of Science Society in collaboration with The American Astronomical Society, The American Mathematical Society, The American Physical Society, The Mathematical Association of America and Various Other Organizations. Baltimore: The Williams & Wilkins Co., 1928.

The Royal Society Newton Tercentenary Celebrations 15–19 July 1946. Cambridge: At the University Press, 1947.

Spargo, Peter E. "Newton's library." *Endeavour* **31** (1972), 29–33.

Strong, E. W. "Newton's 'mathematical way.'" *Journal of the History of Ideas* **12** (1951), 95–110.

Taylor, Frank Sherwood. "An alchemical work of Sir Isaac Newton." *Ambix* **5** (1956), 59–84.

Thackray, Arnold. *Atoms and Powers. An Essay on Newtonian Matter-Theory and the Development of Chemistry.* Harvard Monographs in the History of Science. Cambridge, Massachusetts: Harvard University Press, 1970.

"'Matter in a nut-shell': Newton's *Opticks* and eighteenth-century chemistry." *Ambix* **15** (1968), 29–53.

Turnor, Edmund. *Collections for the History of the Town and Soke of Grantham. Containing Authentic Memoirs of Sir Isaac Newton, Now First Published From the Original MSS. in the Possession of the Earl of Portsmouth.* London: Printed for William Miller, Albemarle-Street, by W. Bulmer and Co. Cleveland-Row, St. James's, 1806.

Westfall, Richard S. *Force in Newton's Physics. The Science of Dynamics in the Seventeenth Century.* London: Macdonald; New York: American Elsevier, 1971.

"The foundations of Newton's philosophy of nature." *The British Journal for the History of Science* **1** (1962–63), 171–82.

"Isaac Newton's coloured circles twixt two contiguous glasses." *Archives of the History of the exact Sciences* **2** (1965), 181–96.

"Newton and the fudge factor." *Science* **179** (1973), 751–58.

B. THE SEVENTEENTH-CENTURY BACKGROUND

Allen, P. "Scientific studies in the English universities of the seventeenth century." *Journal of the History of Ideas* **10** (1949), 219–53.

Andreae, Johann Valentin. *Christianopolis, an Ideal State of the Seventeenth Century.* Translated, with historical introduction, by Felix Emil Held. New York: Oxford University Press, American Branch; London, Toronto, Melbourne, and Bombay: Humphrey Milford, 1916.

Ashmole, Elias. *Elias Ashmole (1617–1692). His Autobiographical and Historical Notes, his Correspondence, and Other Contemporary Sources Relating to his Life and Work.* Edited, with a biographical introduction, by C. H. Josten. 5 Volumes. Oxford: At the Clarendon Press, 1966.

Aubrey, John. *"Brief Lives." Chiefly of Contemporaries, set down by John Aubrey, between the Years 1669 & 1696.* Edited by Andrew Clark. 2 Volumes. Oxford: At the Clarendon Press, 1898.

Miscellanies Upon Various Subjects. 4th Edition. London: John Russell Smith, 1857.

Barnett, Pamela R. *Theodore Haak, F.R.S. (1605–1690): The First German Translator of Paradise Lost.* 's-Gravenhage: Mouton & Co., 1962.

Barrow, Isaac. *The Theological Works of Isaac Barrow, D.D., Master of Trinity College, Cambridge.* Edited by Alexander Napier. 9 Volumes. Cambridge: At the University Press, 1859.

The Works Of the Learned Isaac Barrow, D.D. Late Master of Trinity-College in Cambridge. (Being All his English Works.) Published by His Grace, Dr. John Tillotson, Late Archbishop of Canterbury. 3 Volumes in 2. London: Printed for *Brabazon Aylmer*, at the *Three Pigeons* against the *Royal-Exchange* in *Cornhill*, 1700.

Ben-David, Joseph. *The Scientist's Role in Society. A Comparative Study.* Foundations of Modern Sociology Series. Edited by Alex Inkeles. Englewood Cliffs, New Jersey: Prentice-Hall, 1971.

Birch, Thomas. *The History of the Royal Society of London for Improving of Natural Knowledge, From its First Rise, In which The most considerable of those Papers communicated to the Society, which have hitherto not been published, are inserted in their proper order, As a Supplement to The Philosophical Transactions.* Facsimile reprint of the London edition of 1756–57. 4 Volumes. Bruxelles: Culture et Civilisation, 1968.

Bligh, E. W. *Sir Kenelm Digby and his Venetia*. London: Sampson Low, Marston & Co., 1932.

Block, Ernst. "Die chemischen Theorien bei Descartes und den Cartesianern." *Isis* **1** (1913–14), 590–636.

Boas (Hall), Marie. "The establishment of the mechanical philosophy." *Osiris* **10** (1952), 413–541.

Boyle, Robert. *Certain Physiological Essays, Written at distant Times, and on several Occasions*. London: Printed for *Henry Herringman* at the *Anchor* in the Lower walk in the New-Exchange, 1661.

    Certain Physiological Essays And other Tracts; Written at distant Times, and on several Occasions. The Second Edition. Wherein some of the Tracts are enlarged by Experiments, and the Work is increased by the Addition of a Discourse about the Absolute Rest in Bodies. London: Printed for *Henry Herringman* at the *Blew Anchor* in the Lower Walk of the *New-Exchange*, 1669.

[—] "*Of the Incalescence of* Quick-silver *with* Gold, *generously imparted by* B. R." *Philosophical Transactions* **10** (1675–75/76), 515–33.

    The Origine of Formes and Qualities, (According to the Corpuscular Philosophy), Illustrated by Considerations and Experiments, (Written by way of Notes upon an Essay about Nitre). Oxford: Printed by H. Hall Printer to the University, for Ric: Davis, 1666.

    The Works of the Honourable Robert Boyle. To which is prefixed The Life of the Author. A new edition. 6 Volumes. London: Printed for J. and F. Rivington, L. Davis, W. Johnston, S. Crowder, T. Payne, G. Kearsley, J. Robson, B. White, T. Becket and P. A. De Hondt, T. Davies, T. Cadell, Robinson and Roberts. Richardson and Richardson, J. Knox, W. Woodfall, J. Johnson, and T. Evans, 1772.

Brown, Harcourt. *Scientific Organizations in Seventeenth Century France*. Reissue of the 1934 edition. New York: Russell & Russell, 1967.

Browne, Thomas. *The Works of Sir Thomas Browne*. Edited by Geoffrey Keynes. 4 Volumes. Chicago, London, and Toronto: University of Chicago Press, 1964.

Burtt, Edwin Arthur. *The Metaphysical Foundations of Modern Physical Science*. Revised edition. Doubleday Anchor Books. Garden City, New York: Doubleday & Co., 1954.

Caspar, Max. *Kepler*. Translated and edited by C. Doris Hellman. London and New York: Abelard-Schuman, 1959.

Cassirer, Ernst. *The Platonic Renaissance in England*. Translated by James P. Pettegrove. Austin, Texas: The University of Texas Press, 1953.

Charleton, Walter. *Physiologia-Epicuro-Gassendo-Charltoniana: Or a Fabrick of Science Natural, Upon the Hypothesis of Atoms, Founded by Epicurus, Repaired by Petrus Gassendus, Augmented by Walter Charleton, Dr. in Medicine, and Physician to the late Charles, Monarch of Great Britain*. Indexes and introduction by Robert Hugh Kargon. Reprint of the London edition of 1654. The Sources of Science, no. 31. New York and London: Johnson Reprint Corporation, 1966.

Colie, Rosalie L. *Light and Enlightenment. A Study of the Cambridge Platonists and the Dutch Arminians*. Cambridge: At the University Press, 1957.

A Collection of Letters Illustrative of the Progress of Science in England from the Reign of Queen Elizabeth to that of Charles the Second. Edited by James Orchard Halliwell. London: Printed for the Historical Society of Science, 1841.

Comenius. Edited, with introduction, by John Sadler. Educational Thinkers Series. London: Collier–Macmillan, 1969.

Comenius, John Amos. *A Reformation of Schooles, designed in two excellent Treatises: The first whereof Summarily sheweth, The great necessity of a generall Reformation of Common Learning. What grounds of hope there are for such a Reformation. How it may be brought to passe. The second answers certaine objections ordinarily made against such undertakings, and describes the severall Parts and Titles of Workes which are shortly to follow. Written many yeares agoe in Latine by that Reverend, Godly, Learned, and famous Divine Mr. John Amos Comenius, one of the Seniours of the exiled Church of Moravia: And now upon the request of many translated into English, and published by Samuel Hartlib, for the generall good of this Nation.* London: Printed for Michael Sparke senior, at the Blew Bible in Greene Arbor, 1642.

The Way of Light. Translated, with introduction, by E. T. Campagnac. Liverpool: The University Press; London: Hodder & Stoughton, 1938.

Conway Letters. The Correspondence of Anne, Viscountess Conway, Henry More, and their Friends, 1642–1684. Collected and edited, with a biographical account, by Marjorie Hope Nicolson. New Haven, Connecticut: *Yale University Press,* and to be sold in London by Humphrey Milford, *Oxford University Press,* 1930.

Curtis, Mark H. *Oxford and Cambridge in Transition, 1558–1642. An Essay on Changing Relations between the English Universities and English Society.* Oxford: At the Clarendon Press, 1965.

Debus, Allen G. *Science and Education in the Seventeenth Century: the Webster–Ward Debate.* London: Macdonald; New York: American Elsevier, 1970.

Descartes, René. *Correspondence avec Arnauld et Morus.* Introduction and notes by Geneviève Lewis. Bibliotheque des textes philosophiques, Henry Gouhier, Directeur. Paris: Librairie philosophique J. Vrin, 1953.

[d'Espagnet, Jean.] *Enchyridion Physicae Restitutae; or, The Summary of Physicks Recovered. Wherein the true Harmony of Nature is explained, and many Errours of the Ancient Philosophers, by Canons and certain Demonstrations, are clearly evidenced and evinced.* London: Printed for *W. Bentley,* and are to be sold by *W. Sheares* at the *Bible,* and *Robert Tutchein* at the *Phenix,* in the New-Rents in S. *Pauls* Church-Yard, 1651.

Digby, Kenelm. *Of Bodies, and Of Mans Soul. To Discover the Immortality of Reasonable Souls. With two Discourses Of the Powder of Sympathy, and Of the Vegetation of Plants.* London: Printed by *S. G.* and *B. G.* for *John Williams,* and are to be sold in Little *Britain* over against St. *Buttolphs-Church,* 1669.

Dijksterhuis, E. J. *The Mechanization of the World Picture.* Translated by C. Dikshoorn. London, Oxford, New York: Oxford University Press, 1969.

Evans, Robert John Weston. *Rudolf II and His World. A Study in Intellectual History 1576–1612.* Oxford: At the Clarendon Press, 1973.

Evelyn, John. *Diary.* Edited by E. S. de Beer. 6 Volumes. Oxford: Clarendon Press, 1955.

Fuller, Thomas. *The History of the Worthies of England: Endeavoured by Thomas Fuller, D.D. First Printed in 1662.* A new edition. 2 Volumes. London: Printed for F. C. and J. Rivington; T. Payne; Wilkie and Robinson; Longman, Hurst, Rees, Orme, and Brown; Cadell and Davies; R. H. Evans; J. Mawman; J. Murray; and R. Baldwin, 1811.

Fulton, John F. *A Bibliography of the Honourable Robert Boyle, Fellow of the Royal Society.* 2nd Edition. Oxford: At the Clarendon Press, 1961.

"Robert Boyle and his influence on thought in the seventeenth century." *Isis* **18** (1932), 88–90.

265

Gelbart, Nina Rattner. "The intellectual development of Walter Charleton." *Ambix* 18 (1971), 149–68.

Hooke, Robert. *The Diary of Robert Hooke, M.A., M.D., F.R.S., 1672–1680, Transcribed from the Original in the Possession of the Corporation of the City of London (Guildhall Library)*. Edited by Henry W. Robinson and Walter Adams, with foreword by Sir Frederick Gowland Hopkins. London: Taylor & Francis, 1935.

Houghton, W. E., Jr. "The English virtuoso in the seventeenth century." *Journal of the History of Ideas* 3 (1942), 51–73, 190–219.

Hutin, Serge. *Henry More. Essai sur les doctrines théosophiques chez les Platoniciens de Cambridge*. Studien und Materialien zur Geschichte der Philosophie, Herausgegeben von Heinz Heimsoeth, Dieter Henrich, und Giorgio Tonelli, Band 2. Hildesheim: Georg Olms Verlagsbuchhandlung, 1966.

Jacob, J. R. "The ideological origins of Robert Boyle's natural philosophy." *Journal of European Studies* 2 (1972), 1–21.

Jahn, Melvin E. "A bibliographic history of John Woodward's *An essay toward a natural history of the Earth.*" *The Journal of the Society for the Bibliography of Natural History* 6 (1972), 181–213.

Johnson, Francis R. "Gresham College: precursor of the Royal Society." *Journal of the History of Ideas* 1 (1940), 413–38.

Jones, Richard Foster. *Ancients and Moderns. A Study of the Rise of the Scientific Movement in Seventeenth-Century England*. 2nd Edition. St. Louis: Washington University Studies, 1961.

Koyré, Alexandre. *From the Closed World to the Infinite Universe*. Johns Hopkins Paperbacks edition. Baltimore: The Johns Hopkins Press, 1968.

Kuhn, Thomas S. *The Structure of Scientific Revolutions*. Phoenix Books. Chicago and London: The University of Chicago Press, 1964.

Lenoble, Robert. *Mersenne, ou la naissance du mécanisme*. Paris: Librairie philosophique J. Vrin, 1943.

[Longueville, Thomas.] T. L. *The Life of Sir Kenelm Digby by One of his Descendants*. London, New York and Bombay: Longmans, Green and Co., 1896.

Lyons, Sir Henry. *The Royal Society, 1660–1940. A History of its Administration under its Charters*. Cambridge: University Press, 1944.

Maddison, R. E. W. *The Life of the Honourable Robert Boyle F.R.S.* London: Taylor & Francis; New York: Barnes and Noble, 1969.

Monconys, Balthazar de. *Iovrnal des voyages de Monsievr de Monconys, Conseiller du Roy en ses Conseils d'Estat & Priué, & Lieutenant Criminel au Siege Presidial de Lyon. Oú les Sçauants trouueront vn nombre infini de nouueautez, en Machines de Mathematique, Experiences Physiques, Raisonnemens de la belle Philosophie, curiositez de Chymie, & conuersations des Illustres de ce Siecle; Outre la description de diuers Animaux & Plantes rares, plusieurs Secrets inconnus pour le Plaisir & la Santé, les Ouurages des Peintres fameux, les Coûtumes & Moeurs des Nations, & ce qu'il y a de plus digne de la connoissance d'vn honeste Homme dans les trois Parties du Monde. Enrichi de quantité de Figures en Taille-douce des lieux & des choses principales. Auec des Indices tres-exacts & tres-commodes pour l'vsage. Publié par le Sieur le Liergves son Fils.* 3 Volumes. Lyon: Chez Horace Boissat, & George Remevs, 1665–66.

More, Henry. *Enthusiasmus Triumphatus (1662)*. Introduction by M. V. DePorte. The Augustan Reprint Society, Publication 118. Los Angeles: William Andrews Clark Memorial Library, 1966.

The Immortality of the Soul, So farre forth as it is demonstrable from the Knowledge of Nature and the Light of Reason. London: Printed by *J. Flesher*, for *William Morden* Bookseller in Cambridge, 1659.

Philosophical Writings of Henry More. Edited, with introduction and notes, by Flora Isabel Mackinnon. The Wellesley Semi-Centennial Series. New York, London, Melbourne & Bombay: Oxford University Press, American Branch, 1925.

More, Louis Trenchard. *The Life and Works of the Honourable Robert Boyle*. London, New York, Toronto: Oxford University Press, 1944.

O'Brien, John J. "Samuel Hartlib's influence on Robert Boyle's scientific development. Part I. The Stalbridge period." *Annals of Science* **21** (1965), 1–14.

"Samuel Hartlib's influence on Robert Boyle's scientific development. Part II. Boyle in Oxford." *Annals of Science* **21** (1965), 257–76.

Oldenburg, Henry. *The Correspondence of Henry Oldenburg*. Edited and translated by A. Rupert Hall and Marie Boas Hall. 7 Volumes, continuing. Madison and Milwaukee: The University of Wisconsin Press, 1965– .

Ornstein, Martha. *The Role of Scientific Societies in the Seventeenth Century*. 3rd Edition. Chicago: University of Chicago Press, 1938.

Osmond, Percy H. *Isaac Barrow. His Life and Times*. London: Society for Promoting Christian Knowledge, 1944.

Patterson, T. S. "John Mayow in contemporary setting," *Isis* **15** (1931), 47–96, 504–46.

Petersson, R. T. *Sir Kenelm Digby, The Ornament of England, 1603–1665*. London: Jonathan Cape, 1956.

Power, Henry. *Experimental Philosophy*. The Sources of Science, no. 21. New York and London: Johnson Reprint Corporation, 1966.

Purver, Margery. *The Royal Society: Concept and Creation*. Introduction by H. R. Trevor-Roper. Cambridge, Massachusetts: The M.I.T. Press, 1967.

Raven, Charles E. *John Ray, Naturalist. His Life and Works*. 2nd Edition. Cambridge: At the University Press, 1950.

The Rise of Modern Science. External or Internal Factors? Edited, with introduction, by George Basalla. Problems in European Civilization. Lexington, Massachusetts: D. C. Heath and Co., 1968.

Rowbottom, Margaret E. "The earliest published writing of Robert Boyle." *Annals of Science* **6** (1950), 376–89.

The Royal Society: its Origins and Founders. Edited by Sir Harold Hartley. London: The Royal Society, 1960.

Samuel Hartlib and the Advancement of Learning. Edited, with introduction, by Charles Webster. Cambridge Texts and Studies in the History of Education. General editors, A. C. F. Beales, A. V. Judges and J. P. C. Roach. Cambridge: At the University Press, 1970.

Science, Medicine, and History. Essays on the Evolution of Scientific Thought and Medical Practice Written in Honour of Charles Singer. Collected and edited by E. Ashworth Underwood. 2 Volumes. London, New York, and Toronto: Oxford University Press, 1953.

Society, Medicine and Society in the Renaissance. Essays to honor Walter Pagel. Edited by Allen G. Debus. 2 Volumes. New York: Science History Publications, 1972.

Solomon, Howard M. *Public Welfare, Science, and Propaganda in Seventeenth Century France. The Innovations of Théophraste Renaudot*. Princeton: Princeton University Press, 1972.

Sprat, Thomas. *History of the Royal Society*. Edited with critical apparatus by Jackson I. Cope and Harold Whitmore Jones. Washington University Studies. St. Louis, Missouri: Washington University, 1958.

The History of the Royal-Society of London. For the Improving of Natural Knowledge.

London: Printed by *T. R.* for *J. Martyn* at the *Bell* without *Temple-bar*, and *J. Allestry* at the *Rose* and *Crown* in *Duck-lane*, Printers to the *Royal Society*, 1667.

Stimson, Dorothy. *Scientists and Amateurs. A History of the Royal Society*. New York: Henry Schuman, 1948.

Syfret, R. H. "The origins of the Royal Society." *Notes and Records of the Royal Society* **5** (1948), 75–137.

Trevelyan, G. M. *Trinity College. An Historical Sketch*. Cambridge: At the University Press, 1943.

Walker, D. P. *The Ancient Theology. Studies in Christian Platonism from the Fifteenth to the Eighteenth Century*. London: Duckworth, 1972.

"The *Prisca Theologia* in France." *Journal of the Warburg and Courtauld Institutes* **17** (1954), 204–59.

Ward, Richard. *The Life of the Learned and Pious Dr Henry More Late Fellow of Christ's College in Cambridge By Richard Ward, A.M. Rector of Ingoldsby in Lincolnshire, 1710. To which are annexed Divers Philosophical Poems and Hymns*. Edited, with introduction and notes, by M. F. Howard. London: *Published and Sold by* The Theosophical Publishing Society at 161 New Bond St., 1911.

Webster, Charles. "Henry Power's experimental philosophy." *Ambix* **14** (1967), 150–78.

"Water as the ultimate principle of nature: The background to Boyle's *Sceptical Chymist*." *Ambix* **13** (1966), 96–107.

Westfall, Richard S. *Science and Religion in Seventeenth-Century England*. Yale Historical Publications, edited by David Horne, Miscellany 67. New Haven: Yale University Press, 1958.

Woodward, John. *An Essay toward a Natural History of the Earth: and Terrestrial Bodies, Especially Minerals: As also of the Sea, Rivers, and Springs. With an Account of the Universal Deluge: And of the Effects that it had upon the Earth*. London: Printed for *Ric. Wilkin* at the *Kings-Head* in St. *Paul's* Churchyard, 1695.

Worthington, John. *The Diary and Correspondence of Dr. John Worthington, Master of Jesus College, Cambridge, Vice-Chancellor of the University of Cambridge, etc., etc. From the Baker MSS. in the British Museum and the Cambridge University Library and Other Sources*. Edited by James Crossley and Richard Copley Christie. Remains Historical and Literary Connected with the Palatine Counties of Lancaster and Chester, volumes XIII, XXXVI, and CXIV. 2 Volumes in 3. Manchester: The Chetham Society, 1847–86.

C. THE CHEMICAL BACKGROUND

Boas (Hall), Marie. "Acid and alkali in seventeenth century chemistry." *Archives internationales d'histoire des sciences* **9** (1956), 13–28.

Robert Boyle and Seventeenth-Century Chemistry. Cambridge: At the University Press, 1958.

Coleby, L. J. M. "John Francis Vigani, first Professor of Chemistry in the University of Cambridge." *Annals of Science* **8** (1952), 46–60.

Crosland, Maurice P. *Historical Studies in the Language of Chemistry*. Cambridge, Massachusetts: Harvard University Press, 1962.

Debus, Allen G. "Fire analysis and the elements in the sixteenth and seventeenth centuries." *Annals of Science* **23** (1967), 127–47.

Hales, Stephen. *Vegetable Staticks: Or, An Account of some Statical Experiments on the Sap in Vegetables: Being an Essay towards a Natural History of Vegetation. Also, a Specimen of an Attempt to Analyse the Air, By a great Variety of Chymico-Statical Experiments; Which were read at several Meetings before the Royal Society.* London: Printed for W. and J. Innys, at the West End of St. *Paul's*; and T. Woodward, over-against St. *Dunstan's* Church in *Fleetstreet*, 1727.

John Dalton and the Progress of Science: Papers Presented to a Conference of Historians of Science Held in Manchester September 19–24, 1966, to Mark the Bicentenary of Dalton's Birth. Edited by D. S. L. Cardwell. Manchester: Manchester University Press; New York: Barnes & Noble, 1968.

Knight, D. M. *Atoms and Elements. A Study of Theories of Matter in England in the Nineteenth Century.* London: Hutchinson & Co., 1967.

le Fèvre, Nicolas. *A Compleat Body of Chymistry: Wherein is contained whatsoever is necessary for the attaining to the Curious Knowledge of this Art; Comprehending in General the whole Practice thereof; and Teaching the most exact Preparation of Animals, Vegetables and Minerals, so as to preserve their Essential Vertues. Laid open in two Books, and Dedicated to the Use of all Apothecaries, &c. By Nicasius le Febure, Royal Professor in Chymistry to His Majesty of England, and Apothecary in Ordinary to His Honourable Household. Fellow of the Royal Society. Rendred into English, by P. D. C. Esq; one of the Gentlemen of His Majesties Privy-Chamber. Corrected and amended; with the additions of the late French Copy.* 2 Volumes in 1. London: Printed for O. *Pulleyn* Junior, and are to be sold by *John Wright* at the Sign of the *Globe* in *Little-Brittain*, 1670.

Lémèry, Nicolas. *A Course of Chymistry; containing An easie Method of Preparing those Chymical Medicines which are used in Physick. With Curious Remarks, and Useful Discourses upon Each Preparation, for the Benefit of such as desire to be Instructed in the Knowledge of this Art.* 3rd Edition from the 8th French edition, much enlarged. Translated by James Keill. London: Printed for *W. Kettilby*, and sold by *James Bonwicke* at the *Hat* and *Star* in St. *Paul's* Church-yard, 1698.

Traité de l'antimonie, contenant L'Analyse Chymique de ce Mineral, & un recueil d'un grand nombre d'operations rapportees a l'Academie Royale des science, avec les raison-nemens qu'on a crus necessaires. Paris: Chez Jean Boudot, 1707.

Lewis, William. *A Course of Practical Chemistry. In Which are contained All the Operations Described in Wilson's Complete Course of Chemistry. With Many new, and several uncommon Processes. To each Article is given, The Chemical History, and to most, an Account of the Quantities of Oils, Salts, Spirits, yielded in Distillation, &c. From Lemery, Hoffman, the French Memoirs, Philosophical Transactions, &c. and from the Author's own Experience. With Copper Plates.* London: Printed for J. Nourse, at the *Lamb*, against *Katherine-street*, in the *Strand*, 1746.

Mellor, J. W. *A Comprehensive Treatise on Inorganic and Theoretical Chemistry.* 16 Volumes. New York: John Wiley & Sons, 1960.

Metzger, Hélène. *Les doctrines chimiques en France du début du XVIIe à la fin du XVIIIe siècle.* Paris: Les Presses Universitaires de France, 1923.

Les doctrines chimiques en France du début du XVIIe à la fin du XVIIIe siècle. Nouveau tirage. Paris: Librairie scientifique et technique Albert Blanchard, 1969.

Newton, Stahl, Boerhaave et la doctrine chimique. Paris: Librairie Felix Alcan, 1930.

Paneth, F. A. "The epistemological status of the chemical concept of element (I)." *British Journal for the Philosophy of Science* **13** (1962), 1–14.

"The epistemological status of the chemical concept of element (II)." *The British Journal for the Philosophy of Science* **13** (1962), 144–60.

Partington, J. R. "The concepts of substance and chemical element." *Chymia* 1 (1948), 114–15.

A History of Chemistry. 4 Volumes. London: Macmillan & Co.; New York: St. Martin's Press, 1961–70.

269

Everyday Chemistry. 3rd Edition. London: Macmillan & Co.; New York: St. Martin's Press, 1957.

Peck, E. Saville. "John Francis Vigani, first Professor of Chemistry in the University of Cambridge (1703–12), and his Materia Medica Cabinet in the library of Queens' College." *Proceedings of the Cambridge Antiquarian Society* **34** (1934), 34–49.

Philosophical Transactions and Collections To the End of the Year 1700. *Abridg'd and Dispos'd under General Heads, by John Lowthorp.* 3 Volumes. London: Printed for Thomas Bennet, Robert Knaplock, and Richard Wilkin, 1708. *Vol. III. In Two Parts. The First Containing all the Anatomical, Medical, and Chymical, and the Second all the Philological and Miscellaneous Papers.*

Sennertus, Daniel, Culpeper, Nich., and Cole, Abdiah. *Chymistry Made Easie and Useful. Or, The Agreement and Disagreement Of the Chymists and Galenists.* London: Printed by *Peter Cole,* Printer and Book-seller, at the Sign of the Printing-press in Cornhill, near the Royal Exchange, 1662.

Siegfried, Robert and Dobbs, Betty Jo. "Composition, a neglected aspect of the chemical revolution." *Annals of Science* **24** (1968), 275–93.

Smith, Edgar F. *Old Chemistries.* New York and London: McGraw-Hill, 1927.

Tachenius, Otto. *Otto Tachenius his Hippocrates Chymicus Discovering The Ancient foundation of the late Viperine Salt with his Clavis there unto annexed, Translated by J. W.* London: Printed & are to be sold by Nath: Crouch at the George at the lower end of Cornhill over against ye Stocks Market, 1677.
Otto Tachenius his Hippocrates Chymicus Discovering The Ancient foundation of the late Viperine Salt with his Clavis there unto annexed, Translated by J. W. London: Printed & are to be sold by W. Marshall at the Bible in Newgate Street, 1690.

Vigani, John Francis. *Medulla Chymiae. Notis experientiâ nixis illustrata observationibusque Practicis aucta a Davide Stam.* Lugduni Batavorum: Apud Felicem Lopez, 1693.

Vlastos, G. "Minimal parts in Epicurean atomism." *Isis* **56** (1965), 121–47.

D. THE ALCHEMICAL BACKGROUND

Andreae, John Valentin. *Ioannis Valentini Andreae Theologi Q. Württenbergensis. Vita, ab ipso conscripta. Ex autographo, in Bibl. Guelferbytano recondito, adsumtis Codd. Stuttgartianis, Schorndorfiensi, Tubingensi, nunc primum edidit F. H. Rheinwald, Dr. Cum icone et chirographo Andreano.* Berolino: Apud Herm. Schultzium, 1849.

Armytage, W. H. G. "The early Utopists and science in England." *Annals of Science* **12** (1956), 247–54.

Berthelot, Marcellin Pierre Eugene. *La chimie au moyen âge.* 3 Volumes. Paris: Imprimerie nationale, 1893.
Introduction a l'étude de la chimie des anciens et du moyen âge. Paris: Georges Steinheil, 1889.
Les origines de l'alchimie. Paris: Georges Steinheil, 1885.

Bolton, Henry Carrington. *The Follies of Science at the Court of Rudolph II. 1576–1612.* Milwaukee: Pharmaceutical Review Publishing Co., 1904.

Burckhardt, Titus. *Alchemy, Science of the Cosmos, Science of the Soul.* Translated by W. Stoddart. London: Stuart & Watkins, 1967.

Burland, C. A. *The Arts of the Alchemists.* New York: The Macmillan Co., 1968.

Canseliet, E. *Alchimie, Etudes diverses de Symbolisme hermétique et de pratique Philosophale.* Paris: Jean-Jacques Pauvert, 1964.

Červenka, Jaromír. *Die Naturphilosophie des Johann Amos Comenius.* Praha: Academia Verlag der Tschechoslowakischen Akademie der Wissenschaften, 1970.

"J. A. Komenský, Ladislav Velen ze Žerotina a Alchymie." *Acta Comeniana* **24** (1970), 21–44. Summary in German.

Craven, J. B. *Count Michael Maier, Doctor of Philosophy and of Medicine, Alchemist, Rosicrucian, Mystic, 1568–1622. Life and Writings.* Reprint of the Kirkwall edition of 1910. London: Dawsons, 1968.

Doctor Robert Fludd (Robertus de Fluctibus), The English Rosicrucian. Life and Writings. Reprint of the Kirkwall edition of 1902. N. p.: Occult Research Press, n. d. [196?].

Debus, Allen G. "Alchemy and the historian of science." *History of Science* **6** (1967), 128–38.

The English Paracelsians. New York: Franklin Watts, 1966.

"Gabriel Plattes and his chemical theory of the formation of the earth's crust." *Ambix* **9** (1961), 162–65.

"Harvey and Fludd: The irrational factor in the rational science of the seventeenth century." *Journal of the History of Biology* **3** (1970), 81–105.

"Mathematics and nature in the chemical texts of the Renaissance." *Ambix* **15** (1968), 1–28.

"The Paracelsian aerial nitre." *Isis* **55** (1964), 43–61.

"Renaissance chemistry and the work of Robert Fludd." *Ambix* **14** (1967), 42–59.

"Robert Fludd and the circulation of the blood." *Journal of the History of Medicine and Allied Sciences* **16** (1961), 374–93.

"Robert Fludd and the use of Gilbert's *De Magnete* in the weapon–salve controversy." *Journal of the History of Medicine and Allied Sciences* **19** (1964), 389–417.

Dobbs, Betty Jo. "Studies in the natural philosophy of Sir Kenelm Digby. Part I." *Ambix* **18** (1971), 1–25.

"Studies in the natural philosophy of Sir Kenelm Digby. Part II. Digby and alchemy." *Ambix* **20** (1973) 143–63.

"Studies in the natural philosophy of Sir Kenelm Digby. Part III. Digby's experimental alchemy – the book of *Secrets.*" *Ambix* **21** (1974), 1–28.

Eliade, Mircea. *The Forge and the Crucible.* Translated by Stephen Corrin. Harper Torchbooks. New York and Evanston: Harper & Row, 1971.

Figurevski, N. A. "The alchemist and physician Arthur Dee (Artemii Ivanovich Dii)." *Ambix* **8** (1965), 35–51.

Fordham, Frieda. *An Introduction to Jung's Psychology.* A Pelican Book. Harmondsworth: Penguin Books, 1956.

French, Peter J. *John Dee. The World of an Elizabethan Magus.* London: Routledge & Kegan Paul, 1972.

Fromm, Erich. *Escape from Freedom.* New York: Holt, Rinehart and Winston, 1941.

Geoghegan, D. "Gabriel Plattes' caveat for alchymists." *Ambix* **10** (1962), 97–102.

Gibbs, F. W. "Boerhaave's chemical writings." *Ambix* **6** (1958), 117–35.

Gregory, Joshua C. "Chemistry and alchemy in the natural philosophy of Sir Francis Bacon, 1561–1626." *Ambix* **2** (1938–46), 93–111.

Guerlac, Henry, "John Mayow and the aerial nitre. Studies on the chemistry of John Mayow – I." *Actes du VIIe Congrès International d'Histoire des Sciences. Jérusalem (4–12 Août 1953).* Collection des Travaux de l'Academie internationale d'Histoire des Sciences, No. 8. Publiée avec le Concours financier de L'UNESCO. Jérusalem: F. S. Bodenheimer, n. d.

"The poets' nitre. Studies in the chemistry of John Mayow. II." *Isis* **45** (1954), 243–55.

Hall, Thomas S. "Life, death and the radical moisture. A study of thematic pattern in medieval medical theory." *Clio Medica* **6** (1971), 3–23.

Heym, Gerard. "Review of Jung's *Mysterium conjunctionis.*" *Ambix* **6** (1957–58), 47–51.

"Review of Jung's *Paracelsica.*" *Ambix* **2** (1938–46), 196–98.

"Review of Jung's *Psychologie und Alchemie*," *Ambix* **3** (1948–49), 64–67.

Holmyard, Eric John. *Alchemy*. Harmondsworth: Penguin Books, 1968.

Hopkins, Arthur John. *Alchemy, Child of Greek Philosophy*. New York: Columbia University Press, 1934.

Hubicki, Włodzimierz. "Michael Sendivogius's Theory, its Origin and Significance in the History of Chemistry." In *Proceedings of the Tenth International Congress of the History of Science, Ithaca (Aug. 26–Sept. 2 1962)*. 2 Volumes. Paris: Hermann, 1964.

Ihde, Aaron J. "Alchemy in reverse: Robert Boyle on the degradation of gold." *Chymia* **9** (1964), 47–57.

Jong, H. M. E. de. *Michael Maier's Atalanta Fugiens, Bronnen van een Alchemistisch Emblemenboek*. Utrecht: Schotanus & Jens, 1965.
  Michael Maier's Atalanta fugiens. Sources of an Alchemical Book of Emblems. With 82 Illustrations. Janus: Revue internationale de l'histoire des sciences, de la médecine, de la pharmacie et de la technique. Suppléments, Volume VIII. Redaction: E. M. Bruins, R. J. Forbes, G. A. Lindeboom, D. A. Wittop Koning. Leiden: E. J. Brill, 1969.

Jung, Carl Gustav. *The Collected Works of C. G. Jung*. Edited by Herbert Read, Michael Fordham, Gerhard Adler, and William McGuire. Translated by R. F. C. Hull, Leopold Stein in collaboration with Diana Riviere, and H. G. Baynes. 17 Volumes in 18, continuing. Bollingen Series XX. New York: Pantheon Books; Princeton: Princeton University Press; London: Routledge & Kegan Paul, 1953– . Vol. 9, part 1: *The Archetypes and the Collective Unconscious*. Vol. 9, part 2: *Aion: Researches into the Phenomenology of the Self*. Vol. 12: *Psychology and Alchemy*. Vol. 13: *Alchemical Studies*. Vol. 14: *Mysterium conjunctionis: An Inquiry into the Separation and Synthesis of Psychic Opposites in Alchemy*.

Kopp, Hermann Franz Moritz. *Die Alchemie in Älterer und Neuerer Zeit. Ein Beitrag zur Kulturgeschichte*. 2 Volumes. Heidelberg: Carl Winter's Universitätsbuchhandlung, 1886.

Leicester, Henry M. *The Historical Background of Chemistry*. Reprint of the 1956 edition. New York: Dover Publications, 1971.

Lindeboom, G. A. *Herman Boerhaave. The Man and his Work*. Foreword by E. Ashworth Underwood. London: Methuen & Co., 1968.

Lindsay, Jack. *The Origins of Alchemy in Graeco-Roman Egypt*. New York: Barnes & Noble, 1970.

Marx, Jacques. "Alchimie et Palingénésie." *Isis* **62** (1971), 275–89.

Montgomery, John Warwick. *Cross and Crucible, Johann Valentin Andreae (1586–1654), Phoenix of the Theologians. Vol. I. Andreae's Life, World-view, and Relations with Rosicrucianism and Alchemy. Vol. II. The Chymische Hochzeit with Notes and Commentary*. International Archives of the History of Ideas, no. 55. The Hague: Martinus Nijhoff, 1973.

"Cross, constellation, and crucible: Lutheran astrology and alchemy in the age of the Reformation." *Ambix* **11** (1963), 65–86.

More, Louis Trenchard. "Boyle as alchemist." *Journal of the History of Ideas* **2** (1941), 61–76.

Multhauf, Robert P. *The Origins of Chemistry*. Oldbourne History of Science Library. London: Oldbourne, 1966.

Pachter, Henry M. *Paracelsus: Magic into Science*. New York: Henry Schuman, 1951.

Pagel, Walter. "Jung's views on alchemy." *Isis* **39** (1948), 44–48.

    Das Medizinische Weltbild des Paracelsus seine Zusammenhänge mit Neuplatonismus und Gnosis. Kosmosophie Forschungen und Texte zur Geschichte des Weltbildes, der Naturphilosophie, der Mystick und des Spiritualismus vom Spätmittelalter bis zur Romantik. Im Auftrage der Paracelsus-Kommission und in Verbindung mit der Paracelsus-Ausgabe. Herausgegeben von Kurt Goldammer. Wiesbaden: Franz Steiner Verlag GMBH, 1962.

    "Paracelsus and the neo-Platonic and Gnostic tradition." *Ambix* **8** (1960), 125–66.

    Paracelsus. An Introduction to Philosophical Medicine in the Era of the Renaissance. Basel and New York: S. Karger, 1958.

    "The religious and philosophical aspects of van Helmont's science and medicine." *Bulletin of the History of Medicine Supplements* **1**–**4** (1943–45), no. 2. Baltimore: The Johns Hopkins Press, 1944.

    "Review of Frances A. Yates' *Giordano Bruno and the Hermetic Tradition*. Appendix: Hermetic Alchemy at Bruno's time." *Ambix* **12** (1964), 72–76.

Partington, J. R. "The life and work of John Mayow (1641–1679). Part i." *Isis* **47** (1956), 217–30.

    "The life and work of John Mayow. Part ii." *Isis* **47** (1956), 405–17.

Patterson, T. S. "Jean Beguin and his Tyrocinium Chymicum." *Annals of Science* **2** (1937), 243–98.

Rattansi, P. M. "Alchemy and natural magic in Raleigh's 'History of the World.'" *Ambix* **13** (1966), 122–38.

    "The Helmontian–Galenist controversy in Restoration England." *Ambix* **12** (1964), 1–23.

    "Paracelsus and the puritan revolution." *Ambix* **11** (1963), 24–32.

Read, John. *The Alchemist in Life, Literature, and Art*. London, Edinburgh, Paris, Melbourne, Toronto, and New York: Thomas Nelson and Sons, 1947.

    Prelude to Chemistry: An Outline of Alchemy, its Literature and Relationships. London: G. Bell and Sons, 1936; New York: The Macmillan Co., 1937.

    Through Alchemy to Chemistry. A Procession of Ideas and Personalities. Harper Torchbooks. New York and Evanston: Harper & Row, 1963.

Redgrove, Herbert Stanley. *Alchemy: Ancient and Modern, Being a Brief Account of the Alchemistic Doctrines, and their Relations, to Mysticism on the One Hand, and to Recent Discoveries in Physical Science on the Other Hand; Together with some Particulars Regarding the Lives and Teachings of the Most Noted Alchemists*. 2nd and Revised Edition. London: William Rider & Son, 1922.

Robbins, Rossell Hope. "Alchemical texts in Middle English verse: corrigenda and addenda." *Ambix* **13** (1966), 62–73.

Ross, Percy. *A Professor of Alchemy (Denis Zachaire)*. London: George Redway, 1887.

Schmieder, Karl Christoph. *Geschichte der Alchemie*. Halle: Verlag der Buchhandlung des Waisenhauses, 1832.

Smith, Charlotte Fell. *John Dee (1527–1608)*. London: Constable & Co. 1909.

Stillman, John Maxson. *The Story of Alchemy and Early Chemistry*. Reprint of the 1924 edition. New York: Dover Publications, 1960.

Taylor, Frank Sherwood. *The Alchemists, Founders of Modern Chemistry*. New York: Collier Books, 1962.

The Alchemists, Founders of Modern Chemistry. New York: Henry Schuman, 1949.

Taylor, Frank Sherwood and Josten, C. H. "Johannes Banfi Hunyades 1576–1650." *Ambix* **5** (1953–56), 44–52.

"Johannes Banfi Hunyades. A supplementary note." *Ambix* **5** (1953–56), 115.

Waite, Arthur Edward. *Alchemists Through the Ages. Lives of the Famous Alchemistical Philosophers from the year 850 to the close of the 18th century, together with a Study of the Principles and Practice of Alchemy, including a Bibliography of Alchemical and Hermetic Philosophy*. Introduction by Paul M. Allen. Reprint of the London Edition of 1888. Blauvelt, New York: Rudolf Steiner Publications, 1970.

The Real History of the Rosicrucians, Founded on Their Own Manifestoes, and on Facts and Documents Collected from the Writings of Initiated Brethren. London: George Redway, 1887.

Webster, Charles. "The authorship and significance of *Macaria*." *Past and Present*, no. 56 (August, 1972), 34–48.

"English medical reformers of the puritan revolution: a background to the 'Society of Chymical Physitians.'" *Ambix* **14** (1967), 16–41.

West, Muriel. "Notes on the importance of alchemy to modern science in the writings of Francis Bacon and Robert Boyle." *Ambix* **9** (1961), 102–14.

Wilkinson, Ronald Sterne. "The alchemical library of John Winthrop, Jr. (1606–1676) and his descendants in colonial America. Parts I–III." *Ambix* **11** (1963), 33–51.

"The alchemical library of John Winthrop, Jr. (1606–1676) and his descendants in colonial America. Part IV. The Catalogue of books." *Ambix* **13** (1966), 139–86.

"Further thoughts on the identity of 'Eirenaeus Philalethes.'" *Ambix* **19** (1972), 204–08.

"George Starkey, physician and alchemist." *Ambix* **11** (1963), 121–52.

"The Hartlib Papers and seventeenth century chemistry. Part I." *Ambix* **15** (1968), 54–69.

"The Hartlib Papers and seventeenth century chemistry. Part II." *Ambix* **17** (1970), 85–110.

"The problem of the identity of Eirenaeus Philalethes." *Ambix* **12** (1964), 24–43.

"Some bibliographical puzzles concerning George Starkey." *Ambix* **20** (1973), 235–44.

Wolbarsht, M. L. and Sax, D. S. "Charles II, a royal martyr." *Notes and Records of the Royal Society of London* **16** (1961), 154–57.

Yates, Frances A. *Giordano Bruno and the Hermetic Tradition*. Chicago: The University of Chicago Press; London: Routledge & Kegan Paul; Toronto: The University of Toronto Press, 1964.

Theatre of the World. Chicago: The University of Chicago Press, 1969.

The Rosicrucian Enlightenment. London and Boston: Routledge & Kegan Paul, 1972.

Young, Robert Fitzgibbon. *Comenius in England. The Visit of Jan Amos Komensky (Comenius), the Czech Philosopher and Educationist to London in 1641–42; Its Bearing on the Origins of the Royal Society, on the Development of the Encyclopaedia, and on Plans for the Higher Education of the Indians of New England and Virginia*. London: Oxford University Press, 1932.

E. ALCHEMY

[Andreae, John Valentin.] *The Hermetick Romance: or the Chymical Wedding. Written in high Dutch By Christian Rosencreutz. Translated by E. Foxcroft, late Fellow of Kings Colledge in Cambridge.* N.p. [London]: Printed, by *A. Sowle*, at the *Crooked-Billet* in *Holloway-Lane Shoreditch:* And sold at the *Three-Kyes* in *Nags-Head-Court Grace-Church-street*, 1690.

Artis avriferae, qvam chemiam vocant, Volumina duo, qvae continent Tvrbam Philosophorvm, aliosqúe antiquissimos auctores, quae versa pagina indicat. Accessit nouiter volumen tertium, continens: 1. Lullij vltimum Testamentum. 2. Elucidationem Testam. totius ad R. Odoardum. 3. Potestatem diuitiarum, cum optima expositione Testamenti Hermetis. 4. Compendium Artis Magicae, quoad compositionem Lapidis. 5. De Lapide & oleo Philosophorum. 6. Modum accipiendi aurum potabile. 7. Compendium Alchimiae & naturalis Philosophiae. 8. Lapidarium. Item Alberti Magni secretorum Tractatus. Abbreuiationes quasdam de Secretis Secretorum Ioannis pauperum. Arnaldi Quaest. de Arte Transmut. Metall. eiusqúe Testamentum. Omnia hactenus nunquam visa nec edita. Cum Indicibus rerum & verborum locupletissimis. 3 Volumes in 1. Basileae: Typis Conradi Waldkirchii, 1610.

Atwood, Mary Anne. *Hermetic Philosophy and Alchemy. A suggestive inquiry into " The Hermetic Mystery" with a dissertation on the more celebrated of the alchemical philosophers.* Introduction by Walter Leslie Wilmhurst. Revised Edition. New York: The Julian Press, 1960.

Aurifontina Chymica: or, A Collection Of Fourteen small Treatises Concerning the First Matter of Philosophers. For the discovery of their (hitherto so much concealed) Mercury. Which many have studiously endeavoured to Hide, but these to make Manifest, for the benefit of Mankind in general. Translated by John Frederick Houpreght. London: Printed for *William Cooper*, at the *Pelican* in *Little-Britain*, 1680.

Aurora consurgens: A Document Attributed to Thomas Aquinas on the Problem of Opposites in Alchemy. Edited, with commentary, by Marie-Louis von Franz. Translated by R. F. C. Hull and A. S. B. Glover. Bollingen Series LXXVII. New York: Pantheon Books, 1966.

Basilius Valentinus. *Basilius Valentinus, A Benedictine Monk, Of Natural & Supernatural Things. Also Of the first Tincture, Root, and Spirit of Metals and Minerals, how the same are Conceived, Generated, Brought forth, Changed, and Augmented. Whereunto is added, Frier Roger Bacon, of the Medicine or Tincture of Antimony; Mr. John Isaac Holland, his Work of Saturn, and Alex. Van Suchten, of the Secrets of Antimony.* Translated out of High Dutch by Daniel Cable. London: Printed, and are to be Sold by *Moses Pitt* at the *White Hart* in *Little Britain*, 1671.

Cvrrvs trivmphalis antimonii. Fratris Basillii Valentini Monachi Benedictini. Opvs Antiquioris Medicinae & Philosophiae Hermeticae studiosis dicutum. È Germanico in Latinum versum operâ, studio & sumptibus Petri Ioannis Fabri Doctoris Medici Monspeliensis. Et notis perpetuis ad Marginem appositis ab eodem illustratum. Tolosae: Apud Petrvm Bosc, 1646.

The Triumphal Chariot of Antimony. With the Commentary of Theodore Kerckringius, A Doctor of Medicine. Being the Latin Version Published at Amsterdam in the Year 1685 Translated into English. With a Biographical Preface by Arthur Edward Waite. Reprint of the London edition of 1894. London: Vincent Stuart, 1962.

Birrius, Martinus. *Tres tractatus De metallorum transmutatione. Quid singulis contineatur, sequens pagina indicat. Incognito auctore. Adjuncta est Appendix Medicamentorum Antipodagricorum & Calculifragi. Quae omnia ad bonum publicum promovendum nunc primum in lucem edi curavit Martinus Birrius, Philosophiae & Medicinae*

Doctor, Practicus Amstelodamensis, Apud quem Medicamenta ista reperiunter. Amstel-odami: Apud Johannem Janssonium à Waesberge, & Viduam Elizei Weyerstraet, 1668.

Boerhaave, Herman. *A New Method of Chemistry; including the Theory and Practice of that art: Laid down on Mechanical Principles, and accommodated to the Uses of Life. The whole making a clear and Rational System of Chemical Philosophy. To which is prefix'd A Critical History of Chemistry and Chemists, From the Origin of the Art to the present Time.* Translated by P. Shaw and E. Chambers. London: Printed for J. Osborn and T. Longman at the *Ship* in *Pater-noster-Row*, 1727.

Some Experiments Concerning Mercury. Translated from the Latin, communicated by the Author to the Royal Society. London: Printed for J. Roberts, near the *Oxford-Arms*, in *Warwick-Lane*, 1734.

The Book of Quinte Essence or The Fifth Being; That is to say, Man's Heaven. A tretice in englisch breuely drawne out of ye book of quintis eessencijs in latyn, yat hermys ye prophete and kyng of Egipt, after ye flood of Noe fadir of philosophris, hadde by reuelacioun of an aungil of god to him sende. Edited from the Sloane MS. 73, About 1460–70 A. D. Edited by Frederick J. Furnivall. London: Published for the Early English Text Society, by N. Trubner & Co., 1866, revised 1889.

Chymical, Medicinal, and Chyrurgical Addresses: Made to Samuel Hartlib, Esquire, viz. 1. Whether the Vrim & Thummin were given in the Mount, or perfected by Art. 2. Sir George Ripley's Epistle to King Edward unfolded. 3. Gabriel Plats Caveat for Alchymists. 4. A Conference concerning the Phylosophers Stone. 5. An Invitation to a free and generous Communication of Secrets and Receits in Physick. 6. Whether or no, each Several Disease hath a Particular Remedy? 7. A new and easie Method of Chirurgery, for the curing of all fresh Wounds or other Hurts. 8. A Discourse about the Essence or Existence of Mettals. 9. The New Postilions, pretended Prophetical Prognostication, Of what shall happen to Physitians, Chyrurgeons, Apothecaries, Alchymists, and Miners. London: Printed by G. Dawson for Giles Calvert at the Black-spread Eagle at the west end of *Pauls*, 1655.

Collectanea Chemica: Being Certain Select Treatises on Alchemy and Hermetic Medicine by Eirenaeus Philalethes, George Starkey, Dr. Francis Antony, Sir George Ripley, and Anonymous Unknown. Reprint. London: Vincent Stuart, 1963.

Digby, Kenelm. *A Choice Collection of Rare Secrets and Experiments in Philosophy. As Also Rare and unheard-of Medicines, Menstruums, and Alkahests; with the True Secret of Volatilizing the fixt Salt of Tartar. Collected And Experimented by the Honourable and truly Learned Sir Kenelm Digby, Kt. Chancellour to Her Majesty the Queen-Mother. Hitherto kept Secret since his Decease, but now Published for the good and benefit of the Publick, By George Hartman.* London: Printed for the Author, and are to be sold by *William Cooper*, at the *Pelican* in *Little Britain*; and *Henry Faithorne* and *John Kersey*, at the *Rose* in *St. Paul's* Church-yard, 1682.

Eirenaeus Philalethes. *Enarratio methodica trium gebri medicinarum, In quibus continetur lapidis philosophici vera confectio.* [Londini]: Sumptibus Guilielmi Cooper, Bibliopolae Londinensis, ad Insigne Pelicani, in Vico vulgò dicto Little Britain, 1678.

Ripley Reviv'd: or an Exposition upon Sir George Ripley's Hermetico-Poetical Works. Containing the plainest and most excellent Discoveries of the most hidden Secrets of the Ancient Philosophers, that were ever yet Published. Written by Eirenaeus Philalethes an Englishman, stiling himself Citizen of the World. London: Printed by *Tho. Ratcliff* and *Nat. Thompson*, for *William Cooper* at the *Pelican* in *Little-Britain*, 1678.

Secrets Reveal'd: or, An Open Entrance to the Shut-Palace of the King: Containing,

276

The greatest Treasure in Chymistry, Never yet so plainly Discovered. Composed By a most famous English-man, Styling himself Anonymous, or Eyraeneus Philaletha Cosmopolita: Who, by Inspiration and Reading, attained to the Philosophers Stone at his Age of Twenty three Years, Anno Domini, 1645. Published for the Benefit of all English-men, by W. C. Esq.; a true Lover of Art and Nature. London: Printed by *W. Godbid* for *William Cooper* in Little St. *Bartholomews,* near *Little-Britain,* 1669.

Fasciculus Chemicus: or Chymical Collections. Expressing The Ingress, Progress, and Egress, of the Secret Hermetick Science, out of the choisest and most Famous Authors. Collected and digested in such an order, that it may prove to the advantage, not onely of the Beginners, but Proficients of this high Art, by none hitherto disposed in this Method. Whereunto is added, The Arcanum or Grand Secret of Hermetick Philosophy. Both made English By James Hasolle [Elias Ashmole], Esquire, Qui est Mercurio-philus Anglicus. London: Printed by *J. Flesher* for *Richard Mynne,* at the sign of St. *Paul* in Little *Britain,* 1650.

Ferguson, John. "The marrow of alchemy." *Journal of the Alchemical Society* **3** (1915), 106–29.

Figulus, Benedictus. *A Golden and Blessed Casket of Nature's Marvels.* Translated by Arthur Edward Waite. Reprint of the London Edition of 1893. London: Vincent Stuart, 1963.

Flamel, Nicolas. *Nicholas Flammel, His Exposition of the Hieroglyphicall Figures which he caused to bee painted vpon an Arch in St. Innocents Church-yard, in Paris. Together with The secret Booke of Artephivs, And The Epistle of Iohn Pontanus: Concerning both the Theoricke and the Practicke of the Philosophers Stone. Faithfully, and (as the Maiesty of the thing requireth) religiously done into English out of the French and Latine Copies. By Eirenaevs Orandvs, qui est, Vera veris enodans.* London: Imprinted at *London* by *T. S.* for *Thomas Walkley,* and are to bee solde at his Shop, at the Eagle and Childe in *Britans Bursse,* 1624.

Geber. *The Works of Geber Englished by Richard Russell, 1678.* Edited with introduction by E. J. Holmyard. A new edition. London and Toronto: J. M. Dent & Sons; New York: E. P. Dutton & Co., 1928.

Josten, C. H. "A translation of John Dee's 'Monas Hieroglyphica' (Antwerp, 1564), with an introduction and notes." *Ambix* **12** (1964), 84–221.

Maier, Michael. *Atalanta fugiens hoc est emblemata nova de secretis naturae chymica Authore Michaele Majero.* Herausgegeben, mit Nachwort, von Lucas Heinrich Wüthrich. Faksimile-Druck der Oppenheimer Original-ausgabe von 1618 mit 52 Stichen von Matthaeus Merien d. Ä. Kassel und Basel: Im Bäsenreiter-Verlag, 1964.

Manget, John Jacob. *Bibliotheca chemica curiosa, seu Rerum ad alchemiam pertinentium thesaurus instructissimus: Quo non tantùm artis auriferae, Ac Scriptorum in ea Nobiliorum Historia traditur; Lapidis Veritas Argumentis & Experimentis innumeris, immò & Juris Consultorum Judiciis evincitur; Termini obscuriores explicantur; Cautiones contra Impostores, & Difficultates in Tinctura Universali conficienda occurrentes, declarantur: Verùm etiam tractatus omnes virorum Celebriorum qui in Magno sudarunt Elixyre, quíque ab ipso Hermete, ut dicitur, Trismegisto, ad nostra usque Tempora de Chrysopoea scripserunt, cum praecipuis suis Commentariis, concinno Ordine dispositi exhibentur. Ad quorum omnium Illustrationem additae sunt quamplurimae Figurae aeneae.* 2 Volumes. Genevae: Sumpt. Chouet, G. de Tournes, Cramer, Perachon, Ritter, & S. de Tournes, 1702.

Musaeum Hermeticum reformatum et amplificatum, omnes sopho-spagyricae artis discupulos fidelissimè erudiens, quo pacto Summa illa veraque lapidis philosophici Medicina, qua res omnes qualemcunque defectum patientes, instaurantur, inveniri & haberi queat. Continens tractatus chemicos XXI. Praestantissimos, quorum Nomina & Seriem versa pagella indicabit. Francofurti: Apud Hermannum à Sande, 1678.

Musaeum hermeticum reformatum et amplificatum. Facsimile reprint of the Frankfort edition of 1678. Introduction by Karl R. H. Frick. Introduction translated by C. A. Burland. Graz: Akademische Druck-u. Verlagsanstalt, 1970.

Porta, John Baptista. *Natural Magick.* Edited by Derek J. Price. The Collector's Series in Science. New York: Basic Books, 1957.

Reconditorium ac Reclusorium Opulentiae sapientiaeque Numinis Mundi Magni, Cui deditur in titulum chymica vannus, Obtenta quidem & erecta Auspice Mortale Coepto; Sed Inventa Proauthoribus Immortalibus Adeptis, Quibus Conclusum est, sancitum & decretum, Ut Anno hoc per Mysteriarchum Mercurium, Velut Viocurium, seu Medicurium, statvta oracvla sva exordinè inolescerent, & avrea veritas perspicacioribvs ingeniis undè breviterqve innotesceret. Orbe post Christum natum Millesimo, sexcentesimo, sexagesimo sexto, Idibus Majis. Amstelodami: Apud Joannem Janssonium à Waesberge, & Elizeum Weyerstraet, 1666.

Ripley, George. *The Compound of Alchymy. Or, the ancient hidden Art of Archemie: Conteining the right and perfectest meanes to make the Philosophers Stone, Aurum potabile, with other excellent Experiments. Divided into twelve Gates, First written by the learned and rare Philosopher of our Nation George Ripley, sometime Chanon of Bridlington in Yorkeshire: and Dedicated to K. Edward the 4. Whereunto is adioyned his Epistle to the King, his Vision, his Wheele, and other his Workes, neuer before published: with certaine briefe Additions of other notable Writers concerning the same. Set foorth by Raph Rabbards Gentleman, studious and expert in Archemicall Artes.* London: Imprinted by Thomas Orwin, 1591.

George Ripley, Canon of Bridlington, The Emblematical Scroll. Fitzwilliam MS 276*. Fitzwilliam Museum, Cambridge.

Georgii Riplaei Canonici Angli Opera omnia Chemica, quotquot hactenus visa sunt, quorum aliqua jam primum in lucem prodeunt, aliqua MS. exemplarium collatione à mendis & lacunis repurgata, atque integritati restituta sunt. Horvm Seriem pagina post praefationem prima monstrabit. Cassellis: Typis Jacobi Gentschii, Impensis Sebaldi Kohlers, 1649.

Ruska, Julius. *Turba philosophorum.* Reprint of the Berlin edition of 1931. Berlin, Heidelberg, New York: Springer-Verlag, 1970.

Sendivogius, Michael. *A new Light of Alchymie, taken out of the fountaine of Nature and Manuall Experience. To which is added a Treatise of Sulphur; also Nine Books of the Nature of Things written by Paracelsus, viz., of the Generations, Growths, Conservations, Life, Death, Renewing, Transmutation, Separation, Signatures of Natural Things. Also a Chymical Dictionary explaining hard places and words met withall in the writings of Paracelsus and other obscure Authors. All which are faithfully translated out of the Latin into the English tongue by J. F. [John French] M.D.* London: Printed by Richard Cotes, for Thomas Williams at the Bible in Little-Britain, 1650.

[Starkey, George?] Eirenaeus Philoponos Philalethes. *The Marrow of Alchemy, Being an Experimental Treatise, Discovering The secret and most hidden Mystery of the Philosophers Elixer. Divided into two Parts: The first Containing Four Books Chiefly Illustrating the Theory. The other Containing Three Books, Elucidating the Practique of the Art: In which, The Art is so plainly disclosed as never any before did for the benefit of young Practitioners, And the convincing those who are in Errours Labyrinth.* London: Printed by A. M. for *Edw. Brewster* at the Signe of the Crane in *Pauls* Church-yard, 1654.

Theatrum Chemicum Britannicum, Containing Severall Poeticall Pieces of our Famous

English Philosophers, who have written the Hermetique Mysteries in their owne Ancient Language. Faithfully Collected into one Volume with Annotations thereon by Elias Ashmole, Esq. The First Part. London: Printed by *J. Grismond* for Nath. Brooke, at the Angel in *Cornhill,* 1652.

Theatrum chemicum, praecipuos selectorum auctorum tractatus de chemiae et lapidis philosophici antiquitate, veritate, jure, praestantia, & operationibus, continens: In gratiam Verae Chemiae, & medicinae Chemicae studiosorum (ut qui uberrimam inde optimorum remediorum messem facere poterunt) congestum, & in Sex partes seu volumina digestum; singulis voluminibus, suo auctorum et librorum catalogo primis pagellis: rerum verò & verborum Indice postremis annexo. 6 Volumes. Argentorati: Sumptibus Heredum Eberh. Zetzneri, 1659–61.

The Turba Philosophorum, or Assembly of the Sages. Called also the Book of Truth in the Art and the Third Pythagorical Synod. An ancient alchemical treatise translated from the Latin, the chief readings of the shorter codex, parallels from the Greek alchemists, and explanations of obscure terms. Translated by Arthur Edward Waite. London: George Redway, 1896.

[Vaughan, Thomas.] Eugenius Philalethes. *The Fame and Confession of the Fraternity of R: C: Commonly of the Rosie Cross. With a Praeface annexed thereto, and a short Declaration of their Physicall Work.* Introduction and notes by F. N. Pryce. Facsimile reprint of the London edition of 1652. Margate: W. J. Parrett, 1923.

The Works of Thomas Vaughan: Eugenius Philalethes. Edited, annotated, and introduced by Arthur Edward Waite. London: The Theosophical Publishing House, 1919.

Webster, John. *Metallographia: or, An History of Metals. Wherein is declared the signs of Ores and Minerals both before and after digging, the causes and manner of their generations, their kinds, sorts, and differences; with the description of sundry new Metals, or Semi Metals, and many other things pertaining to Mineral knowledge. As also, The Handling and shewing of their Vegetability, and the discussion of the most difficult Questions belonging to Mystical Chymistry, as of the Philosophers Gold, their Mercury, the Liquor Alkahest, Aurum potabile, and such like. Gathered forth of the most approved Authors that have written in Greek, Latine, or High Dutch; With some Observations and Discoveries of the Author himself.* London: Printed by *A. C.* for *Walter Kettilby* at the *Bishopshead* in St. *Pauls Church-yard,* 1671.

F. CATALOGUES, DICTIONARIES AND ENCYCLOPEDIAS

Alchemy and the Occult. A Catalogue of Books and Manuscripts from the Collection of Paul and Mary Mellon given to Yale University Library. Compiled by Ian Macphail, with essays by R. P. Multhauf and Aniela Jaffé and additional notes by William McGuire. 2 Volumes. New Haven: Yale University Library, 1968.

Alumni cantabrigienses. A Biographical List of all Known Students, Graduates and Holders of Office at the University of Cambridge, From the Earliest Times to 1900. Compiled by John Venn and J. A. Venn. 10 Volumes in 2 parts. Cambridge: At the University Press, 1922–54.

Bibliotheca Alchemica et Chemica: An Annotated Catalogue of Printed Books on Alchemy, Chemistry and Cognate Subjects in the Library of Denis I. Duveen. London: E. Weil, 1949.

Bolton, Henry Carrington. *A Select Bibliography of Chemistry, 1492–1892.* Smithsonian Miscellaneous Collections, vol. 36. Washington: Smithsonian Institution, 1893.

British Museum General Catalogue of Printed Books. Photolithographic edition to 1955. 263 Volumes. London: Trustees of the British Museum, 1965–66.

Caillet, Albert L. *Manuel bibliographique des sciences psychiques ou occultes: Sciences des Mages, Hermétique, Astrologie, Kabbale, Franc-Maçonnerie, Médecine ancienne, Messmérisme, Sorcellerie, Singularités, Aberrations de tout ordre, Curiosités, Sources Bibliographiques et Documentaires sur ces sujets, etc.* 3 Volumes. Paris: Lucien Dorbon, 1912.

Catalogi codicum manuscriptorum bibliothecae Bodleianae, Pars Nona, Codices a Vire Clarissime Kenelm Digby, Eq. Aur., anno 1634 donatos, complectens. Confecit G. D. Macray. Oxonii: e typographeo Clarendoniano, 1883.

Catalogue général des livres imprimés de la Bibliothèque Nationale. Auteurs. 217 Volumes. Paris: Imprimerie Nationale, 1897.

A Catalogue of the Manuscripts in the British Museum Hitherto Undescribed: Consisting of Five Thousand Volumes; including the Collections of Sir Hans Sloan, bart., the Rev. Thomas Birch, D.D..... Compiled by Samuel Ayscough. 2 Volumes. London: Printed for the compiler by J. Rivington, 1782.

Catalogue of the Newton Papers sold by order of The Viscount Lymington to whom they have descended from Catherine Conduitt, Viscountess Lymington, Greatniece of Sir Isaac Newton. London: Sotheby and Co., 1936.

A Catalogue of the Portsmouth Collection of Books and Papers written by or belonging to Sir Isaac Newton, the scientific portion of which has been presented by the Earl of Portsmouth to the University of Cambridge. Drawn up by the Syndicate appointed the 6th November, 1872. Cambridge: At the University Press, 1888.

Chemical, Medical and Pharmaceutical Books Printed before 1800 In the Collections of the University of Wisconsin Libraries. Edited by John Neu, compiled by Samuel Ives, Reese Jenkins, and John Neu. Madison and Milwaukee: The University of Wisconsin Press, 1965.

A Descriptive Catalogue of the Grace K. Babson Collection of the Works of Sir Isaac Newton and the Material relating to him in the Babson Institute Library, Babson Park, Mass. Introduction by Roger Babson Webber. New York: Herbert Reichner, 1950.

Dictionnaire de biographie française. Sous la direction de J. Balteau, M. Barroux, M. Prevost, Roman d'Amat, et R. Limouzin-Lamothe avec le concours de nombreaux collaborateurs. 12 Volumes, continuing. Paris: Librairie Letouzey et Ané, 1933– .

The Dictionary of National Biography From the Earliest Times to 1900. Edited by Sir Leslie Stephen and Sir Sidney Lee. 22 Volumes. London: Oxford University Press, 1949–50.

Ferguson, John. *Bibliotheca Chemica: A Catalogue of the Alchemical, Chemical and Pharmaceutical Books in the Collection of the Late James Young of Kelley and Durris, Esq., LL.D., F.R.S., F.R.S.E.* 2 Volumes. Glasgow: James Maclekose and Sons, 1906.

Harris, John. *Lexicon Technicum or an Universal English Dictionary of Arts and Sciences.* 2 Volumes. Facsimile reprints of the London editions of 1704 (Vol. I) and 1710 (Vol. II). The Sources of Science, no. 28. New York and London: Johnson Reprint Corporation, 1966.

Lüdy, Fritz. *Alchemistische und Chemische Zeichen. Mit 30 Abbildungen und 128 Tafeln.* Berlin: Gesellschaft für Geschichte der Pharmazie, 1928.

Pernety, Antoine Joseph. *Dictionnaire mytho-hermétique, dans lequel on trouve les allégories fabuleuses des poetes, les métaphores, les enigmes et les termes barbares des philosophes hermétiques expliqués.* Paris: Bauche, 1758.

A Philosophicall Epitaph in Hierogliphicall Figures with Explanation, A Briefe of ye golden Calfe (the Worlds Idoll), Glaubers golden Ass well managed Jehior the three Principles or Originall of all things, Published by W. C. Esq. with a Catalogue of

Chymicall Bookes. London: Printed for William Cooper att the Pellican in Litle Britain, 1673.

Ruland, Martin. *A Lexicon of Alchemy. Translated by Arthur Edward Waite.* Facsimile reprint of the London edition of 1892. London: John M. Watkins, 1964.

Schneider, Wolfgang. *Lexikon Alchemistisch-Pharmazeutischer Symbole.* Weinheim/ Bergstr.: Verlag Chemie, 1962.

Singer, Dorothea Waley. *Catalogue of Latin and Vernacular Alchemical Manuscripts in Great Britain and Ireland Dating from Before the XVI Century.* 3 Volumes. Brussels: Maurice Lamertin, 1928–31.

Thorndike, Lynn. *History of Magic and Experimental Science.* 8 Volumes. New York: Columbia University Press, 1941–58.

Index

Abraham: 15, 108
academy
  alchemical: 57, 76
  Florentine: 36–37, 48, 106
Académie Royale des Sciences: 6
  Histoire: 86
  Memoirs: 83n
acid(s): 210, 217–21, 225, 227–30
active principles, *see under* principles
Addresses to Hartlib, see under Hartlib
De aere et aethere, see under Newton,
  Sir Isaac (*Principia*, related writings)
aether: 173, 203–07, 210–12, 230–31
Agricola, George: 8
  De re metallica: 8
Agricola, Johann Christian: 77
agriculture, *see* husbandry
Agrippa von Nettesheim, Cornelius:
  53–54
air: 35, 38–39, 58, 85n, 123–24, 135,
  138, 159–60, 173, 188–90, 204–06,
  210–11, 222–23, 229
  as a chemical element: 35, 135
  philosophical: 159–60, 179, 181,
    183–84, 188–90, 199, 204, 207,
    230; *see also* spirit (universal)
Alazonomastix Philalethes, *see* More,
  Henry
Albertus Magnus: 51
Albineus, Nathan: 50
  Bibliotheca chemica contracta: 50
alcahest, *see* alkahest
alchemist(s): 1, 10–12, 15, 18–19, 28,
  30–32, 34–48, 51–92, 96, 98, 112–22,
  129–96, 204
  Arabic: 19, 40, 51
  Aristotelian: 19, 40–41, 51, 81–83
  Boerhaave as, *see* Boerhaave
  Boyle as: 11, 18, 45, 80, 91, 94–95,
    201–03
  chemical: 83–87
  Digby as, *see* Digby
  empiric: 137
  Greek: 19, 90
  Locke as: 10–12
  mechanical: 19, 44–45, 81–83, 201–04
  medical: 19, 48
  medieval: 19, 35–36, 40–41, 51, 97
  Neoplatonic: 19, 36–39, 48, 85;

  see also Eirenaeus Philalethes,
    Espagnet, *and* Sendivogius
  Newton as: 1, 80, 87 92, 124–93
  Newton's favorite: 67, 174
  Renaissance: 19, 48, 30–32, 35–36,
    40, 51
  Rosicrucian: 19, 53–80
  seventeenth-century: 18–19, 28n,
    31–32, 36–40, 45, 47, 52–92,
    96–97, 112–21, 150–55, 157–93
alchemy: xi–xii, 1–2, 6–95, 97, 99–102,
  105, 111–22, 124–255
  bibliographies of: 21, 26n, 49n, 52,
    113n
  chemical: 43–47, 60–92
  chemicalization of: 47, 63–81, 89–92
  commercial attempts: 18, 72–73
  and dogmatic religion: 48–49
  Great Work of: 1–2, 17–18, 67–68,
    122, 153–55, 157, 180–81, 193,
    222, 230
  history(ies) of: 25, 26n, 41n, 44n,
    48–92, 135n,
  Jungian model of: 26–43, 80–81
  literature of: 1, 10, 14–15, 18–23,
    26–28, 30–36, 40–47, 49–60,
    65–73, 88, 111–13, 125, 129–34
    136–38, 149–55, 179–86, 191–93
  in Newton's science: 194–233
  philosopher's stone: 29n, 31n, 32n,
    34, 45, 51n, 52n, 53n, 54, 67n,
    68n, 69, 71, 78, 97, 117, 130n, 150
  recipes in: 40, 47, 65, 70, 73, 79, 88,
    132, 134–38, 167, 175, 191, 198
  regimen of, in common gold: 1–2, 192–93
  scientific: 89, 91–92, 126–94
  soteriological function of: 26–35,
    40–42, 48–49, 56–57, 80, 91
  spiritual: 35, 42, 53–60, 63–64, 80–81
  terminology of: 23, 28, 30–31, 35–36,
    47, 73, 80–81, 91–92, 111, 127,
    183–85, 219–21; *see also* Hermetic
    terminology
  theories of: 17, 25, 32, 40–41, 43–47,
    71–72, 75, 80–83, 91–92, 128,
    180–84, 186–90, 192–93, 198–203
  transmutation: 8–9, 11, 17–18, 28n,
    32, 34, 40–41, 43–47, 70–73,
    80–87, 91–92, 94–95, 117, 124–26,